Vibration Simulation Using

MATLAB
and ANSYS

Vibration Simulation Using

MATLAB and ANSYS

MICHAEL R. HATCH

Chapman & Hall/CRC
Taylor & Francis Group

Boca Raton   London   New York

Chapman & Hall/CRC is an imprint of the
Taylor & Francis Group, an **informa** business

Library of Congress Cataloging-in-Publication Data

Hatch, Michael R.
  Vibration simulation using MATLAB and ANSYS / Michael R. Hatch.
      p. cm.
  Includes bibliographical references and index.
  ISBN 1-58488-205-0
  1. Vibration--Computer simulation. 2. MATLAB. 3. ANSYS (Computer system) I.
  Title.

TJ177 .H38 2000
620.3'01'13--dc21

00-055517
CIP

Visit the CRC Press Web site at www.crcpress.com

© 2001 by Chapman & Hall/CRC

No claim to original U.S. Government works
International Standard Book Number 1-58488-205-0
Library of Congress Card Number 00-055517
ISBN 978-1-58488-205-3   7 8 9 0

PREFACE

Background

This book resulted from using, documenting and teaching various analysis techniques during a 30-year mechanical engineering career in the disk drive industry. Disk drives use high performance servo systems to control actuator position. Both experimental and analytical techniques are used to understand the dynamic characteristics of the systems being controlled. Constant in-depth communications between mechanical and control engineers are required to bring high performance electro-mechanical systems to market. Having mechanical engineers who can discuss dynamic characteristics of mechanical systems with servo engineers is very valuable in bringing these high-performance systems into production. This book should be useful to both the mechanical and control communities in enhancing their communication.

Purpose of the Book

The book has three main purposes. The first purpose is to collect in one document various methods of constructing and representing dynamic mechanical models. For someone learning dynamics for the first time or for an experienced engineer who uses the tools infrequently, the options available for modeling can be daunting: transfer function form, zpk form, state space form, modal form, state space modal form, etc. Seeing all the methods in one book, with background theory, an example problem and accompanying MATLAB ® (MathWorks, Inc., Natick, MA) code listing for each method, will help put them in perspective and make them readily available for quick reference. (Also, having equation listings with their accompanying MATLAB code is a good way to develop or reinforce MATLAB programming skills.)

The second purpose is to help the reader develop a strong understanding of modal analysis, where the total response of a system can be constructed by combinations of the individual modes of vibration.

The third purpose is to show how to take the results of large dynamic finite element models and build small MATLAB state space dynamic mechanical models for use in mechanical or servo/mechanical system models.

Audience / Prerequisites

This book is meant to be used as a reference book in senior and early graduate-level vibration and servo courses as well as for practicing servo and mechanical engineers. It should be especially useful for engineers who have limited experience with state space. It assumes the reader has a background in basic vibration theory and elementary Laplace transforms.

For those with a strong linear systems background, the first 12 chapters will provide little new information. Chapters 13 and 14, the finite element chapters, may prove interesting for those with little familiarity with finite elements. Chapters 15 to 19 cover methods for creating state space MATLAB models from ANSYS finite element results, then reducing the models.

Programs Used

It is assumed that the reader has access to MATLAB and the Control System Toolbox and is familiar with their basic use. The MATLAB block diagram graphical modeling tool Simulink is used for several examples through the book but is not required. Several excellent texts covering the basics of MATLAB usage can be found on the MathWorks Web page, www.mathworks.com. All the programs were developed using MATLAB Version 5.3.1.

Lumped mass and cantilever examples using the ANSYS® (ANSYS, Inc., Canonsburg, PA) finite element program are used throughout the text. Where ANSYS results are required for input into MATLAB models, they are available by download without having to run the ANSYS code. For those with access to ANSYS, input code is available by download. The last three chapters contain complete ANSYS/MATLAB dynamic analyses of SISO (Single Input Single Output) and MIMO (Multiple Input Multiple Output) disk drive actuator/suspension systems. Revisions 5.5 and 5.6 of ANSYS were used for the examples.

Organization

The unifying theme throughout most of the book is a **three degree of freedom (tdof)** system, simple enough to be solved for all of its dynamic characteristics in closed form, but complex enough to be able to visualize mode shapes and to have interesting dynamics.

Chapters 1 to 16 contain background theoretical material, closed form solutions to the example problem and MATLAB and/or ANSYS code for solving the problems. All closed form solutions are shown in their entirety.

Chapters 17 to 19 analyze complete disk drive actuator/suspension systems using ANSYS and MATLAB. All chapters list and discuss the related MATLAB code, and all but the last three chapters list the related ANSYS code. All the MATLAB and ANSYS input codes, as well as selected output results, are available for downloading from both the MathWorks FTP site and the author's FTP site, both listed at the end of the preface. Reviewers have provided different inputs on the amount and location of MATLAB and ANSYS code in the book. Engineers for whom the material is new have requested that the code be broken up, interspersed with the text and explained, section by section. Others for whom MATLAB code is second nature have suggested either removing the code listings altogether or providing them at the end of the chapters or in an appendix. My apologies to the latter, but I have chosen to intersperse code in the associated text for the new user.

A problem set accompanies the early chapters. A two degree of freedom system, very amenable to hand calculations, is used in the problem sets to allow one to follow through the derivations and codes with less work than the three degree of freedom (tdof) system used in the text. Some of the problems involve modifying the supplied tdof MATLAB code to simulate the two degree of freedom problem, allowing one to become familiar with MATLAB coding techniques and usage.

Following an introductory chapter, Chapter 2 starts with transfer function analysis. A systematic method for creating mass and stiffness matrices is introduced. Laplace transforms and the transfer function matrix are then discussed. The characteristic equation, poles and zeros are defined.

Chapter 3 develops an intuitive method of sketching frequency responses by hand, and the significance of the magnitudes and phases of various frequency ranges are discussed. Following a development of the imaginary plane and plotting of poles and zeros for the various transfer functions, the relationship between the transfer function and poles and zeros is discussed. Finally, mode shapes are defined, calculated and plotted.

Chapter 4 discusses the origin and interpretation of zeros in Single Input and Single Output (SISO) mechanical systems. Various transfer functions are taken for a lumped parameter system to show the origin of the zeros and how they vary depending on where the force is applied and where the output is taken. An ANSYS finite element model of a tip-loaded cantilever is analyzed and the results are converted into a MATLAB modal state space model to show an overlay of the poles of the "constrained" system and their relationship with the zeros of the original model.

Chapter 5, the state space chapter, takes the basic tdof model and uses it to develop the concept of state space representation of equations of motion. A detailed discussion of complex modes of vibration is then presented, including the use of Argand diagrams and individual mode transient responses.

Chapter 6 uses the state space formulation of Chapter 5 to solve for frequency responses and time domain responses. The matrix exponential is introduced both as an inverse Laplace transform and as a power series solution for a single degree of freedom (sdof) mass system. The tdof transient problem is solved using both the MATLAB function ode45 and a MATLAB Simulink model.

Chapter 7, the modal analysis chapter, begins with a definition of principal modes of vibration, then develops the eigenvalue problem. The relationship between the determinant of the coefficient matrix and the characteristic equation is shown. Eigenvectors are calculated and interpreted, and the modal matrix is defined. Next, the relationship between physical and principal coordinate systems is developed and the concept of diagonalizing or uncoupling the equations of motion is shown. Several methods of normalization are developed and compared. The transformation of initial conditions and forces from physical to principal coordinates is developed. Once the solution in principal coordinates is available, the back transformation to physical coordinates is shown. The chapter then goes on to develop various types of damping typically used in simulation and discusses damping requirements for the existence of principal modes. A two degree of freedom model is used to illustrate the form of the damping matrix when proportional damping is assumed, showing that the answer is not intuitive.

In Chapters 8 and 9 the tdof model is solved for both frequency responses and transient responses in closed form and using MATLAB. A description of how individual modes combine to create the overall frequency response is provided, one of several discussions throughout the book which will help to develop a strong mental image of the basics of the modal analysis method.

Chapter 10, the state space modal analysis chapter, shows how to solve the normal mode eigenvalue problem in state space form, discussing the interpretation of the resulting eigenvectors. Equations of motion are developed in the principal coordinates system and again, individual mode contributions to the overall frequency response are discussed. Real modes are discussed in the same context as for complex modes, using Argand diagrams and individual mode transient responses to illustrate.

Chapter 11 continues the modal state space form by solving for the frequency response. Chapter 12 covers time domain response in modal state space form using the MATLAB "ode45" command and "function" files.

Chapters 13 and 14 discuss the basics of static and dynamic analysis using finite elements, the generation of global stiffness and mass matrices from element matrices, mass matrix forms, static condensation and Guyan Reduction. The purpose of the finite element chapters is to familiarize the reader with basic analysis methods used in finite elements. This familiarity should allow a better understanding of how to interpret the results of the models without necessarily becoming a finite element practitioner. A cantilever beam is used as an example in both chapters. In Chapter 14 a complete eigenvalue analysis with Guyan Reduction is carried out by hand for a two-element beam. Then, MATLAB and ANSYS are used to solve the eigenvalue problem with arbitrary cantilever models.

Chapters 15 and 16 use eigenvalue results from ANSYS beam models to develop state space MATLAB models for frequency and time domain analyses. Both chapters discuss simple methods for reducing the size of ANSYS finite element results to generate small, efficient MATLAB state space models which can be used to describe the dynamic mechanical portion of a servo-mechanical model.

Chapter 17 uses an ANSYS model of a single stage SISO disk drive actuator/suspension system to illustrate using dc or peak gains of individual modes to rank modes for elimination when creating a low order state space MATLAB model.

Chapter 18 introduces balanced reduction, another method of ranking modes for elimination, and uses it to produce a reduced model of the SISO disk drive actuator/suspension model from Chapter 17.

In Chapter 19 a complete ANSYS/MATLAB analysis of a two stage MIMO actuator/suspension system is carried out, with balanced reduction used to create a low order model.

Appendix 1 lists the names of all the MATLAB and ANSYS codes used in the book, separated by chapter. It also contains instruction for downloading the MATLAB and ANSYS files from the MathWorks FTP site as well as the author's Web site, www.hatchcon.com.

Appendix 2 contains a short introduction to Laplace transforms.

For MATLAB product information, contact:

> The MathWorks, Inc.
> 3 Apple Hill Drive
> Natick, MA, 01760-2098 U.S.A.
>
> Tel: 508-647-7000
>
> Fax: 508-647-7101
>
> E-mail: info@mathworks.com
>
> Web: www.mathworks.com

For ANSYS product information, contact:

ANSYS, Inc.
Southpointe
275 Technology Drive
Canonsburg, PA 15317

Tel: 724-746-3304

Fax: 724-514-9494

Web: www.ansys.com

Acknowledgments

There are many people whom I would like to thank for their assistance in the creation of this book, some of whom contributed directly and some of whom contributed indirectly.

First, I would like to acknowledge the influence of the late William Weaver, Jr., Professor Emeritus, Civil Engineering Department, Stanford University. I first learned finite elements and modal analysis when taking Professor Weaver's courses in the early 1970s and his teachings have stood me in good stead for the last 30 years.

Dr. Haithum Hindi kindly allowed the use of a portion of his unpublished notes for the Laplace transform presentation in Appendix 2 and provided valuable feedback on the nuances of "modred" and balanced reduction.

I would like to thank my reviewers for their thorough and time-consuming reviews of the document: Stephen Birn, Marianne Crowder, Dr. Y.C. Fu, Dr. Haithum Hindi, Dr. Michael Lu, Dr. Babu Rahman, Kathryn Tao and Yimin Niu. Mark Rodamaker, an ANSYS distributor, kindly reviewed the book from an ANSYS perspective. My daughter-in-law, Stephanie Hatch, provided valuable editing input throughout the book.

I would also like to thank Dr. Wodek Gawronski for his words of encouragement and his helpful suggestions to a new author. Dr. Gawronski's two advanced texts on the subject are highly recommended for those wishing additional information (see References).

TABLE OF CONTENTS

CHAPTER 1

INTRODUCTION

This book has three main purposes. The first purpose is to collect in one document the various methods of constructing and representing dynamic mechanical models. The second purpose is to help the reader develop a strong understanding of the modal analysis technique, where the total response of a system can be constructed by combinations of individual modes of vibration. The third purpose is to show how to take the results of large finite element models and reduce the size of the model (model reduction), extracting lower order state space models for use in MATLAB.

1.1 Representing Dynamic Mechanical Systems

We will see that the nature of damping in the system will determine which representation will be required. In lightly damped structures, where the damping comes from losses at the joints and the material losses, we will be able to use "modal analysis," enabling us to restructure the problem in terms of individual modes of vibration with a particular type of damping called "proportional damping." For systems which have significant damping, as in systems with a specific "damper" element, we will have to use the original, coupled differential equations for solution.

The left-hand block in Figure 1.1 represents a damped dynamic model with coupled equations of motion, a set of initial conditions and a definition of the forcing function to be applied. If damping in the system is significant, then the equations of motion need to be solved in their original form. The option of using the normal modes approach is not feasible. The three methods of solving for time and frequency domain responses for highly damped, coupled equations are shown.

1.2 Modal Analysis

Most practical problems require using the finite element method to define a model. The finite element method can be formulated with specific damping elements in addition to structural elements for highly damped systems, but its most common use is to model lightly damped structures.

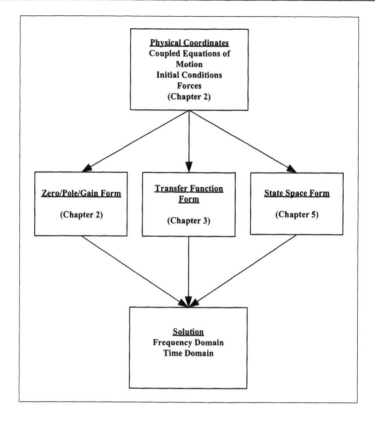

Figure 1.1: Coupled equations of motion flowchart.

The diagram in Figure 1.2 shows the methodology for analyzing a lightly damped structure using normal modes. As with the coupled equation solution above, the solution starts with deriving the undamped equations of motion in physical coordinates.   The next step is solving the eigenvalue problem, yielding eigenvalues (natural frequencies) and eigenvectors (mode shapes). This is the most intuitive part of the problem and gives one considerable insight into the dynamics of the structure by understanding the mode shapes and natural frequencies.

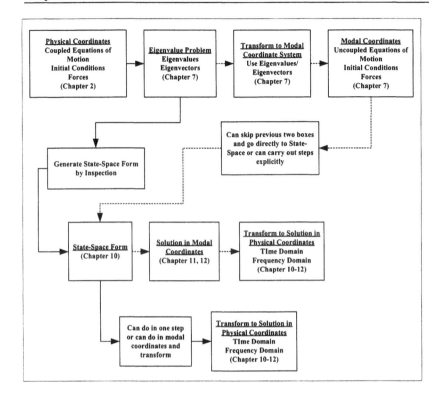

Figure 1.2: Modal analysis method flowchart.

To solve for frequency and time domain responses, it is necessary to transform the model from the original physical coordinate system to a new coordinate system, the modal or principal coordinate system, by operating on the original equations with the eigenvector matrix. In the modal coordinate system the original undamped **coupled** equations of motion are transformed to the same number of undamped **uncoupled** equations. Each uncoupled equation represents the motion of a particular mode of vibration of the system. It is at this step that proportional damping is applied. It is trivial to solve these uncoupled equations for the responses of the modes of vibration to the forcing function and/or initial conditions because each equation is the equation of motion of a simple single degree of freedom system. The desired responses are then back-transformed into the physical coordinate system, again using the eigenvector matrix for conversion, yielding the solution in physical coordinates.

The modal analysis sequence of taking a complicated system, (1) transforming to a simpler coordinate system, (2) solving equations in that coordinate system and then (3) back-transforming into the original coordinate system is analogous to using Laplace transforms to solve differential equations. The original differential equation is (1) transformed to the "s" domain by using a Laplace transform, (2) the algebraic solution is then obtained and is (3) back-transformed using an inverse Laplace transform.

It will be shown that once the eigenvalue problem has been solved, setting up the zero initial condition state space form of the uncoupled equations of motion in principal coordinates can be performed by inspection. The solution and back-transformation to physical coordinates can be performed in one step in the MATLAB solution.

The advantage of the modal solution is the insight developed from understanding the modes of vibration and how each mode contributes to the total solution.

1.3 Model Size Reduction

It is useful to be able to provide a model of the mechanical system to control engineers using the fewest states possible, while still providing a representative model. The mechanical model can then be inserted into the complete mechanical/control system model and be used to define the system dynamics.

Figure 1.3 shows how to convert a large finite element model (and most real finite element models are "large," with thousands to hundreds of thousands of degrees of freedom) to a smaller model which still provides correct responses for the forcing function input and desired output points.

The problem starts out with the finite element model which is solved for its eigenvalues and eigenvectors (resonant frequencies and mode shapes). There are as many eigenvalues and eigenvectors as degrees of freedom for the model, typically too large to be used in a MATLAB model.

Once again, the eigenvalues and eigenvectors provide considerable insight into the system dynamics, but the objective is to provide an efficient, "small" model for inclusion into the mechanical/servo system model. This requires reducing the size of the model while still maintaining the desired input/output relationships.

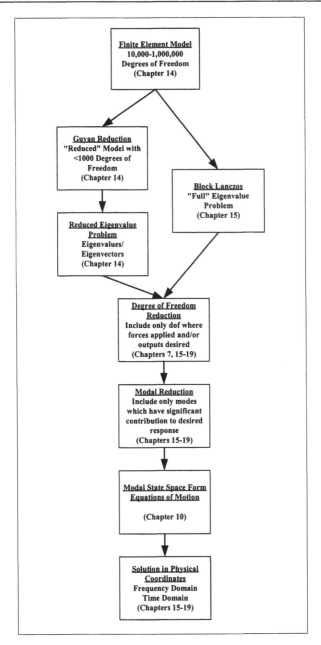

Figure 1.3: Model size reduction flowchart.

The reduction of the size of the model is accomplished in two steps. The first is to reduce the number of degrees of freedom of the model from the original set to a new set which includes only those degrees of freedom where forces are applied and/or where responses are desired.

The second step for Single Input Single Output (SISO) systems is to reduce the number of modes of vibration used for the solution by ranking the relative importance of each mode to the overall response. For Multi Input Multi Output (MIMO) systems, a more sophisticated method of reduction which simultaneously takes into account the controllability and observability of the system is required.

Figure 1.4 shows the overall frequency response for a SISO cantilever beam model discussed in Chapter 15. Superimposed over the overall frequency response is the contribution of each of the individual 10 modes of vibration which make up the overall response.

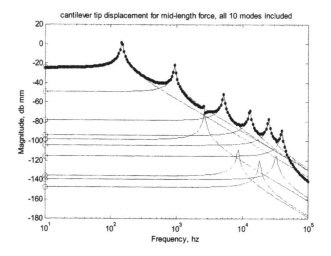

Figure 1.4: Individual mode contribution to overall frequency response.

We will show that modes with little or no displacement at the reduced set of degrees of freedom are candidates for elimination. For example, the three modes which have low frequency magnitudes of less than −120db in Figure 1.4 have no effect on the overall frequency response – their peaks do not show up on the overall frequency response. The less important modes either can be eliminated directly or a more sophisticated method can be used which takes into account the low frequency effects of the removed modes. Both types are discussed in detail, accompanied by examples.

A reduced solution can provide very good results with a significant reduction in number of states – a model which is very amenable to being combined with a servo model for a complete servo mechanical system model.

CHAPTER 2

TRANSFER FUNCTION ANALYSIS

2.1 Introduction

The purpose of this chapter is to illustrate how to derive equations of motion for Multi Degree of Freedom (mdof) systems and how to solve for their transfer functions.

The chapter starts by developing equations of motion for a specific **three degree of freedom** damped system (indicated throughout the book by the acronym "**tdof**"). A systematic method of creating "global" mass, damping and stiffness matrices is borrowed from the stiffness method of matrix structural analysis. The tdof model will be used for the various analysis techniques through most of the book, providing a common thread that links the pieces into a whole.

Two additional examples are used to illustrate the method for building matrix equations of motion. The first is a lumped mass **six degree of freedom (6dof)** system for which the stiffness matrix is developed. The second is a simplified rotary actuator system from a disk drive, for which the complete undamped equations of motion are developed.

Following the equations of motion sections, the chapter continues with a review of the transfer function and frequency response analyses of a **single degree of freedom (sdof)** damped example. After developing the closed form solution of the equations, MATLAB code is used to calculate and plot magnitude and phase versus frequency for a range of damping values.

The tdof model is then reintroduced and Laplace transforms are used to develop its transfer functions. In order to facilitate hand calculations of poles and zeros, damping is set to zero. The characteristic equation, poles and zeros are then defined and calculated in closed form. MATLAB code is used to plot the pole/zero locations for the nine transfer functions using MATLAB's "pzmap" command.

MATLAB is used to calculate and plot poles and zeros for values of damping greater than zero and we will see that additional real values zeros start appearing as damping is increased from zero. The significance of the real axis zeros is discussed.

2.2 Deriving Matrix Equations of Motion

2.2.1 Three Degree of Freedom (tdof) System, Identifying Components and Degrees of Freedom

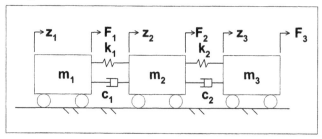

Figure 2.1: tdof system schematic.

The first step in analyzing a mechanical system is to sketch the system, showing the degrees of freedom, the masses, stiffnesses and damping present, and showing applied forces. The tdof system to be followed throughout the book, shown in Figure 2.1, consists of three masses, numbered 1 to 3, two springs between the masses and two dampers also between the masses. The model is purposely not connected to ground to allow a "rigid body" degree of freedom, meaning that at "low" frequencies the set of three masses can all move in one direction or the other as a single rigid body, with no relative motion between them.

The number of **degrees of freedom (dof)** for a model is the number of geometrically independent coordinates required to specify the configuration for the model. For consistency, the notation "z" will be used for degrees of freedom, saving "x" and "y" for state space representations later in the book. For the system shown in Figure 2.1 where each mass can move only along the z axis, a single degree of freedom for each mass is sufficient, hence the degrees of freedom z_1, z_2 and z_3.

2.2.2 Defining the Stiffness, Damping and Mass Matrices

The equations of motion will be derived in matrix form using a method derived from the stiffness method of structural analysis, as follows:

> **Stiffness Matrix: Apply a unit displacement** to each dof, one at a time. Constrain the dof's not displaced and **define the stiffness dependent constraint force** required for all dof's to hold the system in the constrained position.

The row elements of each column of the stiffness matrix are then defined by the constraints associated with each dof that are required to hold the system in the constrained position.

Damping Matrix: Apply a unit velocity to each dof, one at a time. Constrain the dof's not moving and **define the velocity-dependent constraint force** required to keep the system in that state.

The row elements of each column of the damping matrix are then defined by the constraints associated with each dof that are required to keep the system in that state – with one dof moving with constant velocity and all the other dof's not moving.

Mass Matrix: Apply a unit acceleration to each dof, one at a time. Constrain the dof's not being accelerated and **define the acceleration-dependent constraint forces** required.

The row elements of each column of the mass matrix are then defined by the constraints associated with keeping one dof accelerating at a constant rate and the other dof's stationary. Since in this model the only forces transmitted between the masses are proportional to displacement (the springs) and velocity (viscous damping), no forces are transmitted between masses due to one of the masses accelerating. This leads to a diagonal mass matrix in cases where the origin of the coordinate systems are taken through the center of mass of the bodies and the coordinate axes are aligned with the principal moments of inertia of the body.

Table 2.1 shows how the three matrices are filled out. To fill out column 1 of the mass, damping and stiffness matrices, mass 1 is given a unit acceleration, velocity and displacement, respectively. Then the constraining forces required to keep the system in that state are defined for each dof, where row 1 is for dof 1, row 2 is for dof 2 and row 3 is for dof 3.

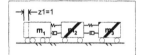 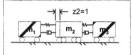 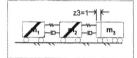

Column 1	Column 2	Column 3
$\text{UNIT} \left\{ \begin{array}{l} \text{accel} \\ \text{vel} \\ \text{disp} \end{array} \right\} \text{dof1}$	$\text{Unit} \left\{ \begin{array}{l} \text{accel} \\ \text{vel} \\ \text{disp} \end{array} \right\} \text{dof2}$	$\text{Unit} \left\{ \begin{array}{l} \text{accel} \\ \text{vel} \\ \text{disp} \end{array} \right\} \text{dof3}$
$\begin{bmatrix} m_1 \\ 0 \\ 0 \end{bmatrix}$	$\begin{matrix} 0 \\ m_2 \\ 0 \end{matrix}$	$\begin{bmatrix} 0 \\ 0 \\ m_3 \end{bmatrix} \begin{matrix} \text{dof1} \\ \text{dof2} \\ \text{dof3} \end{matrix}$
$\begin{bmatrix} c_1 \\ -c_1 \\ 0 \end{bmatrix}$	$\begin{matrix} -c_1 \\ c_1 + c_2 \\ -c_2 \end{matrix}$	$\begin{bmatrix} 0 \\ -c_2 \\ c_2 \end{bmatrix} \begin{matrix} \text{dof1} \\ \text{dof2} \\ \text{dof3} \end{matrix}$
$\begin{bmatrix} k_1 \\ -k_1 \\ 0 \end{bmatrix}$	$\begin{matrix} -k_1 \\ k_1 + k_2 \\ -k_2 \end{matrix}$	$\begin{bmatrix} 0 \\ -k_2 \\ k_2 \end{bmatrix} \begin{matrix} \text{dof1} \\ \text{dof2} \\ \text{dof3} \end{matrix}$

Table 2.1: m, c, k columns and associated dof displacements. The cross-hatched masses in the figures above each column are constrained and non-cross-hatched mass is moved a unit displacement.

The general matrix form for a tdof system is shown below, where the "ij" subscripts in m_{ij}, c_{ij}, k_{ij} are defined as follows: "i" is the row number and "j" is the column number.

$$
\begin{array}{c}
j=1 \ \ j=2 \ \ j=3 \\
\begin{array}{l} i=1 \\ i=2 \\ i=3 \end{array}
\begin{bmatrix} m_{11} & m_{12} & m_{13} \\ m_{21} & m_{22} & m_{23} \\ m_{31} & m_{32} & m_{33} \end{bmatrix}
\begin{bmatrix} \ddot{z}_1 \\ \ddot{z}_2 \\ \ddot{z}_3 \end{bmatrix}
+
\begin{bmatrix} c_{11} & c_{12} & c_{13} \\ c_{21} & c_{22} & c_{23} \\ c_{31} & c_{32} & c_{33} \end{bmatrix}
\begin{bmatrix} \dot{z}_1 \\ \dot{z}_2 \\ \dot{z}_3 \end{bmatrix}
+
\begin{bmatrix} k_{11} & k_{12} & k_{13} \\ k_{21} & k_{22} & k_{23} \\ k_{31} & k_{32} & k_{33} \end{bmatrix}
\begin{bmatrix} z_1 \\ z_2 \\ z_3 \end{bmatrix}
=
\begin{bmatrix} F_1 \\ F_2 \\ F_3 \end{bmatrix}
\end{array} \quad (2.1)
$$

Mass   Damping   Stiffness

Expanding the matrix equations of motion by multiplying across and down:

$$m_{11}\ddot{z}_1 + m_{12}\ddot{z}_2 + m_{13}\ddot{z}_3 + c_{11}\dot{z}_1 + c_{12}\dot{z}_2 + c_{13}\dot{z}_3 + k_{11}z_1 + k_{12}z_2 + k_{13}z_3 = F_1 \quad (2.2)$$

$$m_{21}\ddot{z}_1 + m_{22}\ddot{z}_2 + m_{23}\ddot{z}_3 + c_{21}\dot{z}_1 + c_{22}\dot{z}_2 + c_{23}\dot{z}_3 + k_{21}z_1 + k_{22}z_2 + k_{23}z_3 = F_2 \quad (2.3)$$

$$m_{31}\ddot{z}_1 + m_{32}\ddot{z}_2 + m_{33}\ddot{z}_3 + c_{31}\dot{z}_1 + c_{32}\dot{z}_2 + c_{33}\dot{z}_3 + k_{31}z_1 + k_{32}z_2 + k_{33}z_3 = F_3 \quad (2.4)$$

The matrix equations of motion for our tdof problem, from Table 2.1, is:

$$
\begin{bmatrix} m_1 & 0 & 0 \\ 0 & m_2 & 0 \\ 0 & 0 & m_3 \end{bmatrix}
\begin{bmatrix} \ddot{z}_1 \\ \ddot{z}_2 \\ \ddot{z}_3 \end{bmatrix}
+
\begin{bmatrix} c_1 & -c_1 & 0 \\ -c_1 & (c_1+c_2) & -c_2 \\ 0 & -c_2 & c_2 \end{bmatrix}
\begin{bmatrix} \dot{z}_1 \\ \dot{z}_2 \\ \dot{z}_3 \end{bmatrix}
$$
$$
+
\begin{bmatrix} k_1 & -k_1 & 0 \\ -k_1 & (k_1+k_2) & -k_2 \\ 0 & -k_2 & k_2 \end{bmatrix}
\begin{bmatrix} z_1 \\ z_2 \\ z_3 \end{bmatrix}
=
\begin{bmatrix} F_1 \\ F_2 \\ F_3 \end{bmatrix}
\quad (2.5)
$$

Expanding:

$$m_1\ddot{z}_1 + c_1\dot{z}_1 - c_1\dot{z}_2 + k_1z_1 - k_1z_2 = F_1$$
$$m_2\ddot{z}_2 - c_1\dot{z}_1 + (c_1+c_2)\dot{z}_2 - c_2\dot{z}_3 - k_1z_1 + (k_1+k_2)z_2 - k_2z_3 = F_2 \quad (2.6a,b,c)$$
$$m_3\ddot{z}_3 - c_2\dot{z}_2 + c_2\dot{z}_3 - k_2z_2 + k_2z_3 = F_3$$

2.2.3 Checks on Equations of Motion for Linear Mechanical Systems

Two quick checks which should always be carried out for linear mechanical systems are the following:

1) All diagonal terms must be positive.

2) The mass, damping and stiffness matrices must be symmetrical. For example $k_{ij} = k_{ji}$ for the stiffness matrix.

2.2.4 Six Degree of Freedom (6dof) Model – Stiffness Matrix

The stiffness matrix development for a more complicated model than the tdof model used so far is shown below. The figure below shows a 6dof system with a rigid body mode and no damping.

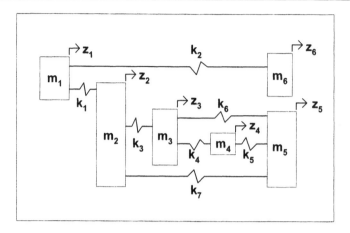

Figure 2.2: 6dof model schematic.

Moving each dof a unit displacement and then writing down the reaction forces to constrain that configuration for each of the column elements, the stiffness matrix for this example can be written by inspection as shown in Table 2.2. Note that the symmetry and positive diagonal checks are satisfied.

$$\begin{bmatrix} (k_1 + k_2) & -k_1 & 0 & 0 & 0 & -k_2 \\ -k_1 & (k_1 + k_3 + k_7) & -k_3 & 0 & -k_7 & 0 \\ 0 & -k_3 & (k_3 + k_4 + k_6) & -k_4 & -k_6 & 0 \\ 0 & 0 & -k_4 & (k_4 + k_5) & -k_5 & 0 \\ 0 & -k_7 & -k_6 & -k_5 & (k_5 + k_6 + k_7) & 0 \\ -k_2 & 0 & 0 & 0 & 0 & k_2 \end{bmatrix}$$

Table 2.2: Stiffness matrix terms for 6dof system.

2.2.5 Rotary Actuator Model – Stiffness and Mass Matrices

The technique is also applicable to systems with rotations combined with translations, as long as rotations are kept small. The system shown below represents a simplified rotary actuator from a disk drive that pivots about its mass center, has force applied at the left-hand end (representing the rotary voice coil motor) and has a "recording head" m_2 at the right-hand end. The "head" is connected to the end of the actuator with a spring and the pivot bearing is connected to ground through the radial stiffness of its bearing.

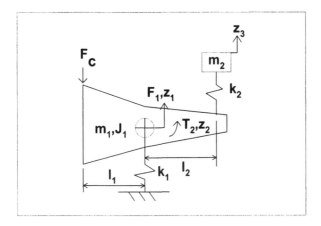

Figure 2.3: Rotary actuator schematic.

Starting off by defining the degrees of freedom, stiffnesses, mass and inertia terms:

dof:

z_1	translation of actuator
z_2	rotation of actuator
z_3	translation of head

Stiffnesses:

k_1	actuator bearing radial stiffness
k_2	"suspension" stiffness

Inertias:

m_1, J_1	actuator mass, inertia
m_2	"head" mass

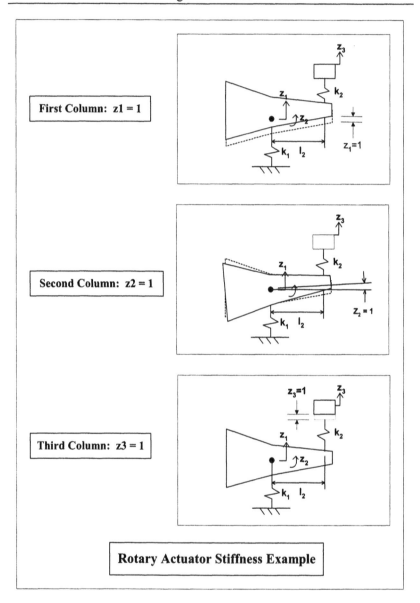

Figure 2.4: Unit displacements to define mass and stiffeness matrices.

See Figure 2.4 to define the entries of each column of (2.7), the forces/moments required to constrain the respective dof in the configuration shown.

$$\begin{bmatrix} m_1 & 0 & 0 \\ 0 & J_1 & 0 \\ 0 & 0 & m_2 \end{bmatrix} \begin{bmatrix} \ddot{z}_1 \\ \ddot{z}_2 \\ \ddot{z}_3 \end{bmatrix} + \begin{bmatrix} (k_1+k_2) & l_2k_2 & -k_2 \\ l_2k_2 & l_2^2k_2 & -l_2k_2 \\ -k_2 & -l_2k_2 & k_2 \end{bmatrix} \begin{bmatrix} z_1 \\ z_2 \\ z_3 \end{bmatrix} = \begin{bmatrix} F_1 \\ T_2 \\ 0 \end{bmatrix} = \begin{bmatrix} -F_c \\ F_c l_1 \\ 0 \end{bmatrix} \qquad (2.7)$$

$$F_1 = -F_c \qquad (2.8)$$

$$T_2 = F_c l_1 \qquad (2.9)$$

2.3 Single Degree of Freedom (sdof) System Transfer Function and Frequency Response

2.3.1 sdof System Definition, Equations of Motion

The sdof system to be analyzed is shown below. The system consists of a mass, m, connected to ground by a spring of stiffness k and a damper with viscous damping coefficient c. Since the mass can only move in the z direction, a single degree of freedom is sufficient to define the system configuration. Force F is applied to the mass.

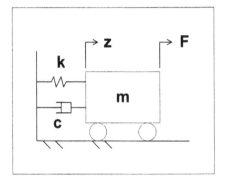

Figure 2.5: Single degree of freedom system.

The equation of motion for this system is given by:

$$m\ddot{z} + c\dot{z} + kz = F \qquad (2.10)$$

2.3.2 Transfer Function

Taking the Laplace transform of a general second order differential equation (DE) with initial conditions is:

$$\text{Second Order DE:} \quad \mathcal{L}\{\ddot{z}(t)\} = s^2 z(s) - sz(0) - \dot{z}(0), \quad (2.11)$$

where $z(0)$ and $\dot{z}(0)$ are position and velocity initial conditions, respectively, and $z(s)$ is the Laplace transform of $z(t)$. See Appendix 2 for more on Laplace transforms.

Because we are taking a transfer function, representing the steady state response of the system to a sinusoidal input, initial conditions are set to zero, leaving

$$\mathcal{L}\{\ddot{z}(t)\} = s^2 z(s) \quad (2.12)$$

The Laplace transform of the sdof equation of motion (2.10), where $F(s)$ represents the Laplace transform of F, is:

$$ms^2 z(s) + csz(s) + kz(s) = F(s) \quad (2.13)$$

Solving for the transfer function:

$$\frac{z(s)}{F(s)} = \frac{1}{ms^2 + cs + k} = \frac{1/m}{s^2 + \dfrac{c}{m}s + \dfrac{k}{m}} \quad (2.14)$$

We can simplify the equation above by applying the following definitions:

1) $\omega_n^2 = \dfrac{k}{m}$, where ω_n is the undamped natural frequency, rad/sec

2) $c_{cr} = 2\sqrt{km}$, where c_{cr} is the "critical" damping value

3) ζ is the amount of proportional damping, typically stated as a percentage of critical damping

4) $2\zeta\omega_n$ is the multiplier of the velocity term, $\dot{z}$, developed below:

$$\frac{c}{m} = 2\zeta\omega_n$$

$$= 2\frac{c}{c_{cr}}\sqrt{\frac{k}{m}}$$

$$= \frac{2c}{2\sqrt{km}}\frac{\sqrt{k}}{\sqrt{m}} \tag{2.15}$$

$$= \frac{c}{m}$$

Rewriting, using the above substitutions:

$$\frac{z(s)}{F(s)} = \frac{1/m}{s^2 + 2\zeta\omega_n s + \omega_n^2} \tag{2.16}$$

2.3.3 Frequency Response

Substituting " $j\omega$ " for "s" to calculate the frequency response, where "j" is the imaginary operator:

$$
\begin{aligned}
\frac{z(j\omega)}{F(j\omega)} &= \frac{1/m}{(j\omega)^2 + 2\zeta\omega_n(j\omega) + \omega_n^2} \\[2mm]
&= \frac{1/m}{-\omega^2 + 2\zeta\omega\omega_n j + \omega_n^2} \\[2mm]
&= \frac{1/(m\omega^2)}{-1 + \dfrac{2\zeta\omega_n j}{\omega} + \dfrac{\omega_n^2}{\omega^2}} \qquad (2.17a,b,c,d,e) \\[2mm]
&= \frac{1/(m\omega^2)}{\left(\dfrac{\omega_n^2}{\omega^2} - 1\right) + \dfrac{2\zeta\omega_n j}{\omega}} \\[2mm]
&= \frac{1/(m\omega^2)}{\left[\left(\dfrac{\omega_n}{\omega}\right)^2 - 1\right] + j2\zeta\left(\dfrac{\omega_n}{\omega}\right)}
\end{aligned}
$$

The frequency response equation above shows how the ratio (z/F) varies as a function of frequency, ω. The ratio is a complex number that has some interesting properties at different values of the ratio (ω_n / ω).

At low frequencies relative to the resonant frequency, $\omega_n^2 \gg \omega\omega_n \gg \omega^2$, and the transfer function is given by:

$$\frac{z(j\omega)}{F(j\omega)} = \frac{1/m}{-\omega^2 + 2\zeta\omega\omega_n j + \omega_n^2}$$

$$\cong \frac{1/m}{\omega_n^2} = \frac{1}{m\omega_n^2} = \frac{1}{m\left(\dfrac{k}{m}\right)} = \frac{1}{k} \qquad (2.18)$$

Since the frequency response value at any frequency is a complex number, we can take the magnitude and phase.

$$\left|\frac{z(j\omega)}{F(j\omega)}\right| = \frac{1}{k}$$

$$(2.19a,b)$$

$$\angle \frac{z(j\omega)}{F(j\omega)} = 0$$

Thus, the gain at low frequencies is a constant, $(1/k)$ or the inverse of the stiffness. Phase is $0°$ because the sign is positive.

At high frequencies, $\omega^2 \gg \omega\omega_n \gg \omega_n^2$, the transfer function is given by:

$$\frac{z(j\omega)}{F(j\omega)} = \frac{1/m}{-\omega^2 + 2\zeta\omega\omega_n j + \omega_n^2}$$

$$\cong \frac{1/m}{-\omega^2} = \frac{-1}{m\omega^2} \qquad (2.20)$$

Once again, taking the magnitude and phase:

$$\left|\frac{z(j\omega)}{F(j\omega)}\right| = \left|\frac{-1}{m\omega^2}\right| = \frac{1}{m\omega^2}$$

$$(2.21a,b)$$

$$\angle \frac{z(j\omega)}{F(j\omega)} = -180°$$

At high frequencies, the gain is given by $1/(m\omega^2)$ and the phase is $-180°$ because the sign is negative.

At resonance, $\omega = \omega_n$, the transfer function is given by:

$$\frac{z(j\omega)}{F(j\omega)} = \frac{1/m}{-\omega^2 + 2\zeta\omega\omega_n j + \omega_n^2} \tag{2.22}$$

$$= \frac{1/m}{2\zeta\omega\omega_n j} = \frac{1/m}{2\zeta\omega_n^2 j} = \frac{1}{\dfrac{2\zeta\omega_n^2 mj}{m}} = \frac{1}{2\zeta kmj} = \frac{1}{2\zeta kj} = \frac{1/k}{2\zeta j} = \frac{-j/k}{2\zeta}$$

Taking magnitude and phase at resonance:

$$\left|\frac{z(j\omega)}{F(j\omega)}\right| = \left|\frac{-j/k}{2\zeta}\right| = \frac{1/k}{2\zeta}$$

$$\angle \frac{z(j\omega)}{F(j\omega)} = -90° \tag{2.23a,b}$$

The magnitude at resonance is seen to be the gain at low frequency, $1/k$, divided by 2ζ. Since ζ is typically a small number, for example 1% of critical damping or 0.01, the magnitude at resonance is seen to be amplified. At resonance the phase angle is $-90°$.

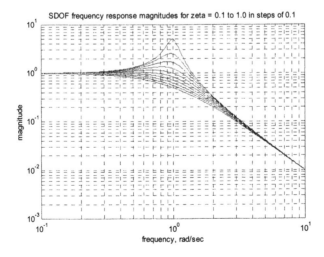

Figure 2.6: sdof magnitude versus frequency for different damping ratios.

The MATLAB code **sdofxfer.m,** listed in the next section, is used to plot the frequency responses from (2.17) for a range of damping values for m = k = 1.0, shown in Figures 2.6 and 2.7. These m and k values give a ω_n value of 1.0 rad/sec.

Since ω_n is 1.0 rad/sec, the resonant peak in Figure 2.6 should occur at that frequency. The low frequency magnitude was shown above to be equal to 1/k = 1.0. The curves for all the damping values approach 1.0 ($10^0 = 1.0$) at low frequencies. At high frequencies the magnitude is given by $1/(m\omega^2)$, and since m = 1, we should have magnitude of $1/\omega^2$. Checking the plot above, at a frequency of 10 rad/sec, the magnitude should be 1/100 or 0.01.

Note that the slope of the low frequency asymptote is zero, meaning it is not changing with frequency. However, the slope of the high frequency asymptote is "-2," meaning that for every decade increase in frequency the magnitude at high frequency decreases by two orders of magnitude by virtue of the ω^2 term in the denominator. The "-2" slope on a log magnitude versus log frequency plot comes from the following:

$$\log|\text{high frequency}| \propto \log\left(\frac{1}{\omega^2}\right) = \log(\omega^{-2}) = -2\log(\omega) \qquad (2.24)$$

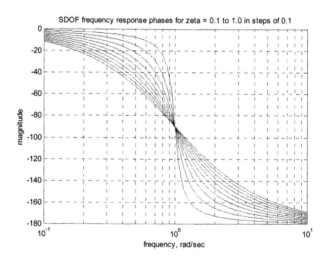

Figure 2.7: sdof phase versus frequency for different damping ratios.

From Figure 2.7, note that at resonance ($\omega_n = 1.0\,\text{rad}/\text{sec}$) the phase for all values of damping is $-90°$. At low frequencies phase is approaching $0°$ and at high frequencies it is approaching $-180°$.

2.3.4 MATLAB Code sdofxfer.m Description

The code uses the transfer function form shown in (2.14) to calculate the complex quantity "xfer," where $s = j\omega$, using a vector of defined ω values. Magnitude and phase of the complex value of the transfer function are then plotted versus frequency.

2.3.5 MATLAB Code sdofxfer.m Listing

```
%        sdofxfer.m plotting frequency responses of sdof model for different damping values

         clf;

         clear all;

%        assign values for mass, percentage of critical damping, and stiffnesses
%        zeta is a vector of damping values from 10% to 100% in steps of 10%

         m = 1;
         zeta = 0.1:0.1:1;                    %  0.1 = 10% of critical
         k = 1;

         wn = sqrt(k/m);

%        Define a vector of frequencies to use, radians/sec.  The logspace command uses
%        the log10 value as limits, i.e. -1 is 10^-1 = 0.1 rad/sec, and 1 is
%        10^1 = 10 rad/sec.  The 400 defines 400 frequency points.

         w = logspace(-1,1,400);

%        pre-calculate the radians to degree conversion

         rad2deg = 180/pi;

%        define s as the imaginary operator times the radian frequency vector

         s = j*w;

%        define a for loop to cycle through all the damping values for calculating
%        magnitude and phase

         for  cnt = 1:length(zeta)

%        define the frequency response to be evaluated

             xfer(cnt,:) = (1/m) ./ (s.^2 + 2*zeta(cnt)*wn*s + wn^2);
```

```
%          calculate the magnitude and phase of each frequency response

              mag(cnt,:) = abs(xfer(cnt,:));

              phs(cnt,:) = angle(xfer(cnt,:))*rad2deg;

end

%          define a for loop to cycle through all the damping values for plotting magnitude

for  cnt = 1:length(zeta)

loglog(w,mag(cnt,:),'k-')
title('SDOF frequency response magnitudes for zeta = 0.1 to 1.0 in steps of 0.1')
xlabel('frequency, rad/sec')
ylabel('magnitude')
grid

hold on

end

hold off

grid on

disp('execution paused to display figure, "enter" to continue'); pause

%          define a for loop to cycle through all the damping values for plotting phase

for  cnt = 1:length(zeta)

semilogx(w,phs(cnt,:),'k-')
title('SDOF frequency response phases for zeta = 0.1 to 1.0 in steps of 0.1')
xlabel('frequency, rad/sec')
ylabel('magnitude')
grid

hold on

end

hold off

grid on

disp('execution paused to display figure, "enter" to continue'); pause
```

2.4 tdof Laplace Transform, Transfer Functions, Characteristic Equation, Poles, Zeros

We now return to the original tdof model as shown in Figure 2.1. In order to define transfer functions and understand poles and zeros of the system, we need to transform from the time domain to the frequency domain. We do this by taking Laplace transforms of the equations of motion.

2.4.1 Laplace Transforms with Zero Initial Conditions

Repeating (2.5) for the tdof system:

$$
\begin{bmatrix} m_1 & 0 & 0 \\ 0 & m_2 & 0 \\ 0 & 0 & m_3 \end{bmatrix} \begin{bmatrix} \ddot{z}_1 \\ \ddot{z}_2 \\ \ddot{z}_3 \end{bmatrix} + \begin{bmatrix} c_1 & -c_1 & 0 \\ -c_1 & (c_1+c_2) & -c_2 \\ 0 & -c_2 & c_2 \end{bmatrix} \begin{bmatrix} \dot{z}_1 \\ \dot{z}_2 \\ \dot{z}_3 \end{bmatrix}
$$
$$
+ \begin{bmatrix} k_1 & -k_1 & 0 \\ -k_1 & (k_1+k_2) & -k_2 \\ 0 & -k_2 & k_2 \end{bmatrix} \begin{bmatrix} z_1 \\ z_2 \\ z_3 \end{bmatrix} = \begin{bmatrix} F_1 \\ F_2 \\ F_3 \end{bmatrix}
$$

(2.25)

Taking Laplace transforms assuming initial conditions of zero, where z_1, z_2, z_3 now represent the Laplace transforms of the original z_1, z_2, z_3:

$$
\begin{bmatrix} m_1 & 0 & 0 \\ 0 & m_2 & 0 \\ 0 & 0 & m_3 \end{bmatrix} \begin{bmatrix} s^2 z_1 \\ s^2 z_2 \\ s^2 z_3 \end{bmatrix} + \begin{bmatrix} c_1 & -c_1 & 0 \\ -c_1 & (c_1+c_2) & -c_2 \\ 0 & -c_2 & c_2 \end{bmatrix} \begin{bmatrix} sz_1 \\ sz_2 \\ sz_3 \end{bmatrix}
$$
$$
+ \begin{bmatrix} k_1 & -k_1 & 0 \\ -k_1 & (k_1+k_2) & -k_2 \\ 0 & -k_2 & k_2 \end{bmatrix} \begin{bmatrix} z_1 \\ z_2 \\ z_3 \end{bmatrix} = \begin{bmatrix} F_1 \\ F_2 \\ F_3 \end{bmatrix}
$$

(2.26)

Rearranging:

$$
\begin{bmatrix} (m_1 s^2 + c_1 s + k_1) & (-c_1 s - k_1) & 0 \\ (-c_1 s - k_1) & (m_2 s^2 + c_1 s + c_2 s + k_1 + k_2) & (-c_2 s - k_2) \\ 0 & (-c_2 s - k_2) & (m_3 s^2 + c_2 s + k_2) \end{bmatrix} \begin{bmatrix} z_1 \\ z_2 \\ z_3 \end{bmatrix} = \begin{bmatrix} F_1 \\ F_2 \\ F_3 \end{bmatrix}
$$

(2.27)

2.4.2 Solving for Transfer Functions

In this section we solve for the nine possible transfer functions for all combinations of degrees of freedom where force is applied and where displacements are taken. Solving for the transfer functions for greater than a 2dof system is a task not to be taken lightly – symbolic algebra programs such as Mathematica, Maple or the MATLAB Symbolic Toolbox should be used.

$$
\begin{array}{ccc}
\dfrac{z_1}{F_1} & \dfrac{z_1}{F_2} & \dfrac{z_1}{F_3} \\[2ex]
\dfrac{z_2}{F_1} & \dfrac{z_2}{F_2} & \dfrac{z_2}{F_3} \\[2ex]
\dfrac{z_3}{F_1} & \dfrac{z_3}{F_2} & \dfrac{z_3}{F_3}
\end{array}
$$

Table 2.3: Nine possible transfer functions for tdof system.

The results below were obtained by use of a symbolic algebra program.

$$
\frac{z_1}{F_1} = \left\{
\begin{aligned}
& s^4 (m_2 m_3) + s^3 (m_3 c_1 + m_3 c_2 + m_2 c_2) \\
& + s^2 (c_1 c_2 + m_2 k_2 + m_3 k_1 + m_3 k_2) + s(c_1 k_2 + c_2 k_1) + k_1 k_2
\end{aligned}
\right\} / \text{Den} \quad (2.28)
$$

$$
\frac{z_1}{F_2} = \left\{ s^3 (m_3 c_1) + s^2 (c_1 c_2 + m_3 k_1) + s(c_1 k_2 + k_1 c_2) + k_1 k_2 \right\} / \text{Den} \quad (2.29)
$$

$$
\frac{z_1}{F_3} = \left\{ s^2 (c_1 c_2) + s(c_1 k_2 + c_2 k_1) + k_1 k_2 \right\} / \text{Den} \quad (2.30)
$$

$$
\frac{z_2}{F_1} = \left\{ s^3 (m_3 c_1) + s^2 (c_1 c_2 + m_3 k_1) + s(c_1 k_2 + c_2 k_1) + k_1 k_2 \right\} / \text{Den} \quad (2.31)
$$

$$
\frac{z_2}{F_2} = \left\{
\begin{aligned}
& s^4 (m_1 m_3) + s^3 (m_1 c_2 + m_3 c_1) \\
& + s^2 (m_1 k_2 + c_1 c_2 + m_3 k_1) \\
& + s(c_1 k_2 + c_2 k_1) + k_1 k_2
\end{aligned}
\right\} / \text{Den} \quad (2.32)
$$

$$
\frac{z_2}{F_3} = \left\{ s^3 (m_1 c_2) + s^3 (m_1 k_2 + c_1 c_2) + s(c_1 k_2 + c_2 k_1) + k_1 k_2 \right\} / \text{Den} \quad (2.33)
$$

$$\frac{z_3}{F_1} = \left\{ s^2 \left(c_1 c_2 \right) + s \left(c_1 k_2 + c_2 k_1 \right) + k_1 k_2 \right\} / \text{Den} \tag{2.34}$$

$$\frac{z_3}{F_2} = \left\{ s^3 \left(m_1 c_2 \right) + s^2 \left(m_1 k_2 + c_1 c_2 \right) + s \left(c_1 k_2 + c_2 k_1 \right) + k_1 k_2 \right\} / \text{Den} \tag{2.35}$$

$$\frac{z_3}{F_3} = \left\{ \begin{array}{l} s^4 \left(m_1 m_2 \right) + s^3 \left(m_1 c_2 + m_1 c_1 + m_2 c_1 \right) \\ + s^2 \left(m_2 k_1 + m_1 k_1 + m_1 k_2 + c_1 c_2 \right) \\ + s \left(c_2 k_1 + c_1 k_2 \right) + \left(k_1 k_2 \right) \end{array} \right\} / \text{Den} \tag{2.36}$$

Where Den is:

$$\text{Den} = s^2 \left\{ \begin{array}{l} s^4 \left(m_1 m_2 m_3 \right) + s^3 \left(m_2 m_3 c_1 + m_1 m_3 c_1 + m_1 m_2 c_2 + m_1 m_3 c_2 \right) \\ + s^2 \left(m_1 m_3 k_1 + m_1 m_3 k_2 + m_1 m_2 k_2 + m_2 c_1 c_2 + m_3 c_1 c_2 + m_1 c_1 c_2 \right. \\ \left. + k_1 m_2 m_3 \right) \\ + s \left(m_3 c_1 k_2 + m_2 c_2 k_1 + m_1 c_2 k_1 + m_1 c_1 k_2 + m_3 c_2 k_1 + m_2 c_1 k_2 \right) \\ + \left(m_1 k_1 k_2 + m_2 k_1 k_2 + m_3 k_1 k_2 \right) \end{array} \right\} \tag{2.37}$$

Note that all the transfer functions have the same denominator, Den, called the **characteristic equation**.

To simplify the system for hand calculations, take:

$$\begin{array}{c} m_1 = m_2 = m_3 = m \\ c_1 = c_2 = c \\ k_1 = k_2 = k \end{array} \tag{2.38}$$

$$z_{11} = \frac{z_1}{F_1} = \left(m^2 s^4 + 3mcs^3 + \left(c^2 + 3mk \right) s^2 + 2cks + k^2 \right) / \text{Den1} \tag{2.39}$$

$$z_{12} = \frac{z_1}{F_2} = \left(mcs^3 + \left(c^2 + mk \right) s^2 + 2cks + k^2 \right) / \text{Den1} \tag{2.40}$$

$$z_{13} = \frac{z_1}{F_3} = \left(c^2 s^2 + 2cks + k^2 \right) / \text{Den1} \tag{2.41}$$

$$z_{21} = \frac{z_2}{F_1} = \left(mcs^3 + \left(c^2 + mk \right) s^2 + \left(2ck \right) s + k^2 \right) / Den1 \qquad (2.42)$$

$$z_{22} = \frac{z_2}{F_2} = \left(m^2 s^4 + 2mcs^3 + \left(2mk + c^2 \right) s^2 + 2cks + k^2 \right) / Den1 \quad (2.43)$$

$$z_{23} = \frac{z_2}{F_3} = \left(mcs^3 + \left(c^2 + mk \right) s^2 + 2cks + k^2 \right) / Den1 \qquad (2.44)$$

$$z_{31} = \frac{z_3}{F_1} = \left(c^2 s^2 + 2cks + k^2 \right) / Den1 \qquad (2.45)$$

$$z_{32} = \frac{z_3}{F_2} = \left(mcs^3 + \left(c^2 + mk \right) s^2 + 2cks + k^2 \right) / Den1 \qquad (2.46)$$

$$z_{33} = \frac{z_3}{F_3} = \left(m^2 s^4 + 3mcs^3 + \left(c^2 + 3mk \right) s^2 + 2cks + k^2 \right) / Den1 \quad (2.47)$$

Where:

$$Den1 = \left\{ m^3 s^4 + 4m^2 cs^3 + \left(4m^2 k + 3mc^2 \right) s^2 + 6mcks + 3mk^2 \right\} s^2 \qquad (2.48)$$

To enable hand calculations of roots, simplify another level by making damping equal to zero:

$$\frac{z_1}{F_1} = \left(m^2 s^4 + 3mks^2 + k^2 \right) / Den2 \qquad (2.49)$$

$$\frac{z_1}{F_2} = \left(mks^2 + k^2 \right) / Den2 \qquad (2.50)$$

$$\frac{z_1}{F_3} = k^2 / Den2 \qquad (2.51)$$

$$\frac{z_2}{F_1} = \left(mks^2 + k^2 \right) / Den2 \qquad (2.52)$$

$$\frac{z_2}{F_2} = \left(m^2s^4 + 2mks^2 + k^2\right)/Den2 \tag{2.53}$$

$$\frac{z_2}{F_3} = \left(mks^2 + k^2\right)/Den2 \tag{2.54}$$

$$\frac{z_3}{F_1} = k^2/Den2 \tag{2.55}$$

$$\frac{z_3}{F_2} = \left(mks^2 + k^2\right)/Den2 \tag{2.56}$$

$$\frac{z_3}{F_3} = \left(m^2s^4 + 3mks^2 + k^2\right)/Den2 \tag{2.57}$$

$$Den2 = s^2\left(m^3s^4 + 4m^2ks^2 + 3mk^2\right) \tag{2.58}$$

2.4.3 Transfer Function Matrix for Undamped Model

A more convenient method of arranging and keeping track of the various transfer functions is to use a matrix form for the transfer function, called the **transfer function matrix**:

$$\begin{bmatrix} z_{11} & z_{12} & z_{13} \\ z_{21} & z_{22} & z_{23} \\ z_{31} & z_{32} & z_{33} \end{bmatrix} \tag{2.59}$$

Where:

$$\begin{bmatrix} z_1 \\ z_2 \\ z_3 \end{bmatrix} = \begin{bmatrix} z_{11} & z_{12} & z_{13} \\ z_{21} & z_{22} & z_{23} \\ z_{31} & z_{32} & z_{33} \end{bmatrix} \begin{bmatrix} F_1 \\ F_2 \\ F_3 \end{bmatrix} \tag{2.60}$$

The transfer function matrix can then be written for the undamped case as follows, where each term of the numerator matrix is divided by the common denominator:

$$\begin{bmatrix} z_1 \\ z_2 \\ z_3 \end{bmatrix} =$$

$$\frac{\begin{bmatrix} (m^2s^4 + 3mks^2 + k^2) & (mks^2 + k^2) & k^2 \\ (mks^2 + k^2) & (m^2s^4 + 2mks^2 + k^2) & (mks^2 + k^2) \\ k^2 & (mks^2 + k^2) & (m^2s^4 + 3mks^2 + k^2) \end{bmatrix}}{s^2 \left(m^3s^4 + 4m^2ks^2 + 3mk^2 \right)} \begin{bmatrix} F_1 \\ F_2 \\ F_3 \end{bmatrix}$$

(2.61)

2.4.4 Four Distinct Transfer Functions

We will be dealing with only Single Input Single Output (SISO) systems until Chapter 19, when a Multi Input Multi Output (MIMO) system is examined. This means that we will be applying only a single force to the system at any time, F_1, F_2 or F_3, and will only be taking the displacement of a single degree of freedom, z_1, z_2 or z_3.

Because there are three inputs and three outputs, there are nine possible SISO transfer functions to investigate. However, because of the symmetry of the system ($z_{ij} = z_{ji}$) there are only four distinct transfer functions. Expanding the denominator into factors and simplifying:

$$\frac{z_1}{F_1} = \frac{m^2s^4 + 3mks^2 + k^2}{s^2 \left(m^3s^4 + 4m^2ks^2 + 3mk^2 \right)}$$

(2.62)

$$\frac{z_2}{F_1} = \frac{(mks^2 + k^2)}{s^2 \left(m^3s^4 + 4m^2ks^2 + 3mk^2 \right)}$$

$$= \frac{k(ms^2 + k)}{s^2 (ms^2 + k)(m^2s^2 + 3km)}$$

$$= \frac{k}{s^2 (m^2s^2 + 3km)} \quad \text{(note cancelling of pole/zero)}$$

(2.63)

$$\frac{z_3}{F_1} = \frac{k^2}{s^2 \left(m^3s^4 + 4m^2ks^2 + 3mk^2 \right)}$$

(2.64)

$$\frac{z_2}{F_2} = \frac{m^2s^4 + 2mks^2 + k^2}{s^2\left(m^3s^4 + 4m^2ks^2 + 3mk^2\right)} \tag{2.65}$$

2.4.5 Poles

The **poles, eigenvalues**, or **resonant frequencies**, are the roots of the characteristic equation. Poles show the frequencies where the system will amplify inputs, and are a basic characteristic of the system. The poles are not a function of which transfer function is used since all the transfer functions for a given system have the same characteristic equation, as shown by the common denominator of (2.61).

The poles for a system depend only on the distribution of mass, stiffness, and damping throughout the system, not on where the forces are applied or where displacements are measured.

Setting the characteristic equation equal to zero and solving for the roots (poles):

$$s^2\left(m^3s^4 + 4m^2ks^2 + 3mk^2\right) = 0 \tag{2.66}$$

$$s^2 = 0 \text{ is a double root at the origin } s_{1,2} = 0 \tag{2.67}$$

Now taking the term in parentheses and setting equal to zero:

$$\left(m^3\right)s^4 + \left(4m^2k\right)s^2 + \left(3mk^2\right) = 0 \tag{2.68}$$

Solving as a quadratic in s^2:

$$s^2 = \frac{-4m^2k \pm \left(16m^4k^2 - 12m^4k^2\right)^{\frac{1}{2}}}{2m^3}$$

$$= \frac{-4m^2k \pm \left(4m^4k^2\right)^{\frac{1}{2}}}{2m^3}$$

$$= \frac{-4m^2k \pm 2m^2k}{2m^3} = \frac{-2m^2k}{m^3}$$

$$= \frac{-2k}{2m}, \frac{-6k}{2m}$$

$$= \frac{-k}{m}, \frac{-3k}{m} \tag{2.69}$$

$$s_{3,4} = \pm j\sqrt{\frac{k}{m}} = \pm j1 \tag{2.70}$$

$$s_{5,6} = \pm j\sqrt{\frac{3k}{m}} = \pm j1.732 \tag{2.71}$$

Because there is no damping, the poles all fall on the s-plane imaginary axis.

2.4.6 Zeros

The **zeros** of each SISO transfer function are defined by the roots of its numerator. Zeros show the frequencies where the system will attenuate inputs. Unlike the poles, which are a characteristic of the system and are the same for every transfer function, zeros can be different for every transfer function and some transfer functions may have no zeros. Chapter 4 will discuss one physical interpretation of zeros, showing how to calculate the number of zeros for various transfer functions for a series-connected lumped mass system.

Calculate the z_1/F_1 zeros:

$$m^2 s^4 + 3mks^2 + k^2 = 0 \tag{2.72}$$

$$s^2 = \frac{-3mk \pm \left(9m^2 k^2 - 4m^2 k^2\right)^{\frac{1}{2}}}{2m^2}$$

$$= \frac{-3mk \pm \sqrt{5}mk}{2m^2} = \frac{-3k \pm \sqrt{5}k}{2m}$$

$$= \left(\frac{k}{m}\right)\left(\frac{-3 \pm \sqrt{5}}{2}\right) = \left(\frac{k}{m}\right)(-0.3820), \ \left(\frac{k}{m}\right)(-2.618) \tag{2.73}$$

Taking the square root of the two values above gives two pair of complex conjugate roots:

$$s_{1,2} = \pm j0.618\sqrt{\frac{k}{m}} = \pm j\,0.618 \qquad (2.74)$$

$$s_{3,4} = \pm j1.618\sqrt{\frac{k}{m}} = \pm j1.618 \qquad (2.75)$$

Calculate the z_2 / F_1 zeros:

$$mks^2 + k^2 = 0 \qquad (2.76)$$

$$s^2 = \frac{-k^2}{mk} = \frac{-k}{m} \qquad (2.77)$$

$$s_{1,2} = \pm j\sqrt{\frac{k}{m}} = \pm j \qquad (2.78)$$

Calculate the z_3 / F_1 zeros:

$$k^2 = 0 \quad \text{there are no zeros.} \qquad (2.79)$$

Calculate the z_2 / F_2 zeros:

$$m^2 s^4 + 2mks^2 + k^2 = 0 \qquad (2.80)$$

$$s^2 = \frac{-2mk \pm \left(4m^2k^2 - 4m^2k^2\right)}{2m^2}$$

$$= \frac{-2mk}{2m^2} = \frac{-k}{m} \pm 0 \qquad (2.81)$$

$$s_{1,2} = \pm j\sqrt{\frac{k}{m}} = \pm j \qquad (2.82)$$

$$s_{3,4} = \pm j \qquad (2.83)$$

As with the poles, since there is no damping in the system, all the zeros are also on the imaginary axis.

2.4.7 Summarizing Poles and Zeros, Matrix Format

$$\frac{\begin{bmatrix} (\pm 0.62, \pm 1.62) & \pm j & \text{none} \\ \pm j & (\pm j, \pm j) & \pm j \\ \text{none} & \pm j & (\pm 0.62, \pm 1.62) \end{bmatrix}}{(\pm 0 j)(\pm 1, \pm 1.732) j} \qquad (2.84)$$

The 3x3 matrix of zero values for the 3x3 transfer function matrix is in the numerator of (2.82) and the pole values are in the denominator.

2.5 MATLAB Code tdofpz3x3.m – Plot Poles and Zeros

2.5.1 Code Description

The program listing below uses the "num/den" form of the transfer function and calculates and plots all nine pole/zero combinations for the nine different transfer functions. It prompts for values of the two dampers, c1 and c2, where the default values (hitting the "enter" key) are set to zero to match the hand-calculated values in (2.82). The "transfer function" forms of the transfer functions are then converted to "zpk - zero/pole/gain" form to enable graphical construction of frequency response in the next chapter.

The values of the poles and zeros as well as the "zpk" forms of the transfer functions are listed in the MATLAB command window.

Note that in most MATLAB code, the critical definitions and calculations take only a few commands while plotting and annotating the plots take the bulk of the space.

2.5.2 Code Listing

```
%        tdofpz3x3.m        plotting poles/zeros of tdof model, all 9 plots

        clf;

        clear all;

%        using MATLAB's pzmap function with the "tf" form using num/den
%        to define the numerator and denominator terms of the different
%        transfer functionx

%        assign values for masses, damping, and stiffnesses

        m1 = 1;
        m2 = 1;
```

```
        m3 = 1;
        k1 = 1;
        k2 = 1;

%       prompt for c1 and c2 values, set to zero to match closed form solution

        c1 = input('enter value for damper c1, default is zero, ... ');

        if isempty(c1)
                c1 = 0;
        end

        c2 = input('enter value for damper c2, default is zero, ... ');

        if isempty(c2)
                c2 = 0;
        end

%       define row vectors of numerator and denominator coefficients

        den = [(m1*m2*m3) (m2*m3*c1 + m1*m3*c1 + m1*m2*c2 + m1*m3*c2) ...
                (m1*m3*k1 + m1*m3*k2 + m1*m2*k2 + m2*c1*c2 + m3*c1*c2 + ...
                m1*c1*c2 + k1*m2*m3) ...
                        (m3*c1*k2 + m2*c2*k1 + m1*c2*k1 + m1*c1*k2 + ...
                        m3*c2*k1 + m2*c1*k2) ...
                        (m1*k1*k2 + m2*k1*k2 + m3*k1*k2) 0 0];

        z11num = [(m2*m3) (m3*c1 + m3*c2 + m2*c2) (c1*c2 + m2*k2 +...
                        m3*k1 + m3*k2)  (c1*k2 + c2*k1) (k1*k2)];

        z21num = [(m3*c1) (c1*c2 + m3*k1) (c1*k2 + c2*k1) (k1*k2)];

        z31num = [(c1*c2) (c1*k2 + c2*k1) (k1*k2)];

        z22num = [(m1*m3) (m1*c2 + m3*c1) (m1*k2 + c1*c2 + m3*k1) ...
                        (c1*k2 + c2*k1) (k1*k2)];

%       use the "tf" function to convert to define "transfer function" systems

        sysz11 = tf(z11num,den)

        sysz21 = tf(z21num,den)

        sysz31 = tf(z31num,den)

        sysz22 = tf(z22num,den)

%       use the "zpk" function to convert from transfer function to zero/pole/gain form

        zpkz11 = zpk(sysz11)

        zpkz21 = zpk(sysz21)

        zpkz31 = zpk(sysz31)
```

```
                zpkz22 = zpk(sysz22)

%               use the "pzmap" function to map the poles and zeros of each transfer function

                [p11,z11] = pzmap(sysz11);

                [p21,z21] = pzmap(sysz21);

                [p31,z31] = pzmap(sysz31);

                [p22,z22] = pzmap(sysz22);

                p11

                z11

                z21

                z31

                z22

%               plot z11 for later use

                subplot(1,1,1)
                plot(real(p11),imag(p11),'k*')
                hold on
                plot(real(z11),imag(z11),'ko')
                title('Poles and Zeros of z11')
                ylabel('Imag')
                axis([-2 2 -2 2])
                axis('square')
                grid
                hold off

                disp('execution paused to display figure, "enter" to continue'); pause

%               plot all 9 plots on a 3x3 grid

                subplot(3,3,1)
                plot(real(p11),imag(p11),'k*')
                hold on
                plot(real(z11),imag(z11),'ko')
                title('Poles and Zeros of z11')
                ylabel('Imag')
                axis([-2 2 -2 2])
                axis('square')
                grid
                hold off

                subplot(3,3,2)
                plot(real(p21),imag(p21),'k*')
                hold on
                plot(real(z21),imag(z21),'ko')
                title('Poles and Zeros of z12')
```

```
ylabel('Imag')
axis([-2 2 -2 2])
axis('square')
grid
hold off

subplot(3,3,3)
plot(real(p31),imag(p31),'k*')
hold on
plot(real(z31),imag(z31),'ko')
title('Poles and Zeros of z13')
ylabel('Imag')
axis([-2 2 -2 2])
axis('square')
grid
hold off

subplot(3,3,4)
plot(real(p21),imag(p21),'k*')
hold on
plot(real(z21),imag(z21),'ko')
title('Poles and Zeros of z21')
ylabel('Imag')
axis([-2 2 -2 2])
axis('square')
grid
hold off

subplot(3,3,5)
plot(real(p22),imag(p22),'k*')
hold on
plot(real(z22),imag(z22),'ko')
title('Poles and Zeros of z22')
ylabel('Imag')
axis([-2 2 -2 2])
axis('square')
grid
hold off

subplot(3,3,6)
plot(real(p21),imag(p21),'k*')
hold on
plot(real(z21),imag(z21),'ko')
title('Poles and Zeros of z23')
ylabel('Imag')
axis([-2 2 -2 2])
axis('square')
grid
hold off

subplot(3,3,7)
plot(real(p31),imag(p31),'k*')
hold on
plot(real(z31),imag(z31),'ko')
title('Poles and Zeros of z31')
```

```
xlabel('Real')
ylabel('Imag')
axis([-2 2 -2 2])
axis('square')
grid
hold off

subplot(3,3,8)
plot(real(p21),imag(p21),'k*')
hold on
plot(real(z21),imag(z21),'ko')
title('Poles and Zeros of z32')
xlabel('Real')
ylabel('Imag')
axis([-2 2 -2 2])
axis('square')
grid
hold off

subplot(3,3,9)
plot(real(p11),imag(p11),'k*')
hold on
plot(real(z11),imag(z11),'ko')
title('Poles and Zeros of z33')
xlabel('Real')
ylabel('Imag')
axis([-2 2 -2 2])
axis('square')
grid
hold off

disp('execution paused to display figure, "enter" to continue'); pause

%       check for real axis values to set plot scale

z11_realmax = max(abs(real(z11)));
z21_realmax = max(abs(real(z21)));
z31_realmax = max(abs(real(z31)));
z22_realmax = max(abs(real(z22)));

maxplot = max([z11_realmax z21_realmax z31_realmax z22_realmax]);

if  maxplot > 2

        maxplot = ceil(maxplot);

else

        maxplot = 2.0;

end

z11_realmax = max(abs(real(z11)));
subplot(1,1,1)
```

```
plot(real(p11),imag(p11),'k*')
hold on
plot(real(z11),imag(z11),'ko')
title('Poles and Zeros of z11, z33')
ylabel('Imag')
axis([-maxplot maxplot -maxplot maxplot])
axis('square')
grid
hold off

disp('execution paused to display figure, "enter" to continue'); pause

plot(real(p21),imag(p21),'k*')
hold on
plot(real(z21),imag(z21),'ko')
title('Poles and Zeros of z21, z12, z23, z32')
ylabel('Imag')
axis([-maxplot maxplot -maxplot maxplot])
axis('square')
grid
hold off

disp('execution paused to display figure, "enter" to continue'); pause

plot(real(p31),imag(p31),'k*')
hold on
plot(real(z31),imag(z31),'ko')
title('Poles and Zeros of z31, z13')
xlabel('Real')
ylabel('Imag')
axis([-maxplot maxplot -maxplot maxplot])
axis('square')
grid
hold off

disp('execution paused to display figure, "enter" to continue'); pause

plot(real(p22),imag(p22),'k*')
hold on
plot(real(z22),imag(z22),'ko')
title('Poles and Zeros of z22')
ylabel('Imag')
axis([-maxplot maxplot -maxplot maxplot])
axis('square')
grid
hold off
```

2.5.3 Code Output – Pole/Zero Plots in Complex Plane

2.5.3.1 Undamped Model – Pole/Zero Plots

The pole/zero plot and pole/zero calculated values for $c1 = c2 = 0$ are shown below. Poles are plotted as asterisks and zeros as circles.

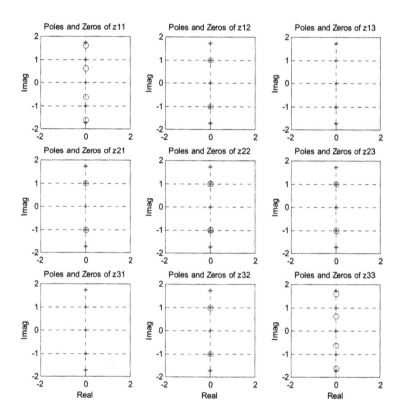

Figure 2.8: Pole/zero plots for nine transfer functions. Poles are indicated by asterisks and zeros by circles.

The first thing to notice about the pole/zero plots is that they all have the same poles. The rigid body mode (resonant frequency = 0 hz) is evident by the pair of zeros at the origin, $\pm 0j$. The zeros of each particular transfer function are seen to be dependent upon which transfer function is taken. Note that with zero damping, all the poles and zeros are on the imaginary axis, indicating that the real portions of their complex values are zero and that there is no damping.

In the next chapter we will discuss frequency responses of transfer functions and will link the pole/zero locations in the complex plane to amplification/attenuation regions of the frequency response plots.

The poles and zeros from the MATLAB output are listed below:

```
poles =

    0
    0
    0 + 1.7321i
    0 - 1.7321i
    0 + 1.0000i
    0 - 1.0000i

zeros_z11 =

    0 + 1.6180i
    0 - 1.6180i
    0 + 0.6180i
    0 - 0.6180i

zeros_z21 =

    0 + 1.0000i
    0 - 1.0000i

zeros_z31 =

  Empty matrix: 0-by-1

zeros_z22 =

  -0.0000 + 1.0000i
  -0.0000 - 1.0000i
   0.0000 + 1.0000i
   0.0000 - 1.0000i
```

Table 2.3: Poles and zeros of tdof transfer functions, undamped.

Repeating the matrix listing of pole/zero locations from previous analysis:

$$\frac{\begin{bmatrix} (\pm 0.62, \pm 1.62) & \pm j & none \\ \pm j & (\pm j, \pm j) & \pm j \\ none & \pm j & (\pm 0.62, \pm 1.62) \end{bmatrix}}{(\pm 0 j)(\pm 1, \pm 1.732)j} \qquad (2.85)$$

Note that MATLAB calculates an "Empty matrix 0 by 1" for the zeros of z31, which matches our calculations which show "none." Also note that several of the plots, z12, z21, z22, z23 and z32, have zeros and poles overlaying each other, where the pole cancels the effect of the zero. We will discuss this cancellation further in the next chapter.

2.5.3.2 Damped Model – Pole/Zero Plots

If damping is not set to zero for c1 and/or c2, the poles (with the exception of the two poles at the origin) and zeros will move from the imaginary axis to the left hand side of the complex plane, with the real parts of the poles and zeros having negative values. The pole/zero plot and MATLAB output listing below are for values of c1 = c2 = 0.1, arbitrarily chosen to illustrate the "damped" case.

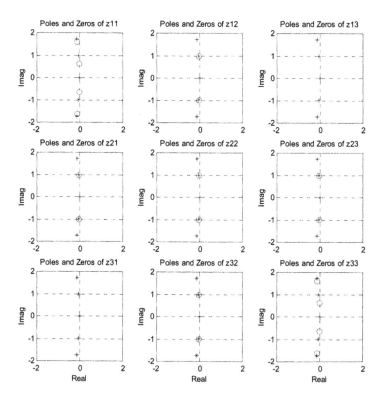

Figure 2.9: Pole/zero plots for nine transfer functions for c1 = c2 = 0.1. Poles are indicated by asterisks and zeros by circles. Negative real axis zeros not shown because of plot scaling.

The limited scale for the nine plots above do not show the real axis zeros, see the figures below for the entire plot. The only poles/zeros that are on the imaginary axis are the two poles at zero, the rigid body mode – which will be described in detail in Chapter 3.

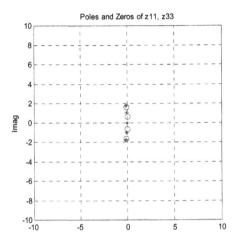

Figure 2.10: Expanded scale pole/zero plots for z11, z33 transfer functions – no real axis zeros.

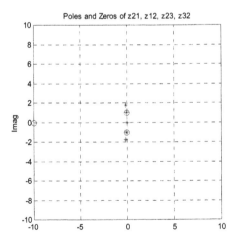

Figure 2.11: Expanded scale pole/zero plots for z21, z12, z23 and z32 transfer functions – one real axis zero at -10.

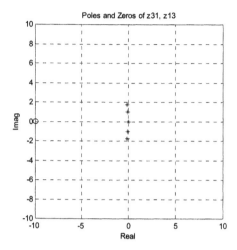

Figure 2.12: Expanded scale pole/zero plots for z31 and z13 transfer functions – two real axis zeros at -10.

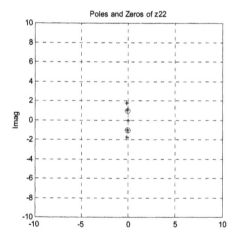

Figure 2.13: Expanded scale pole/zero plots for z31 and z13 transfer functions – no real axis zeros.

The MATLAB calculated values for the poles and zeros for the damped case are below:

p11 =

   0
   0
 -0.1500 + 1.7255i
 -0.1500 - 1.7255i
 -0.0500 + 0.9987i
 -0.0500 - 0.9987i

z11 =

 -0.1309 + 1.6127i
 -0.1309 - 1.6127i
 -0.0191 + 0.6177i
 -0.0191 - 0.6177i

z21 =

-10.0000
 -0.0500 + 0.9987i
 -0.0500 - 0.9987i

z31 =

-10.0000 + 0.0000i
-10.0000 - 0.0000i

z22 =

 -0.0500 + 0.9987i
 -0.0500 - 0.9987i
 -0.0500 + 0.9987i
-0.0500 - 0.9987i

Table 2.4: Poles and zeros of tdof transfer functions, damped.

Several observations can be made about the poles and zeros above. First, all of the poles with the exception of the two rigid body poles p11 = 0 are to the left of the imaginary axis, indicating that the system now has damping. Note that there are several new zeros. The z21 transfer function now has a real zero at −10.0 in addition to the two complex zeros. The z31 transfer function has two zeros now at −10, whereas for the no damping case it had no zeros. These extra zeros do not show up on Figure 2.9 because of plot axis scaling but with the real axis expanded in Figures 2.10 to 2.13 they appear. The reason for these "additional" zeros can be seen if we look at the z21 and z31 transfer functions, repeated from (2.31) and (2.34):

$$\frac{Z_2}{F_1} = \left\{ s^3 \left(m_3 c_1 \right) + s^2 \left(c_1 c_2 + m_3 k_1 \right) + s \left(c_1 k_2 + c_2 k_1 \right) + k_1 k_2 \right\} / \text{Den} \qquad (2.86)$$

$$\frac{Z_3}{F_1} = \left\{ s^2 \left(c_1 c_2 \right) + s \left(c_1 k_2 + c_2 k_1 \right) + k_1 k_2 \right\} / \text{Den} \qquad (2.87)$$

With values for c1 and c2 not equal to zero, the z21 transfer function is third degree, meaning that it should have three roots. With damping equal to zero, only two complex zeros are calculated by MATLAB and by hand. The third root is located at $-\infty$. As damping values for c1 and c2 are increased the root at $-\infty$ moves to the right, towards the origin.

The z31 transfer function has no zeros with zero damping, but is second degree and with infinitely small damping values has two roots at $-\infty$. As the values of c1 and c2 increase, the two zeros at $-\infty$ start moving toward the origin.

2.5.3.3 Root Locus, tdofpz3x3_rlocus.m

In the last two sections we have discussed pole/zero plots for undamped and damped models. For the damped model we chose values of 0.1 for c1 and c2. It would be nice to have a systematic method to display poles and zeros for a range of damping values. There is a MATLAB Control Toolbox function "rlocus" which plots the root locus for an open-loop SISO system. We could use this function if the damping values could be broken out of the system and be treated as a feedback gain. Unfortunately for our tdof system this is not possible, but we can still plot a locus by using a for-loop.

The code listed below, **tdofpz3x3_rlocus.m**, is taken from the initial section of **tdofpz3x3.m**. A for-loop cycles through a vector of damping values, calculating and plotting the poles and zeroes for each damping value.

```
          echo off
%         tdofpz3x3_rlocus.m        plotting locus of poles/zeros of z11 for tdof
%         model for range of damping values.

          clf;

          clear all;

%         assign values for masses, damping, and stiffnesses

          m1 = 1;
          m2 = 1;
          m3 = 1;
```

```
        k1 = 1;
        k2 = 1;

%       define vector of damping values for c1 and c2

        cvec = [0 .2 .4 .6 .8 1.0 1.1 1.05 1.1 1.15 1.16];

        for cnt = 1:length(cvec)

        c1 = cvec(cnt);

        c2 = cvec(cnt);

%       define row vectors of numerator and denominator coefficients

        den = [(m1*m2*m3) (m2*m3*c1 + m1*m3*c1 + m1*m2*c2 + m1*m3*c2) ...
           (m1*m3*k1 + m1*m3*k2 + m1*m2*k2 + m2*c1*c2 + m3*c1*c2 + ...
           m1*c1*c2 + k1*m2*m3) ...
                   (m3*c1*k2 + m2*c2*k1 + m1*c2*k1 + m1*c1*k2 + ...
                   m3*c2*k1 + m2*c1*k2) ...
                   (m1*k1*k2 + m2*k1*k2 + m3*k1*k2) 0 0];

        z11num = [(m2*m3) (m3*c1 + m3*c2 + m2*c2) ...
                (c1*c2 + m2*k2 + m3*k1 + m3*k2) .(c1*k2 + c2*k1) (k1*k2)];

        z21num = [(m3*c1) (c1*c2 + m3*k1) (c1*k2 + c2*k1) (k1*k2)];

        z31num = [(c1*c2) (c1*k2 + c2*k1) (k1*k2)];

        z22num = [(m1*m3) (m1*c2 + m3*c1) (m1*k2 + c1*c2 + m3*k1) ...
                        (c1*k2 + c2*k1) (k1*k2)];

%       use the "tf" function to convert to define "transfer function" systems

        sysz11 = tf(z11num,den);

        sysz21 = tf(z21num,den);

        sysz31 = tf(z31num,den);

        sysz22 = tf(z22num,den);

%       use the "pzmap" function to map the poles and zeros of each transfer function

        [p11,z11] = pzmap(sysz11);

        [p21,z21] = pzmap(sysz21);

        [p31,z31] = pzmap(sysz31);

        [p22,z22] = pzmap(sysz22);

%       plot poles and zeros of z11

        subplot(1,1,1)
```

```
plot(real(p11),imag(p11),'k*')
hold on
plot(real(z11),imag(z11),'ko')
title('Poles and Zeros of z11 for range of damping values c1 and c2')
xlabel('Real')
ylabel('Imag')
axis([-3 1 -2 2])
axis('square')
grid on

end

hold off
```

The root locus plot below is for the following values of damping:

$$cvec = [0\ .2\ .4\ .6\ .8\ 1.0\ 1.1\ 1.05\ 1.1\ 1.15\ 1.16];$$

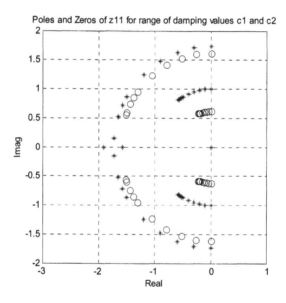

Figure 2.14: Pole zero plot for z11 transfer function.

The plot starts out with damping values of zero for $c1$ and $c2$. The poles and zeros for zero damping are located on the imaginary axis. The poles are located at 0, 0, $\pm 1j$, $\pm 1.732j$. The zeros are located at $\pm 0.62j$ and $\pm 1.62j$.

As damping is increased from zero, the poles and zeros (except the two poles at the origin) start moving to the left, away from the imaginary axis. The poles and zeros move at different rates as damping is increased. The poles at $\pm 1j$

and zeros at $\pm 0.62j$ move to the left less than the poles at $\pm 1.732j$ and the zeros at $\pm 1.62j$. In fact, the two poles at $\pm 1.732j$ move so much that at damping values of 1.16 the poles intercept the real axis and split. One moves to the left and the other to the right along the real axis.

Plotting pole and zero locations as a function of system parameters was introduced in 1949 (Evans 1949), as the Evans root locus technique. The hand plotting originally used has been largely replaced with computer plotting techniques as shown above or by using the "rlocus" function. However, because the ability to hand sketch root loci is such a powerful tool, it is still taught in beginning control theory courses (Franklin 1994).

2.5.3.4 Undamped and Damped Model – tf and zpk Forms

This section is included to start familiarizing the reader with the various forms of transfer functions available with MATLAB and to prepare for issues in the next chapter.

Table 2.6 shows the transfer function form of the four distinct transfer functions for the tdof model for the undamped ($c1 = c2 = 0$) and damped ($c1 = c2 = 0.1$) cases run earlier. The numerator and denominator are both arranged in polynomial form. Table 2.7 shows the zpk form, where the numerator and denominator are both arranged as products of the zeros and poles with a gain term multiplying the numerator.

Note that the denominators of all the undamped transfer functions are the same, as are the denominators of all the damped transfer functions. However, the numerators are all different because of the different number of poles and zeros for each transfer function. For instance the z31 undamped transfer function has no zeros, only a gain term of 1.0, while the z11 undamped transfer function has two sets of complex zeros.

In going from the undamped to damped case, we showed that extra zeros appeared in the z21 and z31 transfer functions. It is easier to see where the extra zeros originate using the zpk form than using the tf form. Comparing the undamped and damped numerators of the z31 zpk transfer function form shows the extra $(s+10)^2$ term, from which the two real axis zeros arise. We will use the zpk form of the transfer functions in the next chapter to calculate frequency response at a specific frequency.

z11 Undamped Transfer function:

$$\frac{s^4 + 3 s^2 + 1}{s^6 + 4 s^4 + 3 s^2}$$

z21 Undamped Transfer function:

$$\frac{s^2 + 1}{s^6 + 4 s^4 + 3 s^2}$$

z31 Undamped Transfer function:

$$\frac{1}{s^6 + 4 s^4 + 3 s^2}$$

z22 Undamped Transfer function:

$$\frac{s^4 + 2 s^2 + 1}{s^6 + 4 s^4 + 3 s^2}$$

z11 Damped Transfer function:

$$\frac{s^4 + 0.3 s^3 + 3.01 s^2 + 0.2 s + 1}{s^6 + 0.4 s^5 + 4.03 s^4 + 0.6 s^3 + 3 s^2}$$

z21 Damped Transfer function:

$$\frac{0.1 s^3 + 1.01 s^2 + 0.2 s + 1}{s^6 + 0.4 s^5 + 4.03 s^4 + 0.6 s^3 + 3 s^2}$$

z31 Damped Transfer function:

$$\frac{0.01 s^2 + 0.2 s + 1}{s^6 + 0.4 s^5 + 4.03 s^4 + 0.6 s^3 + 3 s^2}$$

z22 Damped Transfer function:

$$\frac{s^4 + 0.2 s^3 + 2.01 s^2 + 0.2 s + 1}{s^6 + 0.4 s^5 + 4.03 s^4 + 0.6 s^3 + 3 s^2}$$

Table 2.5: Transfer function (tf) form of undamped and damped tdof transfer functions.

z11 Undamped Zero/pole/gain:

$$\frac{(s^2 + 0.382) (s^2 + 2.618)}{s^2 (s^2 + 1) (s^2 + 3)}$$

z21 Undamped Zero/pole/gain:

$$\frac{(s^2 + 1)}{s^2 (s^2 + 1) (s^2 + 3)}$$

z31 Undamped Zero/pole/gain:

$$\frac{1}{s^2 (s^2 + 1) (s^2 + 3)}$$

z22 Undamped Zero/pole/gain:

$$\frac{(s^2 + 1)^2}{s^2 (s^2 + 1) (s^2 + 3)}$$

z11 Damped Zero/pole/gain:

$$\frac{(s^2 + 0.0382s + 0.382) (s^2 + 0.2618s + 2.618)}{s^2 (s^2 + 0.1s + 1) (s^2 + 0.3s + 3)}$$

z21 Damped Zero/pole/gain:

$$\frac{0.1 (s+10) (s^2 + 0.1s + 1)}{s^2 (s^2 + 0.1s + 1) (s^2 + 0.3s + 3)}$$

z31 Damped Zero/pole/gain:

$$\frac{0.01 (s+10)^2}{s^2 (s^2 + 0.1s + 1) (s^2 + 0.3s + 3)}$$

z22 Damped Zero/pole/gain:

$$\frac{(s^2 + 0.1s + 1)^2}{s^2 (s^2 + 0.1s + 1) (s^2 + 0.3s + 3)}$$

Table 2.6: Zero/Pole/Gain (zpk) for undamped and damped tdof transfer functions.

Problems

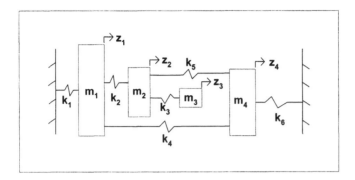

Figure P2.1: four dof system.

P2.1 Derive the global stiffness and mass matrices for the four dof system in Figure P2.1.

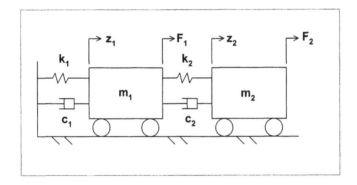

Figure P2.2: two dof problem.

P2.2 Derive the equations of motion in matrix form for the two dof model in Figure P2.2. Check for signs of diagonal terms and symmetry of off-diagonal terms.

P2.3 Solve for the four transfer functions for the two dof problem and define the 2x2 transfer function matrix. Are the denominators of all four transfer functions the same? How many unique transfer functions are there for this problem?

P2.4 Set $m_1 = m_2 = m = 1$, $k_1 = k_2 = k = 1$ and $c_1 = c_2 = 0$ and solve for the eigenvalues for the system. Solve for the zeros of the system and use the form shown in (2.84) to summarize the poles and zeros. Hand sketch the poles and zeros in the s-plane.

P2.5   (MATLAB)   Set $m_1 = m_2 = m = 1$, $k_1 = k_2 = k = 1$.   Modify the **tdofpz3x3.m** file to plot the poles and zeros of the undamped two dof system. Identify the poles and zeros in the MATLAB output listing and compare with the hand-calculated values.

P2.6 (MATLAB) Set $m_1 = m_2 = m = 1$, $k_1 = k_2 = k = 1$, add damping values of $c_1 = c_2 = 0.1$ and plot the poles and zeros in the s-plane. List the poles and zeros from MATLAB and correlate the listed values with the plots. Are there any real axis zeros? How do the real axis zero(s) change with different values of c_1 and c_2 , where $c_1 = c_2$.

CHAPTER 3

FREQUENCY RESPONSE ANALYSIS

3.1 Introduction

In Chapter 2 we calculated the transfer functions and identified the poles and zeros for the undamped system, which are repeated as (3.1) and (3.2) below, respectively. The next step in understanding the system is to plot the frequency domain behavior of each transfer function. Frequency domain behavior means identifying the magnitude and phase characteristics of each transfer function, showing how they change as the frequency of the forcing function is varied over a frequency range. Each transfer function is evaluated in the frequency domain by evaluating it at $s = j\omega$, where ω is the frequency of the forcing function, radians/sec.

$$\begin{bmatrix} z_1 \\ z_2 \\ z_3 \end{bmatrix} = $$

$$\frac{\begin{bmatrix} (m^2s^4 + 3mks^2 + k^2) & (mks^2 + k^2) & k^2 \\ (mks^2 + k^2) & (m^2s^4 + 2mks^2 + k^2) & (mks^2 + k^2) \\ k^2 & (mks^2 + k^2) & (m^2s^4 + 3mks^2 + k^2) \end{bmatrix} \begin{bmatrix} F_1 \\ F_2 \\ F_3 \end{bmatrix}}{s^2 \left(m^3s^4 + 4m^2ks^2 + 3mk^2 \right)}$$

$$(3.1)$$

$$\frac{\begin{bmatrix} (\pm 0.62, \pm 1.62) & \pm j & \text{none} \\ \pm j & (\pm j, \pm j) & \pm j \\ \text{none} & \pm j & (\pm 0.62, \pm 1.62) \end{bmatrix}}{(\pm 0j)(\pm 1, \pm 1.732)j}$$

$$(3.2)$$

Instead of going directly into MATLAB to calculate and plot the frequency responses, we will first sketch them by hand, using information about the low and high frequency asymptotes and the locations of the poles and zeros. We will discuss how to find the gain and phase of a transfer function at a given frequency graphically using the locations of the poles and zeros in the complex plane and then use MATLAB to plot. Finally, mode shapes are defined, then calculated using transfer function information and plotted.

3.2 Low and High Frequency Asymptotic Behavior

It is always good to check either a system's rigid body or spring-like low frequency nature by hand. For this tdof system at very low frequencies there are no spring connections to ground so the system moves as a rigid body, no matter where the force is applied, to F_1, F_2, or F_3.

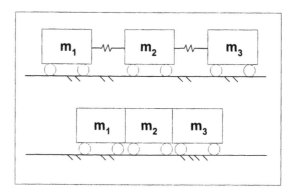

Figure 3.1: Rigid body mode of vibration.

The rigid body equation of motion (where z is the motion of all three masses together) is:

$$(3m)\ddot{z} = F$$

$$\frac{z}{F} = \frac{1}{3ms^2}$$

$$(3.3a,b)$$

Now we can solve for the frequency domain behavior of the system by substituting $j\omega$ for s.

At a radian frequency of 0.1 rad/sec, a frequency taken to be an order of magnitude less than the lowest resonant frequency of 1 rad/sec, the transfer function is:

$$\frac{z}{F} = \frac{1}{3m(j\omega)^2} = \frac{1}{3m[j(0.1)]^2}$$

$$= \frac{-1}{3m(.01)} = \frac{-100}{3m} = \frac{-33.3}{m} = -33.3 \qquad (3.4)$$

Converting from vector (real/imaginary) form to magnitude/phase (polar) form and using the definition of db as follows:

$$db = 20 * \log_{10}(z/F) \tag{3.5}$$

$$\left|\frac{z}{F}\right| = 33.3, \text{ or } 30.45\,db$$

$$\angle\frac{z}{F} = -180° \tag{3.6a,b}$$

These results show that at a frequency of 0.1 rad/sec, the magnitude of the motion of the masses is 33.3*F and the motion is $-180°$ out of phase with the force input.

We will now look at each individual transfer function, checking asymptotic behavior at both low and high frequencies. To do this, the four transfer functions are divided by the mass terms to give coefficients that are proportional to $\omega_n^2 = k/m$:

Starting with the z_1/F_1 transfer function:

$$\frac{z_1}{F_1} = \frac{m^2 s^4 + 3mks^2 + k^2}{s^2(m^3 s^4 + 4m^2 ks^2 + 3mk^2)} \tag{3.7}$$

Dividing numerator and denominator by m^3 allows redefining the equation in terms of ω_n:

$$\left.\frac{z_1}{F_1}\right| = \frac{\dfrac{1}{m}\left(s^4 + \dfrac{3ks^2}{m} + \dfrac{k^2}{m^2}\right)}{s^2\left(s^4 + \dfrac{4ks^2}{m} + \dfrac{3k^2}{m^2}\right)} = \frac{s^4 + 3\omega_n^2 s^2 + \omega_n^4}{ms^2\left(s^4 + 4\omega_n^2 s^2 + 3\omega_n^4\right)} \tag{3.8}$$

$$\div m^3$$

Substituting $s = j\omega$ and looking at low and high frequency behaviors:

$$\left.\frac{z_1}{F_1}\right|_{\omega << \omega_n} = \frac{\left(\omega^4 + 3\omega_n^2\left(-\omega^2\right) + \omega_n^4\right)}{m\left(-\omega^2\right)\left(\omega^4 - 4\omega_n^2\omega^2 + 3\omega_n^4\right)} = \frac{\omega_n^4}{-m\omega^2\left(3\omega_n^4\right)} = \frac{-1}{3m\omega^2} \qquad (3.9)$$

At low frequencies, the rigid body motion of z_1 is falling off at a $\left(-1/\omega^2\right)$ rate, and with a gain of $(1/3m)$. A rate of $\left(-1/\omega^2\right)$ means that every decade of frequency shift, the amplitude drops by a factor of 100. Since a factor of 100 is $-40db$, we should see the low frequency amplitude change 40db/decade.

$$\left.\frac{z_1}{F_1}\right|_{\omega >> \omega_n} = \frac{-\omega^4}{m\omega^2\left(\omega^4\right)} = \frac{-1}{m\omega^2} \qquad (3.10)$$

At high frequencies, the rigid body motion of z_1 is again falling at a $\left(-1/\omega^2\right)$ rate, but the gain is only $(1/m)$ instead of $(1/3m)$. This is because at high frequencies z_1 moves more as a result of F_1; the other two masses do not want to move, as will be seen from the high frequency asymptotes of the z_2/F_1 and z_3/F_1 transfer functions.

Checking $\dfrac{z_2}{F_1}$:

$$\left.\frac{z_2}{F_1}\right| = \frac{\left(\dfrac{mks^2}{m^3} + \dfrac{k^2}{m^3}\right)}{s^2\left(s^4 + \dfrac{4ks^2}{m} + \dfrac{3k^2}{m^2}\right)} = \frac{\left(\omega_n^2 s^2 + \omega_n^4\right)}{s^2 m\left(s^4 + 4\omega_n^2 s^2 + 3\omega_n^4\right)} \qquad (3.11)$$

$\div m^3$

$$\left.\frac{z_2}{F_1}\right|_{\omega << \omega_n} = \frac{-\omega_n^2\omega^2 + \omega_n^4}{-m\omega^2\left(\omega^4 - 4\omega_n^2\omega^2 + 3\omega_n^4\right)}$$

$$= \frac{\omega_n^4}{-m\omega^2 \left(3\omega_n^4\right)} = \frac{-1}{3m\omega^2} \tag{3.12}$$

$$\left. \frac{z_2}{F_1} \right| = \frac{-\omega^2 \omega_n^2}{-m\omega^2 \left(\omega_n^4 - 4\omega_n^2\omega^2 + 3\omega_n^4\right)} = \frac{-\omega^2}{-m\omega^2}\left(\frac{\omega_n^2}{\omega^4}\right) = \frac{\omega_n^2}{m\omega^4} = \frac{k}{\omega^4 m^2} \tag{3.13}$$

$$\omega \gg \omega_n$$

At low frequencies, z_2 looks exactly like z_1. But at high frequencies, z_2 is dropping off at a $(1/\omega^4)$ rate, or 80db/decade, with a gain of (k/m^2).

Checking $\frac{z_3}{F_1}$ now:

$$\left. \frac{z_3}{F_1} \right| = \frac{\dfrac{k^2}{m^3}}{s^2\left(s^4 + \dfrac{4ks^2}{m} + \dfrac{3k^2}{m^2}\right)} = \frac{\omega_n^4}{ms^2\left(s^4 + 4\omega_n^2 s^2 + 3\omega_n^4\right)} \tag{3.14}$$

$$\div m^3$$

$$\left. \frac{z_3}{F_1} \right| = \frac{\omega_n^4}{-m\omega^2\left(\omega^4 - 4\omega_n^2\omega^2 + 3\omega_n^4\right)} = \frac{\omega_n^4}{-m\omega^2\left(3\omega_n^4\right)} = \frac{-1}{3m\omega^2} \tag{3.15}$$

$$\omega \ll \omega_n$$

$$\left. \frac{z_3}{F_1} \right| = \frac{\omega_n^4}{-m\omega^2\left(\omega^4\right)} = \frac{\omega_n^4}{-m\omega^6} = \left(\frac{-k^2}{m^3}\right)\frac{1}{\omega^6} \tag{3.16}$$

$$\omega \gg \omega_n$$

At low frequencies, z_3 looks exactly like z_1 and z_2, but at high frequencies z_3 is dropping at a $(1/\omega^6)$ rate, or 120db/decade, with a gain of $(-k^2/m^3)$.

Checking $\frac{z_2}{F_2}$:

$$\left. \frac{z_2}{F_2} \right| = \frac{\left(\dfrac{m^2 s^4}{m^3} + \dfrac{2mks^2}{m^3} + \dfrac{k^2}{m^3} \right)}{s^2 \left(s^4 + \dfrac{4m^2 ks^2}{m^3} + \dfrac{3mk^2}{m^3} \right)} = \frac{\left(s^4 + 2\omega_n^2 s^2 + \omega_n^4 \right)}{ms^2 \left(s^4 + 4\omega_n^2 s^2 + 3\omega_n^4 \right)} \qquad (3.17)$$

$$\div m^3$$

$$\left. \frac{z_2}{F_2} \right| = \frac{\left(\omega^4 + 2\omega_n^2 \left(-\omega^2 \right) + \omega_n^4 \right)}{-m\omega^2 \left(\omega^4 - 4\omega_n^2 \omega^2 + 3\omega_n^4 \right)} = \frac{\omega_n^4}{-m\omega^2 \left(3\omega_n^4 \right)} = \frac{-1}{3m\omega^2} \qquad (3.18)$$

$$\omega \ll \omega_n$$

$$\left. \frac{z_2}{F_2} \right| = \frac{\omega^4}{-m\omega^2 \left(\omega^4 \right)} = \frac{-1}{m\omega^2} \qquad (3.19)$$

$$\omega \gg \omega_n$$

At low frequencies, z_2 / F_2 looks exactly like z_1 / F_1, z_2 / F_1, and z_3 / F_1. But at high frequencies z_2 / F_2 is dropping at a $(-1/\omega^2)$ rate and has a higher gain of $(1/m)$ instead of $(1/3m)$. Thus, the low and high frequency asymptotes look exactly like z_1 / F_1.

Summarizing the low and high frequency asymptotes, and solving for the gains and phases at $\omega = 0.1$ rad/sec and $\omega = 10$ rad/sec.

$$\left. \frac{z_1}{F_1} \right| = \frac{-1}{3m\omega^2} = \frac{-1}{3m(0.1)^2} = \frac{-1}{3(.01)} = \frac{-100}{3} = -33 = 30.46\,\text{db}, \ 180° \qquad (3.20)$$

$$\omega = 0.1 \frac{\text{rad}}{\text{sec}}$$

$$\left. \frac{z_1}{F_1} \right| = \frac{-1}{m\omega^2} = \frac{-1}{(10)^2} = \frac{-1}{100} = -.01 = -40\,\text{db}, \ 180° \qquad (3.21)$$

$$\omega = 10 \frac{\text{rad}}{\text{sec}}$$

$$\left.\frac{z_2}{F_1}\right|_{\omega=0.1} = \frac{-1}{3m\omega^2} = 30.46\,db,\ 180° \tag{3.22}$$

$$\left.\frac{z_2}{F_1}\right|_{\omega=10} = \frac{k}{m^2\omega^4} = \frac{1}{(10)^4} = 0.0001 = -80\,db,\ 0° \tag{3.23}$$

$$\left.\frac{z_3}{F_1}\right|_{\omega=0.1} = \frac{-1}{3m\omega^2} = 30.46\,db,\ 180° \tag{3.24}$$

$$\left.\frac{z_3}{F_1}\right|_{\omega=10} = \frac{-k^2}{m^3}\frac{1}{\omega^6} = \frac{-1}{1e^6} = -1e^{-6} = -120\,db,\ 180° \tag{3.25}$$

$$\left.\frac{z_2}{F_2}\right|_{\omega=0.1} = \frac{-1}{3m\omega^2} = 30.46\,db,\ 180° \tag{3.26}$$

$$\left.\frac{z_2}{F_2}\right|_{\omega=10} = \frac{-1}{m\omega^2} = \frac{-1}{(10)^2} = -.01 = -40db, 180° \tag{3.27}$$

3.3 Hand Sketching Frequency Responses

Knowing the pole and zero locations and the asymptotes, the complete frequency response can be sketched by hand, as shown in Figure 3.2. We will not worry about the exact magnitudes at the poles and zeros, but will use the hand sketch to get an idea of the overall shape and characteristics of the frequency response. Start by drawing the low and high frequency asymptotes, straight lines with appropriate magnitudes and slopes starting at the 0.1 and 10 rad/sec frequencies. Next, locate the poles and zeros at some distance above and below the asymptote line at the appropriate frequency and start "connecting the dots." Start at the low frequency asymptote and follow it to the first zero or pole encountered. Keep plotting, moving to the next higher frequency pole or zero until all the poles/zeros are passed and move onto the high frequency asymptote. Note that for z21 the pole and zero at 1 rad/sec cancel as do one of the zeros and the pole for z22. Note that z31 has no zeros, only poles. Compare these plots to the MATLAB generated plots in Figure 3.5. Chapter 4 will give a physical interpretation of the zeros.

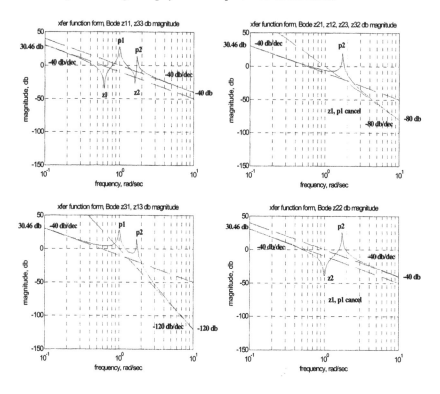

Figure 3.2: Hand sketch of frequency responses using asymptotes and pole/zero locations.

3.4 Interpreting Frequency Response Graphically in Complex Plane

There are many ways to plot frequency responses using MATLAB, as shown in the MATLAB code **tdofxfer.m** in the next section. One method of visualizing graphically what happens in calculating a frequency response is shown below.

In Chapter 2 we defined the four unique transfer functions in both "transfer function" and "zpk" forms. We will use the zpk form to graphically compute the frequency response.

Start by defining a specific frequency for which to calculate the magnitude and phase. Then locate that frequency on the positive imaginary axis.

The gain and phase of the numerator term of a transfer function is the vector product of distances from all the zeros to the frequency of interest times the dc gain. Consider an undamped model, where all the poles and zeros lie on the imaginary axis. If the frequency happens to lie on a zero, that distance is zero, which multiplies all the other zero distances, resulting in a frequency response magnitude of zero. For a damped model the distance will not be zero, as the zeros are to the left of the imaginary axis, but the distance will be small, giving a small multiplier at that frequency and attenuating the response.

The gain and phase of the denominator term is the product of distances from all the poles to the frequency of interest. For an undamped model, if the frequency happens to lie on a pole, that distance is zero, which multiplies all the other pole distances. When the numerator is divided by the zero denominator value, the response goes to ∞. For a damped model the distance will not be zero as the poles are to the left of the imaginary axis; the distance will be small, however, giving a small multiplier at that frequency and amplifying the response.

Once the numerator and denominator are known, a vector division will give the transfer function.

The pole/zero plot, pole/zero values and zpk form for the z11 transfer function are shown below. We will calculate the frequency response for 0.25 rad/sec, where the frequency is indicated in Figure 3.3.

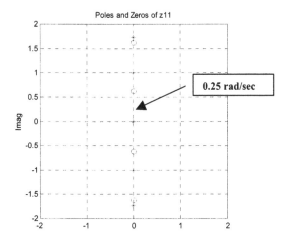

Figure 3.3: Interpreting the frequency response graphically for a frequency of 0.25 rad/sec (tdofpz3x3.m).

poles =

    0
    0
    0 + 1.7321i
    0 - 1.7321i
    0 + 1.0000i
    0 - 1.0000i

zeros_z11 =

    0 + 1.6180i
    0 - 1.6180i
    0 + 0.6180i
    0 - 0.6180i

Table 3.1: Poles and zeros of z11 transfer function, MATLAB listing from tdofpz3x3.m.

z11 Undamped Zero/pole/gain:

(s^2 + 0.382) (s^2 + 2.618)
----
s^2 (s^2 + 1) (s^2 + 3)

Table 3.2: zpk form of z11 transfer function, MATLAB listing from tdofpz3x3.m.

Taking the expression for z11 from the zpk MATLAB listing in Table 3.2, expand the terms to show explicitly the pole and zero values from Table 3.1, substituting $s = 0.25j$ to calculate the frequency response value at 0.25 rad/sec.

$$
\begin{aligned}
z11 &= \frac{(s^2 + 0.382)(s^2 + 2.618)}{s^2(s^2 + 1)(s^2 + 3)} \\[4pt]
&= \frac{(s + 0.618j)(s - 0.618j)(s + 1.618j)(s - 1.618j)}{s^2(s + 1j)(s - 1j)(s + 1.732j)(s - 1.732j)} \\[4pt]
&= \frac{(0.25j + 0.618j)(0.25j - 0.618j)(0.25j + 1.618j)(0.25j - 1.618j)}{(0.25j)^2(0.25j + 1j)(0.25j - 1j)(0.25j + 1.732j)(0.25j - 1.732j)} \\[4pt]
&= \frac{-0.816}{0.172} = -4.74
\end{aligned}
\qquad (3.28)
$$

Taking the magnitude and phase of z11:

$$|z11| = 4.74$$
$$\angle z11 = -180°$$

(3.29)

The frequency response plot from MATLAB code **tdofxfer.m** in Figure 3.4 shows a magnitude of 4.79 (our 4.74 above differs because of rounding errors). The phase plot, not shown here but available by running **tdofxfer.m**, shows −180°.

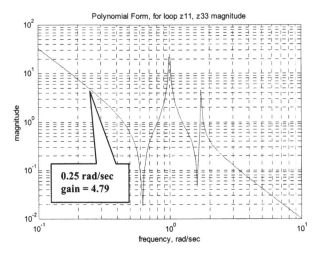

Figure 3.4: z11 frequency response highlighting magnitude at 0.25 rad/sec.

3.5 MATLAB Code tdofxfer.m – Plot Frequency Responses

3.5.1 Code Description

Five different methods of calculating the frequency responses are used in the **tdofxfer.m** code, starting with the simplest and most straightforward method, but not necessarily the most efficient, then going to more sophisticated and efficient methods. The methods are:

1) Polynomial descriptions of the transfer functions: Using a for-loop to cycle through the frequency vector. MATLAB's complex algebra capabilities are used to evaluate the frequency response at each frequency.

2) Polynomial descriptions of the transfer functions: Using MATLAB's vector capabilities instead of a for-loop to

calculate the frequency response at the frequencies in the frequency vector.

3) MATLAB's "transfer function" representations of the transfer functions: MATLAB's automatic bode plotting capability is used, where MATLAB chooses the frequency range to use and automatically plots results.

4) Transfer function representations of the transfer functions: MATLAB's bode plotting capability is used, but this time defining outputs and frequency range with the "bode" command, controlling the output for later plotting.

5) MATLAB's "zero/pole/gain, zpk" form of the input is used.

Because the plotting commands are so lengthy, they will be not be listed. See the downloaded code for the complete code listing.

3.5.2 Polynomial Form, For-Loop Calculation, Code Listing

The "polynomial form" shown below uses (2.28) through (2.36) to define the four distinct frequency responses of the system, allowing the user to specify **any** values of masses, dampers and springs. MATLAB's complex number calculation capabilities are used by defining a vector of radian frequencies "w" and substituting "j*w" for "s." A "for-loop" is then used to cycle through each frequency in the "w" vector and calculate the complex value for the frequency response at that frequency. Because MATLAB does not know how large all of the vectors defined within the "for-loop" are going to be, it resizes each vector during each calculation, a very time-consuming (relatively speaking) operation. We could speed up the operation by defining null vectors of the proper size for each of the "for-loop" variables before the for-loop was entered. This would still require going through the for-loop for every entry in the "w" vector, but would eliminate having to resize the vectors at each calculation. Following the for-loop, magnitudes and phases are calculated using MATLAB's "abs" and "angle" commands and are available for plotting.

```
%        "Polynomial Form, for-loop" frequency response plotting

%        assign values for masses, damping, and stiffnesses

         m1 = 1;
         m2 = 1;
         m3 = 1;
         c1 = 0;
         c2 = 0;
         k1 = 1;
         k2 = 1;
```

```
%         Define a vector of frequencies to use, radians/sec.  The logspace command uses
%         the log10 value as limits, i.e. -1 is 10^-1 = 0.1 rad/sec, and 1 is
%         10^1 = 10 rad/sec.  The 200 defines 200 frequency points.

          w = logspace(-1,1,200);

%         pre-calculate the radians to degree conversion

          rad2deg = 180/pi;

%         Use a for-loop to cycle through all the frequencies, using MATLAB's
%         complex algebra capabilities to evaluate.

          for  cnt = 1:length(w)

%         define s as the imaginary operator times each frequency

                  s = j*w(cnt);

%         define the frequency responses to be evaluated

          den(cnt) = s^2*(s^4*(m1*m2*m3) + s^3*(m2*m3*c1 + m1*m3*c1 + m1*m2*c2
          + m1*m3*c2) + s^2*(m1*m3*k1 + m1*m3*k2 + m1*m2*k2 ...
          + m2*c1*c2 + m3*c1*c2 + m1*c1*c2 + k1*m2*m3) ...
          + s*(m3*c1*k2 + m2*c2*k1 + m1*c2*k1 + m1*c1*k2 ...
          + m3*c2*k1 + m2*c1*k2) + (m1*k1*k2 + m2*k1*k2 + m3*k1*k2));

          z11bf(cnt) = ((m2*m3)*s^4 + (m3*c1 + m3*c2 + m2*c2)*s^3 ...
          + (c1*c2 + m2*k2 + m3*k1 + m3*k2)*s^2 ...
          + (c1*k2 + c2*k1)*s + (k1*k2))/den(cnt);

          z21bf(cnt) = ((m3*c1)*s^3 + (c1*c2 + m3*k1)*s^2 + (c1*k2 + c2*k1)*s ...
          + (k1*k2))/den(cnt);

          z31bf(cnt) = ((c1*c2)*s^2 + (c1*k2 + c2*k1)*s + (k1*k2))/den(cnt);

          z22bf(cnt) = ((m1*m3)*s^4 + (m1*c2 + m3*c1)*s^3 + (m1*k2 + c1*c2 +  ...
                          m3*k1)*s^2
          + (c1*k2 + c2*k1)*s + (k1*k2))/den(cnt);

%         calculate the magnitude and phase of each frequency response

          z11bfmag(cnt) = abs(z11bf(cnt));

          z21bfmag(cnt) = abs(z21bf(cnt));

          z31bfmag(cnt) = abs(z31bf(cnt));

          z22bfmag(cnt) = abs(z22bf(cnt));

          z11bfphs(cnt) = angle(z11bf(cnt))*rad2deg;

          z21bfphs(cnt) = angle(z21bf(cnt))*rad2deg;
```

```
z31bfphs(cnt) = angle(z31bf(cnt))*rad2deg;

z22bfphs(cnt) = angle(z22bf(cnt))*rad2deg;

end                    %         end of for-loop
```

3.5.3 Polynomial Form, Vector Calculation, Code Listing

This section of code defines the transfer functions as in the previous section but instead of using the for-loop for obtaining complex values of the desired quantities at each frequency, this code uses MATLAB's vector calculation capability.   MATLAB can perform operations on vectors directly, very quickly and without having to resize anything as discussed in the previous section. In order to define a vector operation between two vectors, precede the operation symbol (*, /, ^, etc) with a period ("."). This period tells MATLAB to perform an element-by-element operation on or between corresponding elements of the vector(s).   For example, to square every element of a vector, "vec", use the command "vec.^2," and to multiply two elements, "vec1" and "vec2" element by element, use the command "vec1.*vec2." This vector calculation capability will be used wherever appropriate in the balance of the code in the text.

```
%         "Polynomial Form, Vector" method - using MATLAB's vector capabilities instead
%         of the "for" loop.

%         assign values for masses, damping, and stiffnesses

          m1 = 1;
          m2 = 1;
          m3 = 1;
          c1 = 0;
          c2 = 0;
          k1 = 1;
          k2 = 1;

%         Define a vector of frequencies to use, radians/sec.  The logspace command uses
%         the log10 value as limits, i.e. -1 is 10^-1 = 0.1 rad/sec, and 1 is
%         10^1 = 10 rad/sec.  The 200 defines 200 frequency points.

          w = logspace(-1,1,200);

%         pre-calculate the radians to degree conversion

          rad2deg = 180/pi;

%         define s as the imaginary operator times the radian frequency vector

          s = j*w;
```

```
%       define the frequency responses to be evaluated, using the "." prefix
%       in front of each operator to indicate that each

%       define the frequency responses to be evaluated

        den = s.^2.*(s.^4*(m1*m2*m3) + s.^3*(m2*m3*c1 + m1*m3*c1 + m1*m2*c2
        + m1*m3*c2) + s.^2*(m1*m3*k1 + m1*m3*k2 + m1*m2*k2 ...
        + m2*c1*c2 + m3*c1*c2 + m1*c1*c2 + k1*m2*m3) ...
        + s*(m3*c1*k2 + m2*c2*k1 + m1*c2*k1 + m1*c1*k2 ...
        + m3*c2*k1 + m2*c1*k2) + (m1*k1*k2 + m2*k1*k2 + m3*k1*k2));

        z11bfv = ((m2*m3)*s.^4 + (m3*c1 + m3*c2 + m2*c2)*s.^3 ...
        + (c1*c2 + m2*k2 + m3*k1 + m3*k2)*s.^2 ...
        + (c1*k2 + c2*k1)*s + (k1*k2))./den;

        z21bfv = ((m3*c1)*s.^3 + (c1*c2 + m3*k1)*s.^2 + (c1*k2 + c2*k1)*s ...
                + (k1*k2))./den;

        z31bfv = ((c1*c2)*s.^2 + (c1*k2 + c2*k1)*s + (k1*k2))./den;

        z22bfv = ((m1*m3)*s.^4 + (m1*c2 + m3*c1)*s.^3 + (m1*k2 + c1*c2 + m3*k1)*s.^2
                + (c1*k2 + c2*k1)*s + (k1*k2))./den;

%       calculate the magnitude and phase of each frequency response

        z11bfvmag = abs(z11bfv);

        z21bfvmag = abs(z21bfv);

        z31bfvmag = abs(z31bfv);

        z22bfvmag = abs(z22bfv);

        z11bfvphs = angle(z11bfv)*rad2deg;

        z21bfvphs = angle(z21bfv)*rad2deg;

        z31bfvphs = angle(z31bfv)*rad2deg;

        z22bfvphs = angle(z22bfv)*rad2deg;
```

3.5.4 Transfer Function Form – Bode Calculation, Code Listing

This section uses MATLAB's automatic "bode" calculation and plotting capability, as well as the "transfer function" form of input, where the numerator "num" and denominator "den" of each transfer function are input as row vectors in coefficients of descending powers of "s." Using the "bode" command with no left-hand arguments results in MATLAB choosing the frequency range to use and automatically generating plots of magnitude and phase.

```
%          using MATLAB's automatic "bode" plotting capability, defining the transfer
%          functions in "transfer function" form by row vectors of coefficients of "s"

%          assign values for masses, damping, and stiffnesses

           m1 = 1;
           m2 = 1;
           m3 = 1;
           c1 = 0;
           c2 = 0;
           k1 = 1;
           k2 = 1;

%          define row vectors of numerator and denominator coefficients

           den = [(m1*m2*m3) (m2*m3*c1 + m1*m3*c1 + m1*m2*c2 + m1*m3*c2) ...
               (m1*m3*k1 + m1*m3*k2 + m1*m2*k2 + m2*c1*c2 + m3*c1*c2 + ...
               m1*c1*c2 + k1*m2*m3) ...
               (m3*c1*k2 + m2*c2*k1 + m1*c2*k1 + m1*c1*k2 + m3*c2*k1 + m2*c1*k2) ...
               (m1*k1*k2 + m2*k1*k2 + m3*k1*k2) 0 0];

           z11num = [(m2*m3) (m3*c1 + m3*c2 + m2*c2) (c1*c2 + m2*k2 + m3*k1 + m3*k2)
               (c1*k2 + c2*k1) (k1*k2)];

           z21num = [(m3*c1) (c1*c2 + m3*k1) (c1*k2 + c2*k1) (k1*k2)];

           z31num = [(c1*c2) (c1*k2 + c2*k1) (k1*k2)];

           z22num = [(m1*m3) (m1*c2 + m3*c1) (m1*k2 + c1*c2 + m3*k1) ...
               (c1*k2 + c2*k1) (k1*k2)];

%          the bode command with no left hand side arguments automatically chooses
%          frequency limits and plots results

           grid on
           bode(z11num,den);

           disp('execution paused to display figure, "enter" to continue'); pause

           bode(z21num,den);

           disp('execution paused to display figure, "enter" to continue'); pause

           bode(z31num,den);

           disp('execution paused to display figure, "enter" to continue'); pause

           bode(z22num,den);

           disp('execution paused to display figure, "enter" to continue'); pause
```

3.5.5 Transfer Function Form, Bode Calculation with Frequency, Code Listing

This section also uses MATLAB's "bode" plotting capability with the transfer function form of the input but defines magnitude and phase vectors for output and specifies the frequency vector to use. This code also calculates and plots the low and high frequency asymptotes for the four unique transfer functions.

```
%        using MATLAB's "bode" plotting capability, defining the transfer
%        functions in "transfer function" form by row vectors of coefficients of
%        "s"and defining output vectors for magnitude and phase as well as a
%        defined range of radian frequencies

%        assign values for masses, damping, and stiffnesses

         m1 = 1;
         m2 = 1;
         m3 = 1;
         c1 = 0;
         c2 = 0;
         k1 = 1;
         k2 = 1;

%        define row vectors of numerator and denominator coefficients

         den = [(m1*m2*m3) (m2*m3*c1 + m1*m3*c1 + m1*m2*c2 + m1*m3*c2) ...
             (m1*m3*k1 + m1*m3*k2 + m1*m2*k2 + m2*c1*c2 + m3*c1*c2 + ...
             m1*c1*c2 + k1*m2*m3) ...
             (m3*c1*k2 + m2*c2*k1 + m1*c2*k1 + m1*c1*k2 + m3*c2*k1 + m2*c1*k2) ...
             (m1*k1*k2 + m2*k1*k2 + m3*k1*k2) 0 0];

         z11num = [(m2*m3) (m3*c1 + m3*c2 + m2*c2) (c1*c2 + m2*k2 + m3*k1 + m3*k2)
             (c1*k2 + c2*k1) (k1*k2)];

         z21num = [(m3*c1) (c1*c2 + m3*k1) (c1*k2 + c2*k1) (k1*k2)];

         z31num = [(c1*c2) (c1*k2 + c2*k1) (k1*k2)];

         z22num = [(m1*m3) (m1*c2 + m3*c1) (m1*k2 + c1*c2 + m3*k1) ...
             (c1*k2 + c2*k1) (k1*k2)];

%        Define a vector of frequencies to use, radians/sec. The logspace command uses
%        the log10 value as limits, i.e. -1 is 10^-1 = 0.1 rad/sec, and 1 is
%        10^1 = 10 rad/sec. The 200 defines 200 frequency points.

         w = logspace(-1,1,200);

%        calculate the rigid-body motions for low and high frequency portions
%        of all the frequency responses, the denominator entries are vectors with
%        entries being coefficients of the "s" terms in the low and high frequency
%        asymptotes, starting with the highest power of "s" and ending with the
%        "0"th power of "s" or the constant term
```

```
        z11num_lo = [1];

        z11den_lo = [3 0 0];          % -1/(3*w^2)

        z11num_hi = [1];

        z11den_hi = [1 0 0];          % -1/(w^2)

        z21num_lo = [1];

        z21den_lo = [3 0 0];          % -1/(3*w^2)

        z21num_hi = [1];

        z21den_hi = [1 0 0 0 0];      % -1/(3*w^4)

        z31num_lo = [1];

        z31den_lo = [3 0 0];          % -1/(3*w^2)

        z31num_hi = [1];

        z31den_hi = [1 0 0 0 0 0 0];  % -1/(w^2)

        z22num_lo = [1];

        z22den_lo = [3 0 0];          % -1/(3*w^2)

        z22num_hi = [1];

        z22den_hi = [1 0 0];          % -1/(w^2)

%       define the "tf" models from "num, den" combinations

        z11tf = tf(z11num,den);

        z21tf = tf(z21num,den);

        z31tf = tf(z31num,den);

        z22tf = tf(z22num,den);

        z11tf_lo = tf(z11num_lo,z11den_lo);

        z11tf_hi = tf(z11num_hi,z11den_hi);

        z21tf_lo = tf(z21num_lo,z21den_lo);

        z21tf_hi = tf(z21num_hi,z21den_hi);

        z31tf_lo = tf(z31num_lo,z31den_lo);

        z31tf_hi = tf(z31num_hi,z31den_hi);
```

```
          z22tf_lo = tf(z22num_lo,z22den_lo);

          z22tf_hi = tf(z22num_hi,z22den_hi);

%         use the bode command with left hand magnitude and phase vector arguments
%         to provide values for further analysis/plotting

          [z11mag,z11phs] = bode(z11tf,w);

          [z21mag,z21phs] = bode(z21tf,w);

          [z31mag,z31phs] = bode(z31tf,w);

          [z22mag,z22phs] = bode(z22tf,w);

          [z11maglo,z11phslo] = bode(z11tf_lo,w);

          [z21maglo,z21phslo] = bode(z21tf_lo,w);

          [z31maglo,z31phslo] = bode(z31tf_lo,w);

          [z22maglo,z22phslo] = bode(z22tf_lo,w);

          [z11maghi,z11phshi] = bode(z11tf_hi,w);

          [z21maghi,z21phshi] = bode(z21tf_hi,w);

          [z31maghi,z31phshi] = bode(z31tf_hi,w);

          [z22maghi,z22phshi] = bode(z22tf_hi,w);

%         calculate the magnitude in decibels, db

          z11magdb = 20*log10(z11mag);

          z21magdb = 20*log10(z21mag);

          z31magdb = 20*log10(z31mag);

          z22magdb = 20*log10(z22mag);

          z11maglodb = 20*log10(z11maglo);

          z21maglodb = 20*log10(z21maglo);

          z31maglodb = 20*log10(z31maglo);

          z22maglodb = 20*log10(z22maglo);

          z11maghidb = 20*log10(z11maghi);

          z21maghidb = 20*log10(z21maghi);

          z31maghidb = 20*log10(z31maghi);
```

```
z22maghidb = 20*log10(z22maghi);
```

3.5.6 Zero/Pole/Gain Function Form, Bode Calculation with Frequency, Code Listing

This section also uses MATLAB's "bode" plotting capability. This time, with the zero/pole/gain form of the input. It defines magnitude and phase vectors for output and specifies the frequency vector to use.

```
%       using MATLAB's "bode" plotting capability, defining the transfer
%       functions in "zero/pole/gain" form by column vectors of poles and zeros
%       and defining output vectors for magnitude and phase as well as a
%       defined range of radian frequencies

%       assign values for masses, damping, and stiffnesses

        m1 = 1;
        m2 = 1;
        m3 = 1;
        c1 = 0;
        c2 = 0;
        k1 = 1;
        k2 = 1;

        m = m1;
        k = k1;

%       define column vectors of poles and zeros from previous derivation
%
%       there are three ways to make a column vector:
%
%       1)      define a row vector and then transpose it:
%
%               p = [0 0 1*j -1*j sqrt(3*k/m)*j -sqrt(3*k/m)*j]';
%
%       2)      define a column vector by using semi-colons between elements:
%
%               p = [0; 0; 1*j; -1*j; sqrt(3*k/m)*j; -sqrt(3*k/m)*j];
%
%       3)      define a column vector directly:
%
%               p = [      0
%                          0
%                         1*j
%                        -1*j
%                      sqrt(3*k/m)*j
%                     -sqrt(3*k/m)*j ];

%       zeros for z1/f1; quartic so four zeros

        z11_1 = -sqrt((-3*k-sqrt(5)*k)/(2*m));
```

```
            z11_2 = sqrt((-3*k-sqrt(5)*k)/(2*m));

            z11_3 = -sqrt((-3*k+sqrt(5)*k)/(2*m));
            z11_4 = sqrt((-3*k+sqrt(5)*k)/(2*m));

%           zeros for z2/f1; quadratic so two zeros

            z21_1 = -sqrt(-k/m);
            z21_2 = sqrt(-k/m);

%           zeros for z3/f1; no zeros, so use empty brackets

            z31_1 = [];

%           zeros for z2/f2: quadratic so two zeros

            z22_1 = -sqrt(-k/m);
            z22_2 = sqrt(-k/m);

%

            z11 = [z11_1 z11_2 z11_3 z11_4]';

            z21 = [z21_1 z21_2]';

            z31 = z31_1;

            z22 = [z22_1 z22_2]';

            p = [0 0 1*j -1*j sqrt(3*k/m)*j -sqrt(3*k/m)*j]';

            gain = 1;

%           use the zpk command to define the four pole/zero/gain systems

            sys11pz = zpk(z11,p,gain);

            sys21pz = zpk(z21,p,gain);

            sys31pz = zpk(z31,p,gain);

            sys22pz = zpk(z22,p,gain);

%           Define a vector of frequencies to use, radians/sec.  The logspace command uses
%           the log10 value as limits, i.e. -1 is 10^-1 = 0.1 rad/sec, and 1 is
%           10^1 = 10 rad/sec.  The 200 defines 200 frequency points.

            w = logspace(-1,1,200);

%           use the bode command with left hand magnitude and phase vector arguments
%           to provide values for further analysis/plotting

            [z11mag,z11phs] = bode(sys11pz,w);

            [z21mag,z21phs] = bode(sys21pz,w);
```

```
        [z31mag,z31phs] = bode(sys31pz,w);

        [z22mag,z22phs] = bode(sys22pz,w);

%       calculate the magnitude in decibels, db

        z11magdb = 20*log10(z11mag);

        z21magdb = 20*log10(z21mag);

        z31magdb = 20*log10(z31mag);

        z22magdb = 20*log10(z22mag);
```

3.5.7 Code Output – Frequency Response Magnitude and Phase Plots

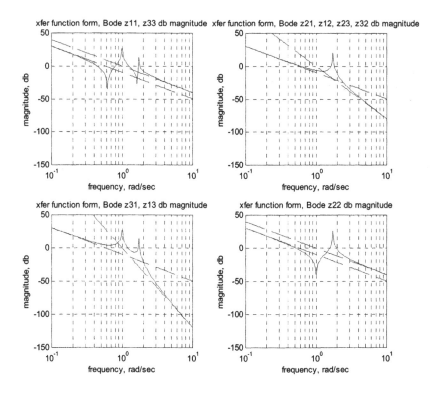

Figure 3.5: Magnitude versus frequency for four distinct frequency responses, including low and high frequency asymptotes.

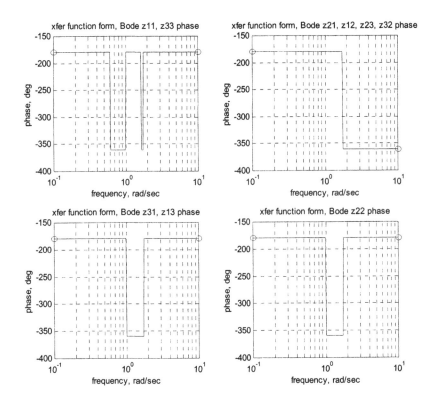

Figure 3.6: Phase versus frequency for four distinct frequency responses, including low
and high frequency asymptotes.

3.6 Other Forms of Frequency Response Plots

Other forms of frequency response plots are shown for a damping value of 2%
of critical damping for each mode. The code used for the plots is from
Chapter 11, **tdofss_modal_xfer_modes.m**.

3.6.1 Log Magnitude versus Log Frequency

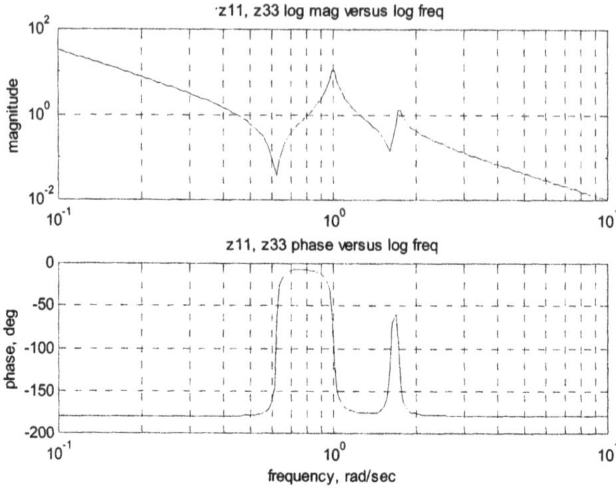

Figure 3.7: Log magnitude versus log frequency.

Comments on the log-log plot:

1) The asymptotic behavior at the low and high frequency ends are clear by checking the slopes.

2) The log frequency scale spreads out the resonances, which otherwise would tend to clump at the lower end of the scale.

3) The log amplitude scale allows reading the gain directly without converting from db.

4) Adding the gain from the mechanics to the gain of the frequency response of the control system allows for definition of the overall series (multiplicative) frequency response.

3.6.2 db Magnitude versus Log Frequency

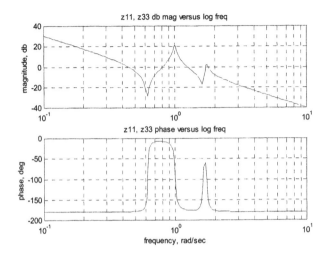

Figure 3.8: db magnitude versus log frequency.

Comments on the db-log plot:

1) The asymptotic behaviors at the low and high frequency
    ends are clear by checking the slopes, i.e.

$$(1/\omega) = -20 \, \text{db/decade}, \, (1/\, \omega^2 \,) = -40 \, \text{db/decade}.$$

2) The log frequency scale spreads out the resonances, which
    otherwise would tend to clump at the lower end of the scale.

3) The db amplitude scale makes it necessary to convert to gain
    if needed.

4) The **product** of two individual frequency response gains can
    be found by **adding** their gains directly on the log scale.

3.6.3 db Magnitude versus Linear Frequency

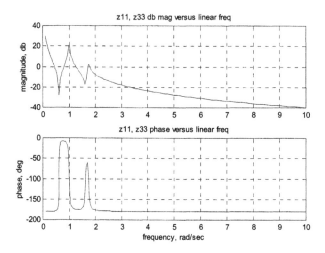

Figure 3.9: db magnitude versus linear frequency.

Comments on the db-linear plot:

1) The asymptotic behaviors at the low and high frequency ends are not clear.

2) The linear frequency scale tends to clump the resonances at the lower end of the scale, although the scale could be shortened since nothing significant is happening at the high end.

3) The db amplitude scale makes it necessary to convert to linear gain if specific gain values are needed.

3.6.4 Linear Magnitude versus Linear Frequency

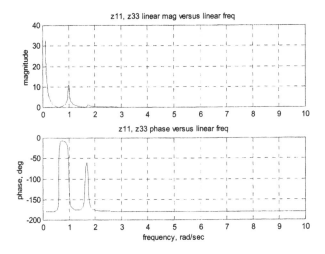

Figure 3.10: Linear magnitude versus linear frequency.

Comments on the linear-linear plot:

1) The asymptotic behaviors at the low and high frequency ends are not clear.

2) The linear frequency scale tends to clump the resonances at the lower end of the scale, although the scale could be shortened since nothing significant is happening at the high end.

3) The linear amplitude scale enables reading gain values directly, but reading values for small gain values is difficult.

4) It is useful for directly adding the individual mode contributions of a frequency response to provide the overall response, shown in Chapter 8, Sections 8.7 and 8.8.

3.6.5 Real and Imaginary Magnitudes versus Log and Linear Frequency

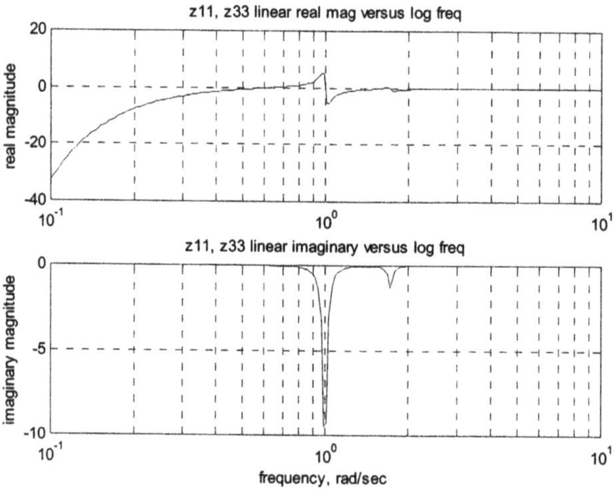

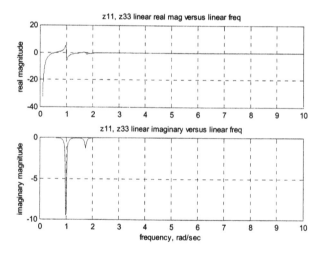

Figure 3.11: Real and imaginary magnitudes versus log frequency.

Figure 3.12: Real and imaginary magnitude versus linear frequency.

Comments on real versus linear frequency, imaginary versus linear frequency:

1)   These plots are useful in understanding the amplitudes of transfer functions at resonance, as the peaks of the imaginary curve represent the amplitude at resonance.

2) While the imaginary plot peaks at each resonance, the real
    plot goes through zero at each resonance.

3.6.6 Real versus Imaginary (Nyquist)

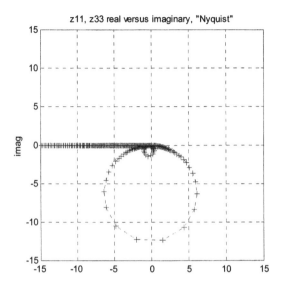

Figure 3.13: Real versus imaginary (Nyquist).

Comments on real versus imaginary:

1) Frequency is not plotted directly on the real/imaginary plot; each
    point on the plot represents a different frequency.

2) Plotting real versus imaginary is a very useful technique when
    identifying resonant characteristics. The two resonances can be
    readily seen, helping in identifying closely spaced resonances.

3) One method of identifying damping in a mode is to use the rate of
    change of amplitude versus frequency (Maia 1997).

3.7 Solving for Eigenvectors (Mode Shapes) Using the Transfer Function Matrix

We have reviewed transfer functions, poles, zeros and frequency responses. The next area we will cover in order to completely define the system is **eigenvectors**, or **mode shapes**. At each natural frequency, the eigenvector defines the relative motion between degrees of freedom. Understanding the distribution of motion in each mode of vibration is essential in order to intelligently modify the system's resonant characteristic to solve resonance problems.

Since eigenvectors define the **relative** motion between degrees of freedom, we need to choose a degree of freedom against which to measure the other motions. We can find the relative motion using **any** column of the transfer function matrix. Choosing z_1 as the reference and solving for z_2 / z_1 and z_3 / z_1 using the first column of the transfer function matrix (we will compare results using the second column later to show that they give the same results):

$$\frac{\dfrac{z_2}{F_1}}{\dfrac{z_1}{F_1}} = \frac{z_{21}}{z_{11}} = \frac{mks^2 + k^2}{m^2 s^4 + 3mks^2 + k^2} \tag{3.29}$$

$$\frac{\dfrac{z_3}{F_1}}{\dfrac{z_1}{F_1}} = \frac{z_{31}}{z_{11}} = \frac{k^2}{m^2 s^4 + 3mks^2 + k^2} \tag{3.30}$$

Now that the ratios are known, we substitute the resonant frequencies (pole values) one at a time to define the mode shape at that frequency, dropping the second index, $z_{21} \rightarrow z_2$.

For mode 1: evaluated at $s = j\omega_1 = 0$

$$\frac{z_2}{z_1} = \frac{mks^2 + k^2}{m^2 s^4 + 3mks^2 + k^2} = \frac{k^2}{k^2} = 1 \tag{3.31}$$

$$z_2 = z_1 \tag{3.32}$$

$$\frac{z_3}{z_1} = \frac{k^2}{m^2s^4 + 3mks^2 + k^2} = \frac{k^2}{k^2} = 1 \tag{3.33}$$

$$z_3 = z_1 \tag{3.34}$$

The interpretation of this mode shape is that at ω_1 the ratios of motion of mass 2 and mass 3 to mass 1 are equal and are equal to 1. This is the rigid body mode at 0 hz.

For mode 2: evaluated at $s = j\omega_2 = j\sqrt{\dfrac{k}{m}}$

$$\frac{z_2}{z_1} = \frac{mks^2 + k^2}{m^2s^4 + 3mks^2 + k^2} = \frac{mk\left(\dfrac{-k}{m}\right) + k^2}{m^2\left(\dfrac{k^2}{m^2}\right) + 3mk\left(\dfrac{-k}{m}\right) + k^2} = \frac{0}{k^2} \tag{3.35}$$

$$z_2 = 0 \tag{3.36}$$

$$\frac{z_3}{z_1} = \frac{k^2}{m^2s^4 + 3mks^2 + k^2} = \frac{k^2}{m^2\left(\dfrac{k^2}{m^2}\right) + 3mk\left(\dfrac{-k}{m}\right) + k^2} = -1 \tag{3.37}$$

$$z_3 = -z_1 \tag{3.38}$$

The interpretation of this mode shape is that at ω_2 mass 2 has zero motion relative to mass 1 (it is stationary). Mass 3 is moving out of phase with mass 1 with equal amplitude.

For mode 3: evaluated at $s = j\omega_3 = j\sqrt{\dfrac{3k}{m}}$

$$\frac{z_2}{z_1} = \frac{mks^2 + k^2}{m^2s^4 + 3mks^2 + k^2} = \frac{mk\left(\dfrac{-3k}{m}\right) + k^2}{m^2\left(\dfrac{9k^2}{m^2}\right) + 3mk\left(\dfrac{-3k}{m}\right) + k^2} \tag{3.39}$$

$$= \frac{-3k^2 + k^2}{9k^2 - 9k^2 + k^2} = \frac{-2k^2}{k^2} = -2$$

$$z_2 = -2z_1 \tag{3.40}$$

$$\frac{z_3}{z_1} = \frac{k^2}{m^2 s^4 + 3mks^2 + k^2} = \frac{k^2}{m^2\left(\dfrac{9k^2}{m^2}\right) + 3mk\left(\dfrac{-3k}{m}\right) + k^2} \tag{3.41}$$

$$= \frac{k^2}{k^2} = 1$$

$$z_3 = z_1 \tag{3.42}$$

The interpretation of this mode shape is that at ω_3 mass 2 is moving with twice the motion of mass 1 and out of phase with it and mass 3 is moving in phase with mass 1 and with the same amplitude.

Showing that the second column of the transfer function matrix could have been used and would have given the same eigenvectors:

$$\frac{\dfrac{z_2}{F_2}}{\dfrac{z_1}{F_2}} = \frac{z_2}{z_1} = \frac{m^2 s^4 + 2mks^2 + k^2}{mks^2 + k^2} \tag{3.43}$$

$$\frac{\dfrac{z_3}{F_2}}{\dfrac{z_1}{F_2}} = \frac{z_3}{z_1} = \frac{mks^2 + k^2}{mks^2 + k^2} = 1 \tag{3.44}$$

For mode 1, $\omega_1 = 0$

$$\frac{z_2}{z_1} = \frac{k^2}{k^2} = 1$$

$$\frac{z_3}{z_1} = 1 \tag{3.45a,b}$$

For mode 2, evaluated at $s = j\omega_2 = j\sqrt{\dfrac{k}{m}}$

$$\frac{z_2}{z_1} = \frac{m^2 s^4 + 2mks^2 + k^2}{mks^2 + k^2} = \frac{m^2 \left(\dfrac{k^2}{m^2}\right) + 2mk\left(\dfrac{-k}{m}\right) + k^2}{mk\left(\dfrac{-k}{m}\right) + k^2} \qquad (3.46)$$

$$= \frac{k^2 - 2k^2 + k^2}{-k^2 + k^2} = \frac{0}{0}$$

$$\frac{z_3}{z_1} = 1 \qquad (3.47)$$

For mode 3, $s = j\omega_3 = j\sqrt{\dfrac{3k}{m}}$

$$\frac{z_2}{z_1} = \frac{m^2 s^4 + 2mks^2 + k^2}{mks^2 + k^2} = \frac{m^2 \left(\dfrac{9k^2}{m^2}\right) + 2mk\left(\dfrac{-3k}{m}\right) + k^2}{mk\left(\dfrac{-3k}{m}\right) + k^2} \qquad (3.48)$$

$$= \frac{9k^2 - 6k^2 + k^2}{-3k^2 + k^2} = \frac{4k^2}{-2k^2} = -2$$

$$\frac{z_3}{z_1} = 1 \qquad (3.49)$$

Summarizing the mode shapes in the **modal matrix**, $\mathbf{z}_m$, where the first through third columns represent mode shapes for the first three modes, respectively, and the first through third rows show the relative motion for the first through third dof's, respectively:

$$\mathbf{z}_m = \begin{bmatrix} 1 & 1 & 1 \\ 1 & 0 & -2 \\ 1 & -1 & 1 \end{bmatrix} \qquad (3.50)$$

Figure 3.14 shows the mode shapes pictorially. There are many different eigenvector scaling, or normalizing techniques, to be discussed later. It is not important which normalization technique is used in visualizing mode shapes. However, in using the modal matrix to calculate responses, the normalization technique used is critical, as we will see in future chapters.

Because there is no damping, these modes are known as "normal" (as opposed to "complex") modes. With a normal mode, if the masses are started with some multiple of the displacements of one of the modes, the system will respond at only that frequency. During that motion, the masses will all reach their maximum and minimum points at the same time. Mode shapes are plotted in Figure 3.14, assuming an arbitrary value of 1 for z_1 :

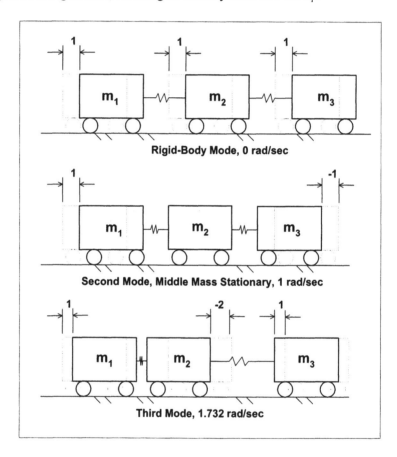

Figure 3.14: Mode shape plots.

Problems

Note: All the problems refer to the two dof system shown in Figure P2.2.

P3.1   Set $m_1 = m_2 = m = 1$, $k_1 = k_2 = k = 1$ and hand sketch the frequency responses for the undamped system.

P3.2   (MATLAB)   Set $m_1 = m_2 = m = 1$, $k_1 = k_2 = k = 1$, modify the **tdofxfer.m** code and plot the frequency responses of the two dof undamped system using the transfer function and zero/pole/gain forms of Sections 3.5.5 and 3.5.6.

P3.3 (MATLAB) Set $m_1 = m_2 = m = 1$, $k_1 = k_2 = k = 1$, add damping to the model from P3.2 and plot the transfer functions in Nyquist form, being careful to use small enough frequency spacing to identify the resonances as shown in Figure 3.13.

P3.4 (MATLAB) Set $m_1 = m_2 = m = 1$, $k_1 = k_2 = k = 1$, choose one of the transfer functions for the undamped system and plot the poles and zeros in the s-plane. Choose a frequency on the positive imaginary axis and hand calculate the gain at that frequency. Correlate with the MATLAB calculated gain.

P3.5 Solve for the two eigenvectors for the system in P3.3 using the transfer function matrix. Hand plot the mode shapes as in Figure 3.14.

CHAPTER 4

ZEROS IN SISO MECHANICAL SYSTEMS

4.1 Introduction

Chapters 2 and 3 discussed poles and zeros of SISO systems and their relationship to transfer functions. The origin and influence of poles are clear. They represent the resonant frequencies of the system, and for each resonant frequency a mode shape can be defined to describe the motion at that frequency. We have seen from our frequency response analyses in Chapter 3 that at the frequencies of the zeros, motions approach or go to zero, depending on the amount of damping present. In Chapters 8 and 11 we will illustrate how all the individual modes of vibration can combine at specific frequencies to create zeros of the overall transfer function.

This chapter will expand on analyses shown in Miu [1993] to develop an intuitive understanding for when to expect zeros in Single Input Single Output (SISO) simple mechanical systems and how to predict the frequencies at which they will occur. We will not cover the theory, but will state the conclusions from Miu and show how the conclusions relate to two example systems.

We will start by defining a series arrangement lumped spring/mass system. We will develop guidelines for defining the number of zeros that should be seen and show how to predict their frequencies. A MATLAB model is used to illustrate the guidelines for various combinations of input and output degrees of freedom. Only the MATLAB code results are discussed; the code itself is not listed or discussed as it uses techniques found later in the book. However, the reader is encouraged to run the code and experiment with various values of the input and number of masses in the model to become familiar with the concept.

Next, an ANSYS finite element model of a tip-excited cantilever is analyzed. The resulting transfer function magnitude is plotted using MATLAB to show an overlay of the poles of the "constrained" system and their relationship with the zeros of the original model.

4.2 "n" dof Example

Figure 4.1 shows a series arrangement of masses and springs, with a total of "n" masses and "n+1" springs. The degrees of freedom are numbered from left to right, z_1 through z_n.

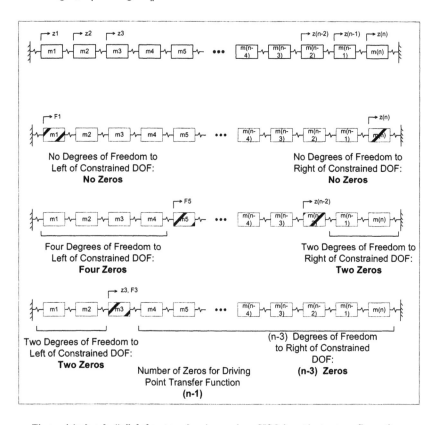

Figure 4.1a,b,c,d: "n" dof system showing various SISO input/output configurations.

Miu [1993] shows that the **zeros** of any particular transfer function are the **poles of the constrained system(s)** to the left and/or right of the system defined by constraining the one or two dof's defining the transfer function. **The resonances of the "overhanging appendages" of the constrained system create the zeros.**

Two limiting cases are immediately available in (1) and (3) below:

1) For the transfer function from one end of the structure to the other, Figure 4.1b, there are no overhanging appendage

structures to the left or right of the constrained structure, so there are no zeros.

2) For an arbitrary transfer function, Figure 4.1c, there will be a structure to the left and/or to the right of the constrained dof's. The total degrees of freedom of the overhanging appendage(s) will give the total number of zeros in the transfer function.

3) For the driving point transfer function, Figure 4.1d, the force and displacement are measured at the same dof, so there are a total of $(n-1)$ degrees of freedom left, hence $(n-1)$ zeros of the transfer function. All but one of the masses are overhanging appendages.

In the analysis that follows, we will calculate frequency responses and pole/zero plots for various transfer functions using the MATLAB code **ndof_numzeros.m**.

4.2.1 MATLAB Code ndof_numzeros.m, Usage Instructions

The MATLAB code is based on the ndof series system in Figure 4.1. The code allows one to choose the total number of masses in the problem and sets the values of the masses and stiffnesses randomly between the values of 1 and 2. The program then allows one to choose which transfer function to calculate, and shows the pole/zero plots for the original system as well as the poles for the two structures to the left and/or right. For now, the reader should not worry about the details of the code, which will be covered in later chapters, but should use the code to study the pole/zero patterns in systems with different numbers of degrees of freedom and for different input/output dof's. Sometimes the random values chosen for stiffnesses and damping will cause the poles and zeros to be so close together that they will cancel each other. If this is the case and the number of poles and zeros do not match the expected number, rerun the code until more widely spaced poles/zeros are randomly chosen and the required poles and zeros are apparent.

4.2.2 Seven dof Model – z7/F1 Frequency Response

Taking a seven-mass model as an example, the resulting frequency responses and pole-zero plots are displayed on the following pages. In all cases, the random distribution of masses and spring stiffnesses is used, resulting in a different set of variables for each run.

Figure 4.2 shows the frequency response for applying a force at the first mass and looking at the output at the last (seventh) mass. Note that in accordance with the prior analysis, there should be no zeros as there are no "overhanging" appendages. Since there are seven masses, there should be seven poles. Since each mass provides an attenuation of –40db/decade, after the last of seven poles the slope of the curve is 7*(–40 db/decade) = –280 db/decade.

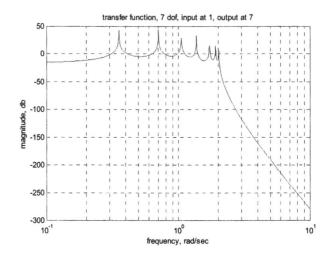

Figure 4.2: z17 transfer function frequency response, seven poles, no zeros.

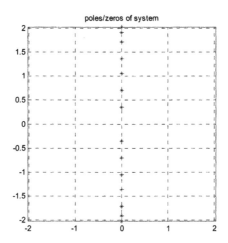

Figure 4.3: z17 pole/zero plot showing only seven poles.

4.2.3 Seven dof Model – z3/F4 Frequency Response

The same seven dof system provides the following frequency response when the force is applied at mass 3 and the output is taken at mass 4. There are two "overhanging" appendages to the left of mass 3, masses 1 and 2, and there are three "overhanging" appendages to the right of mass 4, masses 5, 6 and 7. These masses should combine to give a total of five zeros and once again, seven poles as shown below.

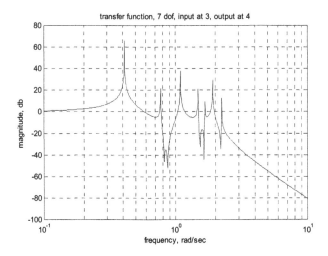

Figure 4.4: z34 transfer function frequency response, seven poles and five zeros.

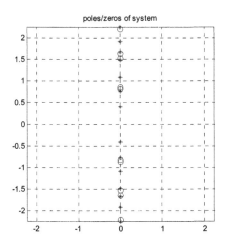

Figure 4.5: z34 pole/zero plot showing seven poles and five zeros.

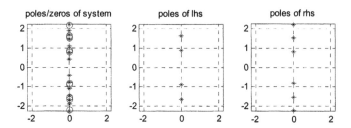

Figure 4.6: z34 poles and zeros; poles of left-hand and right-hand constrained systems are the same as the zeros of the unconstrained system.

The left-hand plot in Figure 4.6 displays the z34 poles and zeros. The middle plot shows the poles of the system to the left of mass 3. The right plot shows the poles of the system to the right of mass 4. It is clear that the poles of the two right plots are the zeros of the z34 system.

4.2.4 Seven dof Model – z3/F3, Driving Point Frequency Response

For the same seven dof system with force and output taken at the same node (driving point transfer function), there should be six "overhanging" masses providing zeros. Therefore the frequency response plot in Figure 4.7 shows six zeros, with alternating pole/zero pairs.

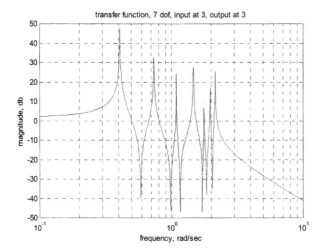

Figure 4.7: z33 transfer function frequency response, seven poles and the expected six zeros.

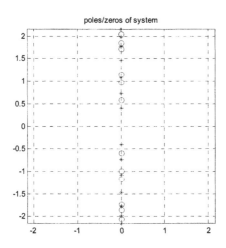

Figure 4.8: z33 pole/zero plot showing seven poles and six zeros.

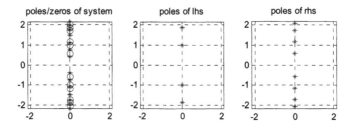

Figure 4.9: z33 poles and zeros. Poles of left-hand and right-hand constrained systems are the same as the zeros of the unconstrained system.

4.3 Cantilever Model – ANSYS

4.3.1 Introduction

Now that we have seen how the "constrained" system artifice works for a simple lumped parameter system, it is interesting to consider how the artifice would work for a continuous system, such as a cantilever beam.

The finite element program ANSYS is used to analyze a cantilever beam with a driving point transfer function at the tip. The transfer function we are interested in is the displacement at the tip, z, due to a vertical force at the tip, F, as shown in Figure 4.10. The "constrained" structure whose poles should define the zero locations for the unconstrained system is the original cantilever with the addition of a simple support at the tip.

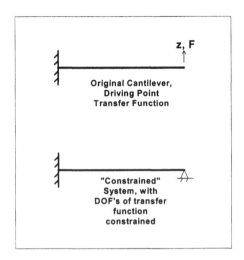

Figure 4.10: Unconstrained and constrained cantilevers used for driving point
transfer function example.

4.3.2 ANSYS Code cantfem.inp Description and Listing

The input listings for the ANSYS models of the cantilever and simply supported tip cantilever are below.   The cantilever input program is **cantfem.inp** and the supported tip input program is **cantzero.inp**.   Both programs can be run if one has access to ANSYS by typing "/input,cantfem,inp" or "/input,cantzero,inp" at the ANSYS program command prompt. The programs will run with no further input and will output graphs of the mode shapes and frequency response. Both programs build the model, and calculate and output the eigenvalues (natural frequencies) and eigenvectors (mode shapes).   **Cantfem.inp** then calculates and outputs the frequency response.   The mode shapes are shown in cantfem.grp and cantzero.grp and the frequency response is shown in cantfem.grp2. They can all be viewed by using the ANSYS Display program and loading the appropriate file.

```
/title, cantfem.inp, 0.05 x 1 x 20mm aluminum cantilever beam, 20 elements

/prep7

et,1,4              ! element type for beam

! aluminum

ex,1,71e6                  ! mN/mm^2
dens,1,2.77e-6             ! kg/mm^3
```

```
nuxy,1,.345

! real value to define beam characteristics

r,1,1,.00001041,.004166,.05,1    ! area, moments of inertia, thickness

! define plotting characteristics

/view,1,1,-1,1    ! iso view
/angle,1,-60      ! iso view
/pnum,mat,1       ! color by material
/num,1            ! numbers off
/type,1,0         ! hidden plot
/pbc,all,1        ! show all boundary conditions

csys,0                         ! define global coordinate system

! nodes

n,1,0,0,0                      ! left-hand node
n,21,20,0,0                    ! right-hand node

fill,1,21                      ! interior nodes

nall
nplo

! elements

type,1
mat,1
real,1
e,1,2
egen,20,1,-1

! constrain left-hand end

nall
d,1,all,0                      ! constrain node 1, all dof's

! constrain all but uz and roty for all other nodes to allow only those dof's
! this will give 20 nodes, node 2 through node 21, each with 2 dof, giving a total of 40 dof
! can calculate a maximum of 40 eigenvalues if don't use Guyan reduction to reduce size of
! eigenvalue problem

nall
nsel,s,node,,2,21
d,all,ux
d,all,uy
d,all,rotx
d,all,rotz

nall
eall
nplo
```

```
eplo

! ****************** eigenvalue run *******************

fini        ! fini just in case not in begin

/solu       ! enters the solution processor, needs to be here to do editing below

allsel      ! default selects all items of specified entity type, typically nodes, elements

antype,modal,new
modopt,reduc,20      ! method - reduced Householder, number of modes to extract
expass,off           ! key = off, no expansion pass, key = on, do expansion
mxpand,20,,,no       ! nummodes to expand
total,20,0           ! total masters, 20 to be used, 1 to exclude rotational dofs

allsel

solve                ! starts the solution of one load step of a solution sequence, modal here

fini

! plot first mode

/post1

set,1,1

pldi,1

! **************** output frequencies *******************

/output,cantfem,frq     ! write out frequency list to ascii file .frq

set,list

/output,term            ! returns output to terminal

! ****************** output eigenvectors *********************

! define nodes for output:  forces applied or output displacements

nsel,s,node,,21         ! cantilever tip

/output,cantfem,eig     ! write out eigenvectors to ascii file .eig

*do,i,1,20
            set,,i
            prdisp
*enddo

/output,term

! **************** plot modes *******************
```

```
! pldi plots

/show,cantfem,grp,0
allsel

/view,1,,-1,,                    ! side view for plotting
/angle,1,0
/auto

*do,i,1,20
          set,1,i
    pldi
*enddo

/show,term

! *************** calculate and plot transfer functions ****************

fini

/assign,rst,junk,rst             ! reassigns a file name to an ANSYS identifier

/solu

dmprat,0.01          ! sets a constant damping ratio for all modes, zeta = 0.01

allsel
eplo                 ! show forces applied

f,21,fz,1            ! 1 mn force applied to node 21, tip node

/title, cantilever with tip load

antype,harmic        ! harmonic (frequency response) analysis

hropt,msup,20        ! mode superposition method, nummodes modes used

harfrq,100,1000000   ! frequency range, hz, for solution, -1 to 10 rad/sec

hrout,off,off        ! amplitude/phase, cluster off

kbc,1

nsubst,10000         ! 10000 frequency points for very fine resolution

outres,nsol,all,     ! controls solution set written to database, nodal dof solution, all
                     ! frequencies, component name for selected set of nodes

solve

fini

/post26

file,,rfrq           ! frequency response results
```

```
xvar,0                   ! display versus frequency

lines,10000              ! specifies the length of a printed page for frequency response listing

nsol,2,21,u,z            ! specifies nodal data to be stored in results file
                         ! u - displacement, z direction
                         ! note that nsol,1 is frequency vector

! plot magnitude

plcplx,0
/grid,1
/axlab,x,frequency, hz
/axlab,y,amplitude, mm
/gropt,logx,1            ! log plot for frequency
/gropt,logy,1            ! log plot for amplitude

/show,cantfem,grp1       ! file name for storing
plvar,2
/show,term

! plot phase

plcplx,1
/grid,1
/axlab,x,freq
/axlab,y,phase, deg      ! label for y axis
/gropt,logx,1            ! log plot for frequency
/gropt,logy,0            ! linear plot for phase

/show,cantfem,grp1
plvar,2
/show,term

! save ascii data to file

prcplx,1                 ! stores phase angle in ascii file .dat

/output,cantfem,dat
prvar,2
/output,term

fini
```

4.3.3 ANSYS Code cantzero.inp Description and Listing

```
/title, cantzero.inp, 0.05 x 1 x 20mm aluminum tip constrained cantilever beam, 20 elements

/prep7

et,1,4                   ! element type for beam

! aluminum
```

```
ex,1,71e6                    ! mN/mm^2
dens,1,2.77e-6               ! kg/mm^3
nuxy,1,.345

! real value to define beam characteristics

r,1,1,.00001041,.004166,.05,1    ! area, moments of inertia, thickness

! define plotting characteristics

/view,1,1,-1,1    ! iso view
/angle,1,-60      ! iso view
/pnum,mat,1        ! color by material
/num,1            ! numbers off
/type,1,0         ! hidden plot
/pbc,all,1        ! show all boundary conditions

csys,0                        ! define global coordinate system

! nodes

n,1,0,0,0                     ! left-hand node
n,21,20,0,0                        ! right-hand node

fill,1,21                     ! interior nodes

nall
nplo

! elements

type,1
mat,1
real,1
e,1,2
egen,20,1,-1

! constrain left-hand end

nall
d,1,all,0                     ! constrain node 1, all dof's
d,21,uz,0                     ! constrain tip

! constrain all but uz and roty for all other nodes to allow only those dof's
! this will give 20 nodes, node 2 through node 21, each with 2 dof, giving a total of 40 dof
! can calculate a maximum of 40 eigenvalues if don't use Guyan reduction to reduce size of
! eigenvalue problem

nall
nsel,s,node,,2,21
d,all,ux
d,all,uy
d,all,rotx
d,all,rotz
```

```
nall
eall
nplo
eplo

! ***************** eigenvalue run *******************

fini                 ! fini just in case not in begin

/solu                ! enters the solution processor, needs to be here to do editing below

allsel               ! default selects all items of specified entity type, typically nodes, elements

antype,modal,new
modopt,reduc,20      ! method - reduced Householder, number of modes to extract
expass,off           ! key = off, no expansion pass, key = on, do expansion
mxpand,20,,,no       ! nummodes to expand
total,20             ! total masters, 20 to be used, exclude rotational dofs

allsel

solve                ! starts the solution of one load step of a solution sequence, modal here

fini

! plot first mode

/post1

set,1,1

pldi,1

! ******************* output frequencies *********************

/output,cantzero,frq     ! write out frequency list to ascii file .frq

set,list

/output,term         ! returns output to terminal

! **************** output eigenvectors ********************

! define nodes for output:  forces applied or output displacements

nsel,s,node,,10      ! cantilever midpoint

/output,cantzero,eig     ! write out eigenvectors to ascii file .eig

*do,i,1,20
        set,,i
        prdisp
*enddo
```

```
/output,term

! ****************** plot modes ********************

! pldi plots

/show,cantzero,grp,0
allsel

/view,1,,-1,,          ! side view for plotting
/angle,1,0
/auto

*do,i,1,20
          set,1,i
   pldi
*enddo

/show,term
```

4.3.4 ANSYS Results, cantzero.m

The driving point frequency response for cantfem.inp is shown in Figure 4.11. The ANSYS frequency and magnitude output results are read into MATLAB and plotted in order to be able to overlay the resonances from the cantzero.inp ANSYS run. The MATLAB code to plot the overlay is **cantzero.m**, which reads in two input programs, **cantfem_magphs.m** and **cantzero_freq.m**.

The resonant frequencies (poles) of the cantilever and constrained tip cantilever models are listed in Table 4.1.

According to the guidelines for zeros discussed earlier in the chapter, the poles of the frequency response plot should be the same frequencies as shown in the "cantfem freq" column above. The zeros of the frequency response should be the same frequencies as shown in the "cantzero freq" column above.

mode	cantfem freq, hz	cantzero freq, hz
1	457.14	2004.6
2	2864.4	6495.0
3	8018.8	13548.
4	15709.	23162.
5	25961.	35336.
6	38771.	50071.
7	54147.	67380.
8	72102.	87291.
9	92672.	0.10985E+06
10	0.11592E+06	0.13520E+06
11	0.14196E+06	0.16337E+06
12	0.17098E+06	0.19495E+06
13	0.20323E+06	0.22951E+06
14	0.23907E+06	0.26909E+06
15	0.27885E+06	0.31129E+06
16	0.32274E+06	0.35968E+06
17	0.37012E+06	0.40928E+06
18	0.41860E+06	0.45602E+06
19	0.46289E+06	0.49344E+06
20	0.49490E+06	0.89212E+06

Table 4.1: Unconstrained (cantfem) and constrained tip (cantzero) cantilever resonances.

The constrained system poles in Figure 4.11 are shown below the curve with "o" symbols. Note that the "o's" align with the zeros of the unconstrained system.

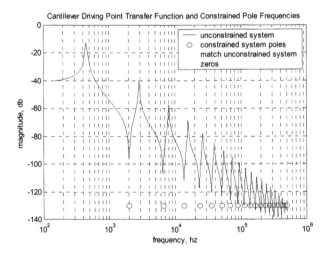

Figure 4.11: Cantilever driving point transfer function frequency response plot with overlaid frequencies of contrained-tip cantilever poles – which should match the unconstrained system zeros.

Problem

Note: The problem refers to the two dof system shown in Figure P2.2.

P4.1 Use the MATLAB code **ndof_numzeros.m** to identify the number of poles and zeros for a five dof system for the following: z11, z23, z33. Correlate the poles of the constrained system with the zeros of the original system.

CHAPTER 5

STATE SPACE ANALYSIS

5.1 Introduction

In Chapter 2 we derived the equations of motion for the tdof system shown in Figure 5.1, and showed how to solve the coupled differential equations for various transfer functions. In order to solve time domain problems using a computer, it is desirable to change the form of the equations for an n dof system with n second order differential equations to 2n first order differential equations. The first order form of equations of motion is known as **state space** form.

This chapter will develop the state space formulation for the tdof example. Once the state space formulation is completed, the subject of complex eigenvalues and eigenvectors, resulting in **complex modes** of vibration, will be covered in some detail. Once complex modes are understood, comprehending **real modes** which arise from the undamped case in the modal analysis section (Chapter 7) is simple.

Having an understanding of complex modes is especially helpful in working with experimental modal analysis. There are some very powerful experimental techniques available for testing and then visualizing the modes of vibration of structures. Frequency response data is taken at a number of selected positions on the structure and software is available to fit the data and define modes of vibration. The software identifies the resonant frequencies of the system and defines a damping value for each mode. It is then possible to create a model of the geometry of the test point locations and build a virtual model which can be animated to display the shape of motion of each mode.

The software has options which allow one to view the mode as either "real" or "complex." When the mode is viewed as "real," all the points on the structure move such that they all reach their maximum or minimum positions at the same point in time, which is consistent with our definition of "principal" or "real" modes defined in Chapter 7.

When the mode is viewed as "complex," the structure does not move such that all points reach either their minimum or maximum positions at the same point in time. Instead there appears to be a wave that moves along the structure as the different points reach their minimum or maximum positions at different times. For lightly damped mechanical structures, the assumption is often made that the modes are "real," allowing use of modal analysis methods and efficient finite element models. For structures that are not "lightly damped," the modal analysis method cannot be used and the state space formulation is the only practical method of solving the problem.

It is difficult to visualize complex modes without an animated structure model, but we will use a graphical method called an **Argand diagram** to explain how modes described by complex eigenvectors and complex eigenvalues combine to create physical motion of the system. We will find that if the unforced system is started from a set of initial conditions that match the complex eigenvector then only a single mode is excited. We will show how to calculate the transient response of the system for that specific initial condition case and illustrate how only a single mode is excited.

Chapter 6 will cover how to use the state space formulation to obtain both frequency and time domain results with MATLAB.

5.2 State Space Formulation

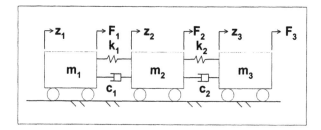

Figure 5.1: Original damped tdof system model.

Repeating the matrix equations of motion from (2.25):

$$\begin{bmatrix} m_1 & 0 & 0 \\ 0 & m_2 & 0 \\ 0 & 0 & m_3 \end{bmatrix}\begin{bmatrix} \ddot{z}_1 \\ \ddot{z}_2 \\ \ddot{z}_3 \end{bmatrix} + \begin{bmatrix} c_1 & -c_1 & 0 \\ -c_1 & (c_1+c_2) & -c_2 \\ 0 & -c_2 & c_2 \end{bmatrix}\begin{bmatrix} \dot{z}_1 \\ \dot{z}_2 \\ \dot{z}_3 \end{bmatrix}$$
$$+ \begin{bmatrix} k_1 & -k_1 & 0 \\ -k_1 & (k_1+k_2) & -k_2 \\ 0 & -k_2 & k_2 \end{bmatrix}\begin{bmatrix} z_1 \\ z_2 \\ z_3 \end{bmatrix} = \begin{bmatrix} F_1 \\ F_2 \\ F_3 \end{bmatrix} \tag{5.1}$$

Expanding the equations:

$$m_1\ddot{z}_1 + c_1\dot{z}_1 - c_1\dot{z}_2 + k_1 z_1 - k_1 z_2 = F_1$$
$$m_2\ddot{z}_2 - c_1\dot{z}_1 + (c_1+c_2)\dot{z}_2 - c_2\dot{z}_3 - k_1 z_1 + (k_1+k_2)z_2 - k_2 z_3 = F_2 \tag{5.2a,b,c}$$
$$m_3\ddot{z}_3 - c_2\dot{z}_2 + c_2\dot{z}_3 - k_2 z_2 + k_2 z_3 = F_3$$

The three equations above are second order differential equations which require knowledge of the initial states of position and velocity for all three degrees of freedom in order to solve for the transient response.

In the state space formulation, the three second order differential equations are converted to six first order differential equations. Following typical state space notation, we will refer to the states as "x" and the output as "y."

Start by solving (5.2) for the three equations for the highest derivatives, in this case the three second derivatives, $\ddot{z}_1$, $\ddot{z}_2$, $\ddot{z}_3$:

$$\ddot{z}_1 = (F_1 - c_1\dot{z}_1 + c_1\dot{z}_2 - k_1z_1 + k_1z_2)/m_1$$
$$\ddot{z}_2 = (F_2 + c_1\dot{z}_1 - (c_1 + c_2)\dot{z}_2 + c_2\dot{z}_3 + k_1z_1 - (k_1 + k_2)z_2 + k\ z_3)/m_2$$
$$\ddot{z}_3 = (F_3 + c_2\dot{z}_2 - c_2\dot{z}_3 + k_2z_2 - k_2z_3)/m_3$$

$$(5.3a,b,c)$$

We now change notation, using "x" to define the six states; three positions and three velocities:

$$x_1 = z_1 \quad \text{Position of Mass 1} \qquad (5.4)$$
$$x_2 = \dot{z}_1 \quad \text{Velocity of Mass 1} \qquad (5.5)$$
$$x_3 = z_2 \quad \text{Position of Mass 2} \qquad (5.6)$$
$$x_4 = \dot{z}_2 \quad \text{Velocity of Mass 2} \qquad (5.7)$$
$$x_5 = z_3 \quad \text{Position of Mass 3} \qquad (5.8)$$
$$x_6 = \dot{z}_3 \quad \text{Velocity of Mass 3} \qquad (5.9)$$

By using this notation, we observe the relationship between the state and its first derivatives:

$$\dot{z}_1 = x_2 = \dot{x}_1 \qquad (5.10)$$
$$\dot{z}_2 = x_4 = \dot{x}_3 \qquad (5.11)$$
$$\dot{z}_3 = x_6 = \dot{x}_5 \qquad (5.12)$$

Also between the first and second derivatives:

$$\ddot{z}_1 = \dot{x}_2 \qquad (5.13)$$
$$\ddot{z}_2 = \dot{x}_4 \qquad (5.14)$$
$$\ddot{z}_3 = \dot{x}_6 \qquad (5.15)$$

Rewriting the three equations for $\ddot{z}_1$, $\ddot{z}_2$, $\ddot{z}_3$ in terms of the six states x_1 through x_6 and adding the three equations defining the position and velocity relationships:

$$\dot{x}_1 = x_2$$
$$\dot{x}_2 = (F_1 - c_1 x_2 + c_1 x_4 - k_1 x_1 + k_1 x_3)/m_1$$
$$\dot{x}_3 = x_4$$
$$\dot{x}_4 = (F_2 + c_1 x_2 - (c_1 + c_2)x_4 + c_2 x_6 + k_1 x_1 - (k_1 + k_2)x_3 + k_2 x_5)/m_2 \qquad (5.16\text{a-f})$$
$$\dot{x}_5 = x_6$$
$$\dot{x}_6 = (F_3 + c_2 x_4 - c_2 x_6 + k_2 x_3 - k_2 x_5)/m_3$$

Rewriting the equations above in matrix form as:

$$
\begin{bmatrix} \dot{x}_1 \\ \dot{x}_2 \\ \dot{x}_3 \\ \dot{x}_4 \\ \dot{x}_5 \\ \dot{x}_6 \end{bmatrix}
=
\begin{bmatrix}
0 & 1 & 0 & 0 & 0 & 0 \\
\dfrac{-k_1}{m_1} & \dfrac{-c_1}{m_1} & \dfrac{k_1}{m_1} & \dfrac{c_1}{m_1} & 0 & 0 \\
0 & 0 & 0 & 1 & 0 & 0 \\
\dfrac{k_1}{m_2} & \dfrac{c_1}{m_2} & \dfrac{-(k_1+k_2)}{m_2} & \dfrac{-(c_1+c_2)}{m_2} & \dfrac{k_2}{m_2} & \dfrac{c_2}{m_2} \\
0 & 0 & 0 & 0 & 0 & 1 \\
0 & 0 & \dfrac{k_2}{m_3} & \dfrac{c_2}{m_3} & \dfrac{-k_2}{m_3} & \dfrac{-c_2}{m_3}
\end{bmatrix}
\begin{bmatrix} x_1 \\ x_2 \\ x_3 \\ x_4 \\ x_5 \\ x_6 \end{bmatrix}
+
\begin{bmatrix} 0 \\ \dfrac{F_1}{m_1} \\ 0 \\ \dfrac{F_2}{m_2} \\ 0 \\ \dfrac{F_3}{m_3} \end{bmatrix} \qquad (1)
$$

$$\dot{\mathbf{x}} = \qquad\qquad \mathbf{A} \qquad\qquad \mathbf{x} + \mathbf{B}\ \mathbf{u}$$
$$(5.17\text{a,b})$$

5.3 Definition of State Space Equations of Motion

Schematically, a SISO state space system is represented as shown in Figure 5.2. We will define the blocks in the following sections. The scalar input u(t) is fed into both the input matrix **B** and the direct transmission matrix **D**. The output of the input matrix is an nx1 vector, where "n" is the number of states. For a SISO system, the direct transmission matrix is a scalar, and its output is fed into a summing junction to be added to the output of the **C** matrix.

The output of the **B** matrix is added to the feedback term coming from the system matrix and is fed into an integrator block, where "I" is an nxn identify matrix. The output matrix has as many rows as outputs, a single row for a

SISO system, and has as many columns as states, n. The output y(t) is the
sum of the output of the **C** and **D** matrices.

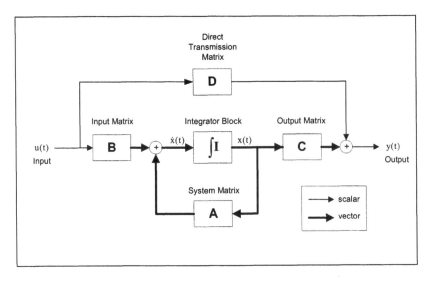

Figure 5.2: State space system block diagram.

Notation for equations of motion in state space form is:

$$\dot{x} = Ax + Bu \qquad\qquad (5.18)$$

where the **A** and **B** matrices are shown in (5.17a). Matrix **A** is known as
the system matrix, matrix **B** is the input matrix, and scalar u is the input. The
column vector **x** is the state of the system.

5.4 Input Matrix Forms

Because "u" is a scalar, the nature of the input matrix **B** changes depending
on what input is used. If the system is a Single Input (SI) system with a force
either at mass 1, 2 or 3, the **B** matrix changes as follows:

$$F_1 : \quad B = \begin{bmatrix} 0 \\ \dfrac{F_1}{m_1} \\ 0 \\ 0 \\ 0 \\ 0 \end{bmatrix}, \quad F_2 : \quad B = \begin{bmatrix} 0 \\ 0 \\ 0 \\ \dfrac{F_2}{m_2} \\ 0 \\ 0 \end{bmatrix}, \quad F_3 : \quad B = \begin{bmatrix} 0 \\ 0 \\ 0 \\ 0 \\ 0 \\ \dfrac{F_3}{m_3} \end{bmatrix} \qquad (5.19a,b,c)$$

If the same forcing function u (for example, a step function or sine function) is applied to several degrees of freedom simultaneously (for example, a force of magnitude F_1 to mass 1 and a force of magnitude F_3 to mass 3) the input matrix would become:

$$B = \begin{bmatrix} 0 \\ \dfrac{F_1}{m_1} \\ 0 \\ 0 \\ 0 \\ \dfrac{F_3}{m_3} \end{bmatrix} \qquad (5.20)$$

For a Multi Input (MI) system, where forces are applied independent of one another to the separate masses, a multiple column input matrix is appropriate. For example, for different inputs at mass 1 and mass 2, none at mass 3, the input matrix would become:

$$B = \begin{bmatrix} 0 & 0 \\ \dfrac{F_1}{m_1} & 0 \\ 0 & F_2 \\ 0 & m_2 \\ 0 & 0 \\ 0 & 0 \end{bmatrix} \qquad (5.21)$$

5.5 Output Matrix Forms

To account for the case where the desired output is not just the states but is some linear combination of the states, an output matrix **C** is defined to relate the outputs to the states. Also, a matrix **D**, known as the direct transmission matrix, is multiplied by the input "u" to account for outputs that are related to the inputs but that bypass the states.

$$\mathbf{y} = \mathbf{C}\mathbf{x} + \mathbf{D}\mathbf{u} \qquad (5.22)$$

The output matrix **C** has as many rows as outputs required and as many columns as states. The direct transmission matrix **D** has the same number of columns as the input matrix **B** and as many rows as the output matrix **C**.

In our example, we are interested in all six of the states, displacements and velocities, so the matrix output equation becomes, where **C** is the identity matrix and **D** is assumed to be zero:

$$
\begin{bmatrix} y_1 \\ y_2 \\ y_3 \\ y_4 \\ y_5 \\ y_6 \end{bmatrix}
=
\begin{bmatrix}
1 & 0 & 0 & 0 & 0 & 0 \\
0 & 1 & 0 & 0 & 0 & 0 \\
0 & 0 & 1 & 0 & 0 & 0 \\
0 & 0 & 0 & 1 & 0 & 0 \\
0 & 0 & 0 & 0 & 1 & 0 \\
0 & 0 & 0 & 0 & 0 & 1
\end{bmatrix}
\begin{bmatrix} x_1 \\ x_2 \\ x_3 \\ x_4 \\ x_5 \\ x_6 \end{bmatrix}
+
\begin{bmatrix} 0 \\ 0 \\ 0 \\ 0 \\ 0 \\ 0 \end{bmatrix}(1) \qquad (5.23)
$$

Expanding, the matrix equations become:

$$y_1 = x_1 \qquad (= z_1) \qquad\qquad (5.24)$$

$$y_2 = x_2 \qquad (= \dot{z}_1) \qquad\qquad (5.25)$$

$$y_3 = x_3 \qquad (= z_2) \qquad\qquad (5.26)$$

$$y_4 = x_4 \qquad (= \dot{z}_2) \qquad\qquad (5.27)$$

$$y_5 = x_5 \qquad (= z_3) \qquad\qquad (5.28)$$

$$y_6 = x_6 \qquad (= \dot{z}_3) \qquad\qquad (5.29)$$

If we were only interested in the three displacements and not the three velocities, the output equation would be, assuming **D** is zero:

$$\begin{bmatrix} y_1 \\ y_2 \\ y_3 \end{bmatrix} = \begin{bmatrix} 1 & 0 & 0 & 0 & 0 & 0 \\ 0 & 0 & 1 & 0 & 0 & 0 \\ 0 & 0 & 0 & 0 & 1 & 0 \end{bmatrix} \begin{bmatrix} x_1 \\ x_2 \\ x_3 \\ x_4 \\ x_5 \\ x_6 \end{bmatrix} + (0)(1) \qquad (5.30)$$

Expanding:

$$y_1 = x_1 \qquad (= z_1) \qquad (5.31)$$

$$y_2 = x_3 \qquad (= z_2) \qquad (5.32)$$

$$y_3 = x_5 \qquad (= z_3) \qquad (5.33)$$

On the other hand, if the outputs are linear combinations of the states, as in a control system problem, the output equation could look like (where a, b and c are scalars), assuming **D** is zero:

$$\begin{bmatrix} y_1 \\ y_2 \\ y_3 \\ y_4 \end{bmatrix} = \begin{bmatrix} 0 & 0 & a & 0 & b & 0 \\ c & 0 & 1 & 0 & 0 & 0 \\ 1 & 0 & 0 & 0 & 0 & 0 \\ 0 & 0 & 0 & 1 & 0 & 0 \end{bmatrix} \begin{bmatrix} x_1 \\ x_2 \\ x_3 \\ x_4 \\ x_5 \\ x_6 \end{bmatrix} + (0)(1) \qquad (5.34)$$

Expanding:

$$y_1 = ax_3 + bx_5 \qquad (= az_2 + bz_3) \qquad (5.35)$$

$$y_2 = cx_1 + x_3 \qquad (= cz_1 + z_2) \qquad (5.36)$$

$$y_3 = x_1 \qquad (= z_1) \qquad (5.37)$$

$$y_4 = x_4 \qquad (= \dot{z}_2) \qquad (5.38)$$

If a single force is applied and a single output is desired (SISO), for example, a force applied at mass 1 and the output displacement at mass 3, assuming **D** is zero:

$$y = \begin{bmatrix} 0 & 0 & 0 & 0 & 1 & 0 \end{bmatrix} \begin{bmatrix} x_1 \\ x_2 \\ x_3 \\ x_4 \\ x_5 \\ x_6 \end{bmatrix} + (0)(1) \qquad (5.39)$$

With all the possible variations of the output equation, the state equation never changes; it is always:

$$\dot{x} = Ax + Bu \qquad (5.40)$$

5.6 Complex Eigenvalues and Eigenvectors – State Space Form

The most basic analysis one can perform on a dynamic system is to solve for its eigenvalues (natural frequencies) and eigenvectors (mode shapes). In this section we will develop the most general case where there are no limitations on the presence or magnitude of the two damping terms, which could result in complex eigenvalues and eigenvectors.

Start by postulating that there is a set of initial conditions such that if the system is released with that set, the system will respond in one of its natural modes of vibration. To that end, we set the forcing function to zero and write the homogeneous state space equations of motion:

$$\dot{x} = Ax \qquad (5.41)$$

We define motion in a principal mode as:

$$x_i = x_{mi}\, e^{\lambda_i t} \qquad (5.42)$$

Where:

λ_i is the ith eigenvalue, the natural frequency of the ith mode of vibration

x_i is the vector of states at the ith frequency

x_{mi} is the ith eigenvector, the mode shape for the ith mode

For our tdof (z_1 to z_3), six state (x_1 to x_6) system, for the ith eigenvalue and eigenvector, the equation would appear as:

$$
\begin{bmatrix} z_{1i} \\ \dot{z}_{1i} \\ z_{2i} \\ \dot{z}_{2i} \\ z_{3i} \\ \dot{z}_{3i} \end{bmatrix} = \begin{bmatrix} x_{1i} \\ x_{2i} \\ x_{3i} \\ x_{4i} \\ x_{5i} \\ x_{6i} \end{bmatrix} = \mathbf{x}_{mi} e^{\lambda_i t} = \begin{bmatrix} x_{m1i} \\ x_{m2i} \\ x_{m3i} \\ x_{m4i} \\ x_{m5i} \\ x_{m6i} \end{bmatrix} e^{\lambda_i t}
\tag{5.43}
$$

Differentiating the modal displacement equation above to get the modal velocity equation:

$$
\dot{\mathbf{x}}_{mi} = \frac{d}{dt}\left[\mathbf{x}_{mi} e^{\lambda t} \right] = \lambda \mathbf{x}_{mi} e^{\lambda t}
\tag{5.44}
$$

Substituting into the state equation and canceling the exponential terms leads to:

$$
\dot{\mathbf{x}} = \mathbf{A}\mathbf{x}
$$
$$
\lambda \mathbf{x}_{mi} e^{\lambda t} = \mathbf{A}\mathbf{x}_{mi} e^{\lambda t}
$$
$$
\lambda \mathbf{x}_{mi} = \mathbf{A}\mathbf{x}_{mi}
$$
$$
(\lambda \mathbf{I} - \mathbf{A})\mathbf{x}_{mi} = 0
\tag{5.45a-d}
$$

Equation (5.45c) is the classic "eigenvalue problem." If $\mathbf{x}_{mi}$ is not equal to zero in (5.45d), a solution exists only if the determinant below is zero (Strang 1998):

$$
\left| (\lambda \mathbf{I} - \mathbf{A}) \right| = 0
\tag{5.46}
$$

Taking the system matrix $\mathbf{A}$ from (5.17a) and inserting in (5.45):

$$(\lambda\mathbf{I} - \mathbf{A}) = \lambda\ \mathbf{I} - \begin{bmatrix} 0 & 1 & 0 & 0 & 0 & 0 \\ \dfrac{-k_1}{m_1} & \dfrac{-c_1}{m_1} & \dfrac{k_1}{m_1} & \dfrac{c_1}{m_1} & 0 & 0 \\ 0 & 0 & 0 & 1 & 0 & 0 \\ \dfrac{k_1}{m_2} & \dfrac{c_1}{m_2} & \dfrac{-(k_1+k_2)}{m_2} & \dfrac{-(c_1+c_2)}{m_2} & \dfrac{k_2}{m_2} & \dfrac{c_2}{m_2} \\ 0 & 0 & 0 & 0 & 0 & 1 \\ 0 & 0 & \dfrac{k_2}{m_3} & \dfrac{c_2}{m_3} & \dfrac{-k_2}{m_3} & \dfrac{-c_2}{m_3} \end{bmatrix}$$

$$(5.47)$$

In Chapter 10 we will use the undamped version of (5.46) with $c1 = c2 = 0$ to discuss "normal" modes, where we will find that taking the determinant in closed form is practical. For the tdof damped system matrix, taking the closed form determinant is far too complicated so we will use MATLAB's "eig" function to solve the eigenvalue problem numerically, using specific values of m, c and k. We will use the MATLAB code **tdof_non_prop_damped.m** as we continue our exploration of complex modes.

5.7 MATLAB Code tdof_non_prop_damped.m: Methodology, Model Setup, Eigenvalue Calculation Listing

The sequence of development of complex modes is as follows:

1) solve original damped system equation for complex eigenvalues and eigenvectors

2) normalize the eigenvector entries to unity

3) calculate magnitude and phase angle of each of the eigenvector entries

4) use the Argand diagram to visualize the motion of a complex mode

5) calculate the percentage of critical damping (damping ratio) for each mode

6) calculate the motions of the three masses for all three modes

7) plot the real and imaginary displacements of each
    of the degrees of freedom separately

We have explored how to calculate the eigenvectors or mode shapes for an
undamped problem using the transfer function matrix (Chapter 3). The modes
for the undamped problem were real modes, meaning that the position
elements of the eigenvectors were real, not complex, and we were able to plot
diagrams showing the shape of the modes. For complex modes, it is not
possible to draw a picture of the deformed mode shape because there are
phase differences between the various degrees of freedom which prevent them
from reaching their maximum/minimum points at the same point in time.
This leads to the apparent "traveling wave" in an animated mode.

The first section of **tdof_non_prop_damped.m** sets up the state space
equations of motion and solves the eigenvalue problem for damping values of
$c_1 = 0.1$, $c_2 = 0.2$:

```
%        tdof_non_prop_damped.m        non-proportionally damped tdof model

         clf;

         legend off;

         subplot(1,1,1);

         clear all;

%        define the values of masses, springs, dampers

         m1 = 1;
         m2 = 1;
         m3 = 1;

         k1 = 1;
         k2 = 1;

%        define arbitrary damping values

         c1 = input('input value for c1, default 0.1, ... ');

         if (isempty(c1))
                 c1 = 0.1;
         else
         end

         c2 = input('input value for c1, default 0.2, ... ');

         if (isempty(c2))
                 c2 = 0.2;
         else
         end
```

```
%         define the system matrix, aphys, in physical coordinates

          aphys = [    0        1         0           0          0        0
                    -k1/m1    -c1/m1      k1/m1       c1/m1       0        0
                       0        0         0           1          0        0
                     k1/m2    c1/m2    -(k1+k2)/m2  -(c1+c2)/m2  k2/m2    c2/m2
                       0        0         0           0          0        1
                       0        0         k2/m3       c2/m3     -k2/m3   -c2/m3];

%         solve for the eigenvalues of the system matrix

          [xm,lambda] = eig(aphys);

%         take the diagonal elements of the generalized eigenvalue matrix lambda

          lambdad = diag(lambda);
```

The six eigenvalues, lambda values, are listed below. Since we have three
degrees of freedom, there should be three sets of complex conjugate
eigenvalues.

```
xm =

Columns 1 through 4

-0.0567 - 0.1940i  -0.0567 + 0.1940i   0.2886 - 0.4085i   0.2886 + 0.4085i
 0.3452 - 0.0535i   0.3452 + 0.0535i   0.3865 + 0.3190i   0.3865 - 0.3190i
 0.0624 + 0.4029i   0.0624 - 0.4029i  -0.0218 - 0.0123i  -0.0218 + 0.0123i
-0.7046 + 0.0162i  -0.7046 - 0.0162i   0.0139 - 0.0209i   0.0139 + 0.0209i
-0.0057 - 0.2089i  -0.0057 + 0.2089i  -0.2668 + 0.4208i  -0.2668 - 0.4208i
 0.3593 + 0.0373i   0.3593 - 0.0373i  -0.4004 - 0.2981i  -0.4004 + 0.2981i

Columns 5 through 6

 0.0000 - 0.5774i   0.0000 + 0.5774i
 0.0000 + 0.0000i   0.0000 - 0.0000i
 0.0000 - 0.5774i   0.0000 + 0.5774i
 0.0000 + 0.0000i   0.0000 - 0.0000i
 0.0000 - 0.5774i   0.0000 + 0.5774i
 0.0000 + 0.0000i   0.0000 - 0.0000i


lambda =

Columns 1 through 4

-0.2250 + 1.7141i      0              0              0
      0          -0.2250 - 1.7141i    0              0
      0                0         -0.0750 + 0.9991i    0
      0                0              0         -0.0750 - 0.9991i
      0                0              0              0
```

```
    0        0        0        0

Columns 5 through 6

    0        0
    0        0
    0        0
    0        0
 -0.0000 + 0.0000i    0
    0        -0.0000 - 0.0000i


lambdad =

 -0.2250 + 1.7141i
 -0.2250 - 1.7141i
 -0.0750 + 0.9991i
 -0.0750 - 0.9991i
 -0.0000 + 0.0000i
 -0.0000 - 0.0000i
```

Note that the two eigenvalues which correspond to each of the three modes are complex conjugates of each other, and that the real parts of the second and third mode eigenvalues are all negative.

We did not specify the form of the eigenvalues, which in the most general case can be complex, as in the second and third modes above. We will now discuss the components of complex eigenvalues. We use the term λ_{n1} to describe the first complex eigenvalue of any of the three sets of eigenvalues above. The term λ_{n2} is used to describe the second complex eigenvalue of the set, and the complex conjugacy of the two is stated as: $\lambda_{n2} = \lambda_{n1}^*$, where the "*" indicates a complex conjugate. The real and imaginary parts will be defined using σ_{nx} and ω_{nx}, respectively:

$$\lambda_{n1} = \sigma_{n1} + j\omega_{n1}$$
$$\lambda_{n2} = \lambda_{n1}^* = \sigma_{n1} - j\omega_{n1}$$

(5.48)

See Figure 5.3 for graphical descriptions of the components of a complex eigenvalue. The figure shows two complex conjugate eigenvalues (poles) in the left half plane as "x" symbols. The real parts of the two eigenvalues are the same and are given the symbol σ, with the imaginary parts both having a distance from the origin of ω, referred to as the damped natural frequency. The radial distance from the origin to the poles is given by ω_n and is referred to as the undamped natural frequency. The angle between the imaginary axis and the line from the origin to the pole is used to define the amount of damping of the mode, referred to as ζ, the damping ratio or percentage of critical damping.   If $\sigma = 0$, $\theta = 0$ and there is no damping, therefore $\omega = \omega_n$.

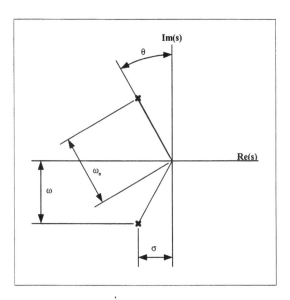

Figure 5.3: Complex eigenvalue (pole) nomenclature in complex plane.

Referring to Figure 5.3 for the definition of θ, the equation for calculating ζ for a mode from the real and imaginary components of the eigenvalue is:

$$
\begin{aligned}
\zeta &= \sin\theta \\
&= \sin\left(\tan^{-1}\left(\frac{\text{Re}(\lambda)}{\text{Im}(\lambda)} \right) \right) \\
&= \sin\left(\tan^{-1}\left(\frac{\sigma}{\omega} \right) \right)
\end{aligned}
\qquad (5.49)
$$

5.8 Eigenvectors – Normalized to Unity

The section of code below reorders the eigenvectors from low to high frequency and normalizes them. The normalization procedure is to divide each eigenvector by its position state for mass 1, the first term in each eigenvector.

```
%         now reorder the eigenvalues and eigenvectors from low to high frequency,
```

```
%          keeping track of how the eigenvalues are ordered in reorder the
%          eigenvectors to match, using indexhz

[lambdaorder,indexhz] = sort(abs(imag(lambdad)));

for  cnt = 1:length(lambdad)

      lambdao(cnt,1) = lambdad(indexhz(cnt));          % reorder eigenvalues

      xmo(:,cnt) = xm(:,indexhz(cnt));            % reorder eigenvector columns

end

%          now normalize the eigenvectors with respect to the position of mass 1, which
%          will be set to 1.0

for  cnt = 1:length(lambdad)

            xmon1(:,cnt) = xmo(:,cnt)/xmo(1,cnt);

end
```

The eigenvectors, normalized such that the displacements of mass 1 are set to 1.0 are shown below as xmon1.

```
lambdao =

 -0.0000 + 0.0000i
 -0.0000 - 0.0000i
 -0.0750 + 0.9991i
 -0.0750 - 0.9991i
 -0.2250 + 1.7141i
 -0.2250 - 1.7141i


xmo =

Columns 1 through 4

 0.0000 - 0.5774i   0.0000 + 0.5774i   0.2886 - 0.4085i   0.2886 + 0.4085i
 0.0000 + 0.0000i   0.0000 - 0.0000i   0.3865 + 0.3190i   0.3865 - 0.3190i
 0.0000 - 0.5774i   0.0000 + 0.5774i  -0.0218 - 0.0123i  -0.0218 + 0.0123i
 0.0000 + 0.0000i   0.0000 - 0.0000i   0.0139 - 0.0209i   0.0139 + 0.0209i
 0.0000 - 0.5774i   0.0000 + 0.5774i  -0.2668 + 0.4208i  -0.2668 - 0.4208i
 0.0000 + 0.0000i   0.0000 - 0.0000i  -0.4004 - 0.2981i  -0.4004 + 0.2981i

Columns 5 through 6

 -0.0567 - 0.1940i  -0.0567 + 0.1940i
  0.3452 - 0.0535i   0.3452 + 0.0535i
  0.0624 + 0.4029i   0.0624 - 0.4029i
 -0.7046 + 0.0162i  -0.7046 - 0.0162i
 -0.0057 - 0.2089i  -0.0057 + 0.2089i
```

```
0.3593 + 0.0373i   0.3593 - 0.0373i


xmon1 =

Columns 1 through 4

 1.0000 - 0.0000i  1.0000 + 0.0000i  1.0000         1.0000
 0.0000 + 0.0000i  0.0000 - 0.0000i -0.0750 + 0.9991i -0.0750 - 0.9991i
 1.0000 - 0.0000i  1.0000 + 0.0000i -0.0050 - 0.0498i -0.0050 + 0.0498i
-0.0000 + 0.0000i -0.0000 - 0.0000i  0.0502 - 0.0013i  0.0502 + 0.0013i
 1.0000 - 0.0000i  1.0000 + 0.0000i -0.9950 + 0.0498i -0.9950 - 0.0498i
 0.0000 + 0.0000i  0.0000 - 0.0000i  0.0248 - 0.9978i  0.0248 + 0.9978i

Columns 5 through 6

 1.0000 - 0.0000i  1.0000 + 0.0000i
-0.2250 + 1.7141i -0.2250 - 1.7141i
-2.0001 - 0.2630i -2.0001 + 0.2630i
 0.9009 - 3.3691i  0.9009 + 3.3691i
 1.0001 + 0.2630i  1.0001 - 0.2630i
-0.6759 + 1.6550i -0.6759 - 1.6550i
```

The six rows of each eigenvector are related to the six states, x_1 to x_6, where x_1, x_3, x_5 are the displacement states and x_2, x_4, x_6 are the velocity states. Each velocity row is equal to the displacement row associated with it times its eigenvector, as can be seen by repeating (5.41) and differentiating it.

$$
\begin{aligned}
\mathbf{x}_i &= \mathbf{x}_{mi}\, e^{\lambda_i t} \\
\dot{\mathbf{x}}_i &= \lambda_i (\mathbf{x}_{mi} e^{\lambda_i t})
\end{aligned}
\tag{5.50}
$$

The tdof model has three degrees of freedom, so we should have three modes of vibration. The first two columns of the eigenvector matrix define mode 1, the third and fourth define mode 2 and the fifth and sixth columns define mode 3. Like the two complex conjugate eigenvalues for each mode, the two eigenvector columns for each of the modes are complex conjugates of each other.

5.9 Eigenvectors – Magnitude and Phase Angle Representation

Another way of looking at the eigenvectors is to calculate the magnitude and phase angle for each entry. The code for doing this follows.

```
%        now calculate the magnitude and phase angle of each of the eigenvector
%        entries

         for  row = 1:length(lambdad)
```

```
                    for  col = 1:length(lambdad)

                            xmon1mag(row,col) = abs(xmon1(row,col));

                            xmon1ang(row,col) = (180/pi)*angle(xmon1(row,col));

                    end

            end

            lambdao

            xmo

            xmon1

            xmon1mag

            xmon1ang
```

The magnitude and phase angles are:

```
xmon1mag =
   1.0000   1.0000   1.0000   1.0000   1.0000   1.0000
   0.0000   0.0000   1.0019   1.0019   1.7288   1.7288
   1.0000   1.0000   0.0501   0.0501   2.0173   2.0173
   0.0000   0.0000   0.0502   0.0502   3.4875   3.4875
   1.0000   1.0000   0.9962   0.9962   1.0341   1.0341
   0.0000   0.0000   0.9981   0.9981   1.7877   1.7877


xmon1ang =
        0         0         0         0         0         0
  90.0000  -90.0000   94.2930  -94.2930   97.4782  -97.4782
   0.0000    0.0000  -95.7723   95.7723 -172.5081  172.5081
  90.0000  -90.0000   -1.4793    1.4793  -75.0299   75.0299
   0.0000    0.0000  177.1334 -177.1334   14.7356  -14.7356
  90.0000  -90.0000  -88.5736   88.5736  112.2138 -112.2138
```

We will see in Chapter 7 that undamped eigenvector oscillatory modes have phases that are multiples of $90°$. For the damped complex eigenvectors the phases are slightly offset from being $90°$ multiples of each other.

5.10 Complex Eigenvectors Combining to Give Real Motions

Now that we have solved for the complex eigenvalues and eigenvectors, we will discuss how we can have the system respond in only a single mode of vibration by releasing the system with a particular set of initial conditions. We will answer the following question:

How does a mode that is described by complex eigenvalues and eigenvectors give "real," physically observable motions (Newland 1989)?

For the nth mode, the motion in that mode is defined as the sum of the motions due to the two conjugate eigenvalues/eigenvectors for that mode, as shown in (5.51). Substituting the complex conjugate value and collecting exponential terms:

$$
\begin{aligned}
\mathbf{x}(t) &= e^{\lambda_{n1}t}\mathbf{x}_{n1} + e^{\lambda_{n2}t}\mathbf{x}_{n2} \\
&= e^{\lambda_{n1}t}\mathbf{x}_{n1} + e^{\lambda_{n1}^{*}t}\mathbf{x}_{n1}^{*} \\
&= e^{(\sigma_{n1}+j\omega_{n1})t}\mathbf{x}_{n1} + e^{(\sigma_{n1}-j\omega_{n1})t}\mathbf{x}_{n1}^{*} \\
&= e^{\sigma_{n1}t}(e^{j\omega_{n1}t}\mathbf{x}_{n1} + e^{-j\omega_{n1}t}\mathbf{x}_{n1}^{*}) \\
&= 2e^{\sigma_{n1}t}\,\mathrm{Re}(\mathbf{x}_{n1})
\end{aligned}
\tag{5.51}
$$

The $e^{j\omega_{n1}t}\mathbf{x}_{n1}$ term represents a vector of magnitude $\left|\mathbf{x}_{n1}\right|$ which is rotating counter-clockwise at the rate of ω_{n1} radians/sec. The $e^{-j\omega_{n1}t}\mathbf{x}_{n1}^{*}$ term represents a vector of magnitude $\left|\mathbf{x}_{n1}^{*}\right|$ which is rotating clockwise at the rate of ω_{n1} radians/sec. This counter-rotation is the key to understanding how the sum of two complex numbers becomes real.   Since the two counter-rotating eigenvector terms are complex conjugates, their imaginary portions are of opposite sign and as they rotate, the sum of the two results in only a real component as the two imaginary portions cancel each other. See the Argand diagram in the next section for a graphical representation.

The $e^{\sigma_{n1}t}$ term is an exponentially decreasing scalar which multiplies the sum of the two counter-rotating vectors. The σ_{n1} term is the real value of the eigenvalue, and for a stable mode, with the poles in the left half of the s-plane, the value is always negative. Thus, $e^{\sigma_{n1}t}$ is exponentially decreasing with a time constant of $1/\sigma_{n1}$.

For real modes, the poles are on the imaginary axis, so $\sigma_{n1}=0$ and $e^{(0)t}=1$. The two counter-rotating vectors are not attenuated in amplitude with time, so the motion is undamped.

If the initial conditions for the system are set at one of the eigenvectors, the system will respond in only that mode. **For systems with complex modes, initial conditions of both displacements and velocities of all the masses must be set simultaneously in order for the system to respond only in that mode.** If the initial conditions for the system are set at any other value, the resulting motion will be composed of a superposition of the motions of several modes.

For undamped systems with normal modes, either the displacement or velocity initial conditions can be set and the system will respond only in that mode (see Chapter 7 for more details).

Equation (5.51) will be used in the MATLAB code for plotting the motion of the system for the two oscillatory modes.

5.11 Argand Diagram Introduction

Since we are dealing with complex modes where different parts of the structure reach their maximum and minimum positions at different times, we cannot plot deformed mode shape plots as we did for the undamped model in Chapter 3. The best way to visualize complex modes is by animating the mode shape, allowing one to see the different parts of the structure moving in time.

The use of an Argand or Phasor diagram is another way to visualize the motion. It plots rotating eigenvectors of position and velocity in the complex plane for each degree of freedom in the eigenvector and shows how the complex conjugate eigenvector components add to create the "real" motion.

The normalized eigenvector matrix, xmon1, is repeated below. The first two states, position and velocity of mass 1, dof z1, are highlighted in bold type for the second mode of vibration.

Figure 5.4 shows Argand diagrams for the highlighted mode and states in the eigenvector matrix below. All three plots are in the complex plane. The upper left-hand plot shows the position and velocity eigenvector components for the third column of the eigenvector matrix, where the position component is 1+0j and the velocity component is –0.075+0.999j. The position component plots from 0 to 1 on the real axis. Notice that the tip of the velocity vector is slightly to the left of the imaginary axis. The $e^{j\omega_2 t}$ term indicates that the position and velocity vectors are both rotating in the counter-clockwise direction at a speed of ω radians/sec, starting from the initial locations defined by the eigenvector components.

```
xmon1 =
   1.0000            1.0000             1.0000             1.0000
   0.0000 + 0.0000i  0.0000 - 0.0000i  -0.0750 + 0.9991i  -0.0750 - 0.9991i
   1.0000 + 0.0000i  1.0000 - 0.0000i  -0.0050 - 0.0498i  -0.0050 + 0.0498i
   0.0000 + 0.0000i  0.0000 - 0.0000i   0.0502 - 0.0013i   0.0502 + 0.0013i
   1.0000 + 0.0000i  1.0000 - 0.0000i  -0.9950 + 0.0498i  -0.9950 - 0.0498i
   0.0000 + 0.0000i  0.0000 - 0.0000i   0.0248 - 0.9978i   0.0248 + 0.9978i


   1.0000            1.0000
  -0.2250 + 1.7141i  -0.2250 - 1.7141i
  -2.0001 - 0.2630i  -2.0001 + 0.2630i
   0.9009 - 3.3691i   0.9009 + 3.3691i
   1.0001 + 0.2630i   1.0001 - 0.2630i
  -0.6759 + 1.6550i  -0.6759 - 1.6550i
```

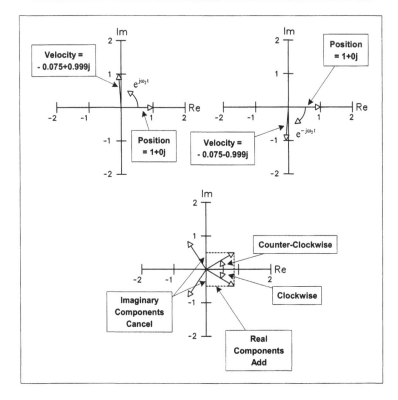

Figure 5.4: Argand diagram explanation.

The upper right-hand plot is similar to the left-hand plot except that the fourth column entries of the eigenvector matrix for the first two states are plotted and the two vectors are rotating in the clockwise direction. Note that the real components of the position and velocity components are the same as the third column, but that the imaginary components are complex conjugates of each other.

The lower plot illustrates the complex plane with both third and fourth eigenvectors shown on the same plot after rotating through the angle $\omega_2 t$. At any time "t," the two counter-rotating position vectors can be added to give the current position. At any time, the two imaginary components cancel out, leaving only the sum of the two real axis components as the "real" position. The same vector addition of the two counter-rotating velocity vectors will give the "real" velocity.

For an undamped model, the lengths of the two original eigenvector components stay the same. For the damped model, the lengths of all the vectors decrease continuously with a time constant of $1/\sigma_2$.

Looking at the Argand diagram above, which shows the "real" motion as twice the real axis component of the vector, it is clear that the motion as a function of time can also be written as:

$$\begin{aligned} \mathbf{x}(t) &= 2\,e^{\sigma_{ni}t}\,|\mathbf{x}_{nl}|\cos(\omega t + \phi_{ni}) \\ &= 2\,e^{\sigma_{ni}t}\,\mathrm{Re}(\mathbf{x}_{nl}) \end{aligned} \qquad (5.52)$$

where the phase angle ϕ_{ni} is given by:

$$\tan(\phi_{ni}) = \mathrm{Im}(z_{ni})/\mathrm{Re}(z_{ni}) \qquad (5.53)$$

5.12   Calculating ζ, Plotting Eigenvalues in Complex Plane, Frequency Response

This section of code calculates the percentage of critical damping for each of the three modes, ζ_i using (5.49).

```
%        calculate the percentage of critical damping for each mode

        zeta1 = 0

        theta2 = atan(real(lambdao(3))/imag(lambdao(3)));
        zeta2 = abs(sin(theta2))

        theta3 = atan(real(lambdao(5))/imag(lambdao(5)));
        zeta3 = abs(sin(theta3))

        plot(lambda,'k*')
        grid on
        axis([-3 1 -2 2])
        axis('square')
        title('Damped Eigenvalues')
        xlabel('real')
```

```
ylabel('imaginary')
text(real(lambdao(3))-1,imag(lambdao(3))+0.1,['zeta = ',num2str(zeta2)])
text(real(lambdao(5))-1,imag(lambdao(5))+0.1,['zeta = ',num2str(zeta3)])
disp('execution paused to display figure, "enter" to continue'); pause
```

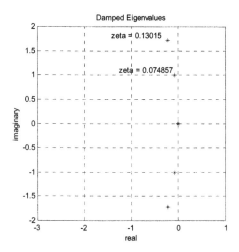

Figure 5.5: Plot of eigenvalues in complex plane for tdof model with c1 = 0.1, c2 = 0.2.

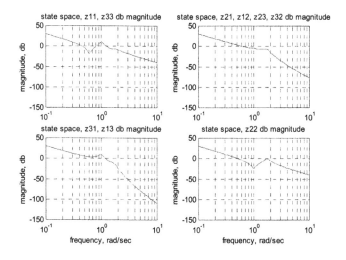

Figure 5.6: Frequency response magnitude plots.

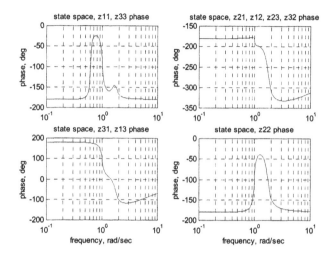

Figure 5.7: Frequency response phase plots.

The magnitude and phase frequency response plots for the system with $c1 = 0.1$ and $c2 = 0.2$ are shown above, using **tdofss.m** to plot. Note the significant attenuation of the resonances with zetas of 7.5% and 13% for modes 1 and 2, respectively. (Note: This amount of damping is very difficult to obtain in most practical structures without the use of additive damping.)

5.13 Initial Condition Responses of Individual Modes

The code below calculates the initial condition response for the oscillatory (not rigid body) second and third modes of the system when started with initial conditions defined by the appropriate eigenvector. Equation (5.51) is repeated below to show the form of the equation for $x(t)$ that is used in the code.

$$
\begin{aligned}
\mathbf{x}(t) &= e^{\sigma_{n1}t} \left(e^{j\omega_{n1}t}\mathbf{x}_{n1} + e^{-j\omega_{n2}t}\mathbf{x}_{n2} \right) \\
&= e^{\sigma_{n1}t} \left(e^{j\omega_{n1}t}\mathbf{x}_{n1} \right) + e^{\sigma_{n1}t} \left(e^{-j\omega_{n2}t}\mathbf{x}_{n2} \right)
\end{aligned}
\tag{5.54}
$$

The real and imaginary components of the eigenvalues are calculated to give σ and ω in the equation above. The real and imaginary displacements of each of the three masses are then calculated for both oscillatory modes for a time period of 15 seconds.

```
%        calculate the motions of the three masses for all three modes - damped case

t = 0:.12:15;
```

```
          sigma11 = real(lambdao(1));      % sigma for first eigenvalue for mode 1
          omega11 = imag(lambdao(1));      % omega for first eigenvalue for mode 1

          sigma12 = real(lambdao(2));      % sigma for second eigenvalue for mode 1
          omega12 = imag(lambdao(2));      % omega for second eigenvalue for mode 1

          sigma21 = real(lambdao(3));      % sigma for first eigenvalue for mode 2
          omega21 = imag(lambdao(3));      % omega for first eigenvalue for mode 2

          sigma22 = real(lambdao(4));      % sigma for second eigenvalue for mode 2
          omega22 = imag(lambdao(4));      % omega for second eigenvalue for mode 2

          sigma31 = real(lambdao(5));      % sigma for first eigenvalue for mode 3
          omega31 = imag(lambdao(5));      % omega for first eigenvalue for mode 3

          sigma32 = real(lambdao(6));      % sigma for second eigenvalue for mode 3
          omega32 = imag(lambdao(6));      % omega for second eigenvalue for mode 3

%         motion of three masses for mode 1

          z111r = exp(sigma11*t).*(exp(i*omega11*t)*xmon1(1,1));      % mass 1
          z112r = exp(sigma12*t).*(exp(i*omega12*t)*xmon1(1,2));      % mass 1

          z121r = exp(sigma11*t).*(exp(i*omega11*t)*xmon1(3,1));      % mass 2
          z122r = exp(sigma12*t).*(exp(i*omega12*t)*xmon1(3,2));      % mass 2

          z131r = exp(sigma11*t).*(exp(i*omega11*t)*xmon1(5,1));      % mass 3
          z132r = exp(sigma12*t).*(exp(i*omega12*t)*xmon1(5,2));      % mass 3

%         motion of three masses for mode 2

          z211r = exp(sigma21*t).*(exp(i*omega21*t)*xmon1(1,3));      % mass 1
          z212r = exp(sigma22*t).*(exp(i*omega22*t)*xmon1(1,4));      % mass 1

          z221r = exp(sigma21*t).*(exp(i*omega21*t)*xmon1(3,3));      % mass 2
          z222r = exp(sigma22*t).*(exp(i*omega22*t)*xmon1(3,4));      % mass 2

          z231r = exp(sigma21*t).*(exp(i*omega21*t)*xmon1(5,3));      % mass 3
          z232r = exp(sigma22*t).*(exp(i*omega22*t)*xmon1(5,4));      % mass 3

%         motion of three masses for mode 3

          z311r = exp(sigma31*t).*(exp(i*omega31*t)*xmon1(1,5));      % mass 1
          z312r = exp(sigma32*t).*(exp(i*omega32*t)*xmon1(1,6));      % mass 1

          z321r = exp(sigma31*t).*(exp(i*omega31*t)*xmon1(3,5));      % mass 2
          z322r = exp(sigma32*t).*(exp(i*omega32*t)*xmon1(3,6));      % mass 2

          z331r = exp(sigma31*t).*(exp(i*omega31*t)*xmon1(5,5));      % mass 3
          z332r = exp(sigma32*t).*(exp(i*omega32*t)*xmon1(5,6));      % mass 3
```

5.14 Plotting Initial Condition Response, Listing

The code listing below is to plot various combinations of real and imaginary components of the displacements of the three masses when released in states which match the eigenvectors.

```
%       plot real and imaginary motions of each mass for the two complex conjugate
%       eigenvectors of mode 2

        plot(t,real(z211),'k-',t,real(z212),'k+-',t,imag(z211),'k.-',t,imag(z212),'ko-')
        title('non-prop damped real and imag for z1, mode 2')
        legend('real','real','imag','imag')
        xlabel('time, sec')
        axis([0 max(t) -1 1])
        grid on

        disp('execution paused to display figure, "enter" to continue'); pause

        plot(t,real(z221),'k-',t,real(z222),'k+-',t,imag(z221),'k.-',t,imag(z222),'ko-')
        title('non-prop damped real and imag for z2 mode 2')
        legend('real','real','imag','imag')
        xlabel('time, sec')
        axis([0 max(t) -1 1])
        grid on

        disp('execution paused to display figure, "enter" to continue'); pause

        plot(t,real(z231),'k-',t,real(z232),'k+-',t,imag(z231),'k.-',t,imag(z232),'ko-')
        title('non-prop damped real and imag for z3 mode 2')
        legend('real','real','imag','imag')
        xlabel('time, sec')
        axis([0 max(t) -1 1])
        grid on

        disp('execution paused to display figure, "enter" to continue'); pause

        plot(t,real(z211+z212),'k-',t,real(z221+z222),'k+-',t,real(z231+z232),'k.-')
        title('non-prop damped, z1, z2, z3 mode 2')
        legend('mass 1','mass 2','mass 3')
        xlabel('time, sec')
        axis([0 max(t) -2 2])
        grid on

        disp('execution paused to display figure, "enter" to continue'); pause

%       plot subplots for notes

        subplot(2,2,1)
        plot(t,real(z211),'k-',t,real(z212),'k+',t,imag(z211),'k.-',t,imag(z212),'ko-')
        title('non-prop damped real and imag for z1, mode 2')
        legend('real','real','imag','imag')
        axis([0 max(t) -1 1])
        grid on
```

```
        subplot(2,2,2)
        plot(t,real(z221),'k-',t,real(z222),'k+',t,imag(z221),'k.-',t,imag(z222),'ko-')
        title('non-prop damped real and imag for z2 mode 2')
        legend('real','real','imag','imag')
        axis([0 max(t) -1 1])
        grid on

        subplot(2,2,3)
        plot(t,real(z231),'k-',t,real(z232),'k+',t,imag(z231),'k.-',t,imag(z232),'ko-')
        title('non-prop damped real and imag for z3 mode 2')
        legend('real','real','imag','imag')
        xlabel('time, sec')
        axis([0 max(t) -1 1])
        grid on

        subplot(2,2,4)
        plot(t,real(z211+z212),'k-',t,real(z221+z222),'k+-',t,real(z231+z232),'k.-')
        title('non-prop damped, z1, z2, z3 mode 2')
        legend('mass 1','mass 2','mass 3')
        grid on
        xlabel('time, sec')
        axis([0 max(t) -2 2])

        disp('execution paused to display figure, "enter" to continue'); pause

        subplot(1,1,1)

%       plot mode 3

        plot(t,real(z311),'k-',t,real(z312),'k+-',t,imag(z311),'k.-',t,imag(z312),'ko-')
        title('non-prop damped real and imag for z1, mode 3')
        legend('real','real','imag','imag')
        xlabel('time, sec')
        axis([0 max(t) -1 1])
        grid on

        disp('execution paused to display figure, "enter" to continue'); pause

        plot(t,real(z321),'k-',t,real(z322),'k+-',t,imag(z321),'k.-',t,imag(z322),'ko-')
        title('non-prop damped real and imag for z2 mode 3')
        legend('real','real','imag','imag')
        xlabel('time, sec')
        axis([0 max(t) -2 2])
        grid on

        disp('execution paused to display figure, "enter" to continue'); pause

        plot(t,real(z331),'k-',t,real(z332),'k+-',t,imag(z331),'k.-',t,imag(z332),'ko-')
        title('non-prop damped real and imag for z3 mode 3')
        legend('real','real','imag','imag')
        xlabel('time, sec')
        axis([0 max(t) -1 1])
        grid on

        disp('execution paused to display figure, "enter" to continue'); pause

        plot(t,real(z311+z312),'k-',t,real(z321+z322),'k+-',t,real(z331+z332),'k.-')
```

```
                title('non-prop damped, z1, z2, z3 mode 3')
                legend('mass 1','mass 2','mass 3')
                xlabel('time, sec')
                axis([0 max(t) -4 4])
                grid on

                disp('execution paused to display figure, "enter" to continue'); pause

%               plot subplots for notes

                subplot(2,2,1)
                plot(t,real(z311),'k-',t,real(z312),'k+-',t,imag(z311),'k.-',t,imag(z312),'ko-')
                title('non-prop damped real and imag for z1, mode 3')
                legend('real','real','imag','imag')
                axis([0 max(t) -1 1])
                grid on

                subplot(2,2,2)
                plot(t,real(z321),'k-',t,real(z322),'k+-',t,imag(z321),'k.-',t,imag(z322),'ko-')
                title('non-prop damped real and imag for z2 mode 3')
                legend('real','real','imag','imag')
                axis([0 max(t) -2 2])
                grid on

                subplot(2,2,3)
                plot(t,real(z331),'k-',t,real(z332),'k+-',t,imag(z331),'k.-',t,imag(z332),'ko-')
                title('non-prop damped real and imag for z3 mode 3')
                legend('real','real','imag','imag')
                xlabel('time, sec')
                axis([0 max(t) -1 1])
                grid on

                subplot(2,2,4)
                plot(t,real(z311+z312),'k-',t,real(z321+z322),'k+-',t,real(z331+z332),'k.-')
                title('non-prop damped, z1, z2, z3 mode 3')
                legend('mass 1','mass 2','mass 3')
                xlabel('time, sec')
                axis([0 max(t) -4 4])
                grid on

                disp('execution paused to display figure, "enter" to continue'); pause
```

5.15 Plotted Results: Argand and Initial Condition Responses

The next four sections plot Argand and initial condition transient responses
for the two oscillatory modes, illustrating the canceling of the imaginary
components and the doubling of the real components.

5.15.1 Argand Diagram, Mode 2

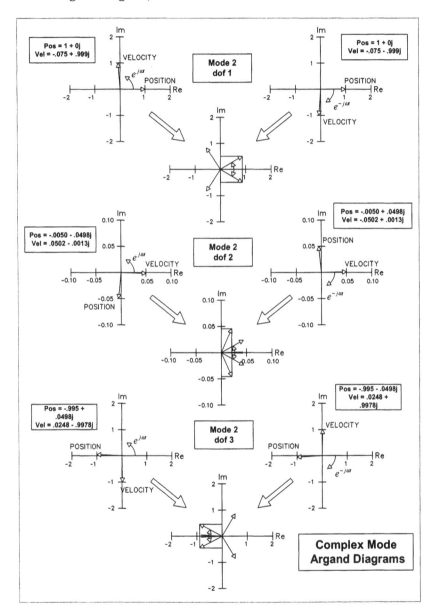

Figure 5.8 Argand diagram for three degrees of freedom for mode 2, complex damping.

5.15.2 Time Domain Responses, Mode 2

The plots below show the motions of the masses decreasing due to the damping. Once again, the imaginary components are out of phase and cancel each other, leaving only twice the real component as the final motion. Unlike the undamped case, the three masses do not reach their maximum or minimum positions at the same time. Since the damping is quite small, it is hard to see on the plots the small differences in times at which the maxima and minima are reached. Note that the unequal damping values for the two dampers make the center mass have a small motion in mode 2. We showed in Chapter 3 that for the undamped case mass 2 has no motion for mode 2.

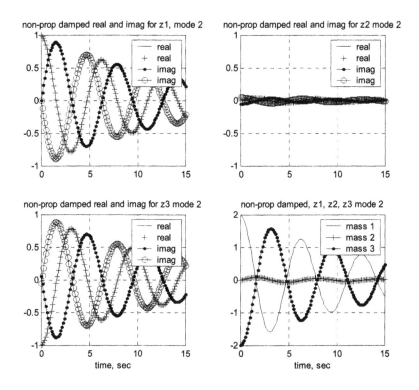

Figure 5.9: Initial condition transient response for mode 2.

5.15.3 Argand Diagram, Mode 3

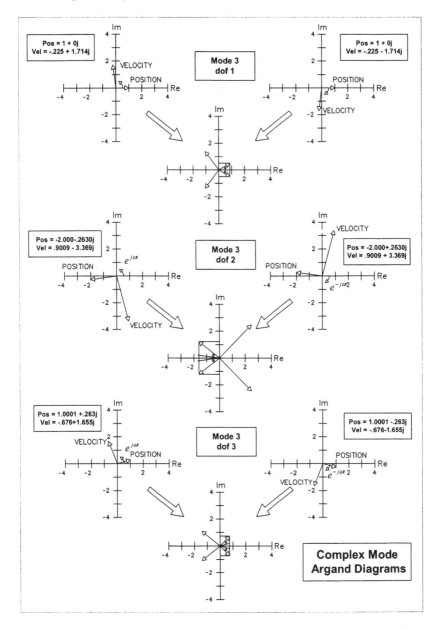

Figure 5.10: Argand diagram for three degrees of freedom for mode 3, complex damping.

5.15.4 Time Domain Responses, Mode 3

Compared to the responses for the mode 2 in Figure 5.9, the response for mode 3 damps out faster for two reasons. First, it has higher damping, 13% versus 7.5%, as shown in Figure 5.5. Secondly, even if zeta were the same for the two modes, the higher frequency of mode 3 will create higher velocities, hence higher damping from the velocity-dependent damping term.

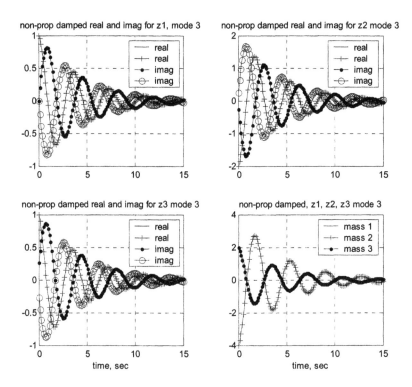

Figure 5.11: Initial condition transient response for mode 3.

Problems

Note: All the problems refer to the two dof system shown in Figure P2.2.

P5.1 Write the damped equations for the two dof system in state space form, both expanded and matrix. Show the input matrix **B** for a step force of magnitude 1 to mass 1 and magnitude –2 for mass 2. Show the output matrix **C** for the following outputs:

a) Position of masses 1 and 2

b) Position and velocity of mass 1

c) 2 times velocity of mass 1 plus 3 times the position of mass 2

P5.2 Set up the eigenvalue problem for the damped two dof problem as in (5.46).

P5.3 (MATLAB) With $m_1 = m_2 = m = 1$, $k_1 = k_2 = k = 1$, modify the code in **tdof_non_prop_damped.m** for the two dof damped model with $c_1 = c_2 = 0.1$ and:

a) list the complex eigenvalues, real and imaginary form

b) list the complex eigenvalues, magnitude and phase angle form

c) normalize the eigenvectors for unity values of the position of mass 1 and hand plot the Argand diagrams for the system

d) list the percentage of critical damping for each mode

e) plot the complex eigenvalues in the s-plane and correlate the three different descriptions in (a), (b) and (d)

P5.4 (MATLAB) Set $m_1 = m_2 = m = 1$, $k_1 = k_2 = k = 1$ and plot the initial condition responses for the system in initial conditions which match the two damped eigenvectors.

P5.5   Set $m_1 = m_2 = m = 1$, $k_1 = k_2 = k = 1$ and hand plot the Argand diagrams for modes 1 and 2.

CHAPTER 6

STATE SPACE:

FREQUENCY RESPONSE, TIME DOMAIN

6.1 Introduction – Frequency Response

This chapter will begin with the state space form of the equations of motion. We will use Laplace transforms to define the transfer function matrix. Next we will solve for the closed form transfer function matrix of the undamped tdof model using a symbolic algebra program and compare the answer with the solution presented in Chapter 2. MATLAB code will be used to set up frequency response calculations, using the full system matrix which allows the user to define damping values.

6.2 Solving for Transfer Functions in State Space Form Using Laplace Transforms

Starting with the complete set of state space equations:

$$\begin{aligned} \dot{\mathbf{x}} &= \mathbf{A}\mathbf{x} + \mathbf{B}\mathbf{u} \\ \mathbf{y} &= \mathbf{C}\mathbf{x} + \mathbf{D}\mathbf{u} \end{aligned} \tag{6.1}$$

Ignoring initial conditions to solve for steady state frequency response, take the matrix Laplace transform of the state equation and solve for $\mathbf{x}(s)$ (Appendix 2):

$$s\mathbf{I}\mathbf{x}(s) = \mathbf{A}\mathbf{x}(s) + \mathbf{B}\mathbf{u}(s) \tag{6.2}$$

$$(s\mathbf{I} - \mathbf{A})\mathbf{x}(s) = \mathbf{B}\mathbf{u}(s) \tag{6.3}$$

$$\mathbf{x}(s) = (s\mathbf{I} - \mathbf{A})^{-1}\,\mathbf{B}\mathbf{u}(s) \tag{6.4}$$

Substituting into the Laplace transform of the output equation:

$$\mathbf{y}(s) = \mathbf{C}\,(s\mathbf{I} - \mathbf{A})^{-1}\,\mathbf{B}\mathbf{u}(s) + \mathbf{D}\mathbf{u}(s) \tag{6.5}$$

Solving for the transfer function $\dfrac{\mathbf{y}(s)}{\mathbf{u}(s)}$:

$$\frac{y(s)}{u(s)} = \mathbf{C} \, (s\mathbf{I} - \mathbf{A})^{-1} \, \mathbf{B} + \mathbf{D} \tag{6.6}$$

Checking consistency of sizes

$$\begin{aligned} nx1 &= (nxn)x(nxn)x(nx1) + (nx1) \\ &= nx1 \end{aligned} \tag{6.7}$$

Letting $m_1 = m_2 = m_3 = m$, $k_1 = k_2 = k_3 = k$, $c_1 = c_2 = 0$ and rewriting the matrix equations of motion to match the original undamped problem used in Section 2.4.3 allows calculation of results by hand. The MATLAB code which follows, however, will allow **any** values to be used for the individual masses, dampers and stiffnesses.

$$(s\mathbf{I} - \mathbf{A}) = s \begin{bmatrix} 1 & 0 & 0 & 0 & 0 & 0 \\ 0 & 1 & 0 & 0 & 0 & 0 \\ 0 & 0 & 1 & 0 & 0 & 0 \\ 0 & 0 & 0 & 1 & 0 & 0 \\ 0 & 0 & 0 & 0 & 1 & 0 \\ 0 & 0 & 0 & 0 & 0 & 1 \end{bmatrix}$$

$$- \begin{bmatrix} 0 & 1 & 0 & 0 & 0 & 0 \\ \dfrac{-k_1}{m_1} & \dfrac{-c_1}{m_1} & \dfrac{k_1}{m_1} & \dfrac{c_1}{m_1} & 0 & 0 \\ 0 & 0 & 0 & 1 & 0 & 0 \\ \dfrac{k_1}{m_2} & \dfrac{c_1}{m_2} & \dfrac{-(k_1+k_2)}{m_2} & \dfrac{-(c_1+c_2)}{m_2} & \dfrac{k_2}{m_2} & \dfrac{c_2}{m_2} \\ 0 & 0 & 0 & 0 & 0 & 1 \\ 0 & 0 & \dfrac{k_2}{m_3} & \dfrac{c_2}{m_3} & \dfrac{-k_2}{m_3} & \dfrac{-c_2}{m_3} \end{bmatrix}$$

$$= \begin{bmatrix} s & -1 & 0 & 0 & 0 & 0 \\ \dfrac{k_1}{m_1} & s+\dfrac{c_1}{m_1} & \dfrac{-k_1}{m_1} & \dfrac{-c_1}{m_1} & 0 & 0 \\ 0 & 0 & s & -1 & 0 & 0 \\ \dfrac{-k_1}{m_2} & \dfrac{-c_1}{m_2} & \dfrac{(k_1+k_2)}{m_2} & s+\dfrac{(c_1+c_2)}{m_2} & \dfrac{-k_2}{m_2} & \dfrac{-c_2}{m_2} \\ 0 & 0 & 0 & 0 & s & -1 \\ 0 & 0 & \dfrac{-k_2}{m_3} & \dfrac{-c_2}{m_3} & \dfrac{k_2}{m_3} & s+\dfrac{c_2}{m_3} \end{bmatrix}$$

$$= \begin{bmatrix} s & -1 & 0 & 0 & 0 & 0 \\ \dfrac{k}{m} & s & \dfrac{-k}{m} & 0 & 0 & 0 \\ 0 & 0 & s & -1 & 0 & 0 \\ \dfrac{-k}{m} & 0 & \dfrac{2k}{m} & s & \dfrac{-k}{m} & 0 \\ 0 & 0 & 0 & 0 & s & -1 \\ 0 & 0 & \dfrac{-k}{m} & 0 & \dfrac{k}{m} & s \end{bmatrix} \tag{6.8}$$

Here, in order to develop the entire 3x3 transfer function matrix, we will use a MIMO representation of **B** and **C**.

Taking **B** equal to the 6x3 matrix gives transfer functions for all three forces:

$$\mathbf{B} = \begin{bmatrix} 0 & 0 & 0 \\ 1/m_1 & 0 & 0 \\ 0 & 0 & 0 \\ 0 & 1/m_2 & 0 \\ 0 & 0 & 0 \\ 0 & 0 & 1/m_3 \end{bmatrix} \tag{6.9}$$

Taking **C** equal to the 3x6 matrix below gives the three displacement transfer functions as outputs:

$$\mathbf{C} = \begin{bmatrix} 1 & 0 & 0 & 0 & 0 & 0 \\ 0 & 0 & 1 & 0 & 0 & 0 \\ 0 & 0 & 0 & 0 & 1 & 0 \end{bmatrix} \tag{6.10}$$

6.3 Transfer Function Matrix

Now that we have the terms required, we can substitute into the equation for the transfer function matrix:

$$\frac{y(s)}{u(s)} = \mathbf{C}\,(s\mathbf{I} - \mathbf{A})^{-1}\mathbf{B} + \mathbf{D} \tag{6.11}$$

We have an expression for $(s\mathbf{I} - \mathbf{A})$ above, but need to have its inverse. Using a symbolic algebra program to calculate the inverse even for this relatively small 3x3 problem yields a result which is too lengthy to be listed here in its entirety. To show that the calculation by hand really works, however, we will expand the equation above symbolically and then substitute the appropriate terms from the inverse to give the results for several of the transfer functions. We will refer to the $(s\mathbf{I} - \mathbf{A})^{-1}$ matrix by the notation "sia" and expand it as follows:

$$\frac{y(s)}{u(s)} = \mathbf{C}\,(s\mathbf{I} - \mathbf{A})^{-1}\mathbf{B}$$

$$= \begin{bmatrix} 1 & 0 & 0 & 0 & 0 & 0 \\ 0 & 0 & 1 & 0 & 0 & 0 \\ 0 & 0 & 0 & 0 & 1 & 0 \end{bmatrix} \begin{bmatrix} siai_{11} & siai_{12} & siai_{13} & siai_{14} & siai_{15} & siai_{16} \\ siai_{21} & siai_{22} & siai_{23} & siai_{24} & siai_{25} & siai_{26} \\ siai_{31} & siai_{32} & siai_{33} & siai_{34} & siai_{35} & siai_{36} \\ siai_{41} & siai_{42} & siai_{43} & siai_{44} & siai_{45} & siai_{46} \\ siai_{51} & siai_{52} & siai_{53} & siai_{54} & siai_{55} & siai_{56} \\ siai_{61} & siai_{62} & siai_{63} & siai_{64} & siai_{65} & siai_{66} \end{bmatrix} \begin{bmatrix} 0 & 0 & 0 \\ 1/m & 0 & 0 \\ 0 & 0 & 0 \\ 0 & 1/m & 0 \\ 0 & 0 & 0 \\ 0 & 0 & 1/m \end{bmatrix}$$

$$
=\begin{bmatrix} siai_{11} & siai_{12} & siai_{13} & siai_{14} & siai_{15} & siai_{16} \\ siai_{31} & siai_{32} & siai_{33} & siai_{34} & siai_{35} & siai_{36} \\ siai_{51} & siai_{52} & siai_{53} & siai_{54} & siai_{55} & siai_{56} \end{bmatrix}
\begin{bmatrix} 0 & 0 & 0 \\ 1/m & 0 & 0 \\ 0 & 0 & 0 \\ 0 & 1/m & 0 \\ 0 & 0 & 0 \\ 0 & 0 & 1/m \end{bmatrix}
$$

$$
=\begin{bmatrix} siai_{12}/m & siai_{14}/m & siai_{16}/m \\ siai_{32}/m & siai_{34}/m & siai_{36}/m \\ siai_{52}/m & siai_{54}/m & siai_{56}/m \end{bmatrix} \tag{6.12}
$$

Listing the values for the $siai_{xx}$ terms used above from the symbolic algebra solution:

$$
siai_{12} = siai_{56} = (m^3s^4 + 3m^2ks^2 + mk^2)/Den
$$
$$
siai_{32} = siai_{14} = siai_{54} = siai_{36} = (m^2ks^2 + mk^2)/Den
$$
$$
siai_{34} = (m^3s^4 + 2m^2ks^2 + mk^2)/Den
$$
$$
siai_{52} = siai_{16} = mk^2/Den
$$

$$
\text{where Den} = s^2(m^3s^4 + 4m^2ks^2 + 3mk^2)
$$

$$
\text{(6.13a-e)}
$$

Dividing each of the above terms by "m" and presenting in the transfer function matrix form of (2.61):

$$
\begin{bmatrix} z_1 \\ z_2 \\ z_3 \end{bmatrix} = \frac{\begin{bmatrix} (m^2s^4 + 3mks^2 + k^2) & (mks^2 + k^2) & k^2 \\ (mks^2 + k^2) & (m^2s^4 + 2mks^2 + k^2) & (mks^2 + k^2) \\ k^2 & (mks^2 + k^2) & (m^2s^4 + 3mks^2 + k^2) \end{bmatrix}}{s^2(m^3s^4 + 4m^2ks^2 + 3mk^2)} \begin{bmatrix} F_1 \\ F_2 \\ F_3 \end{bmatrix}
$$

$$
\text{(6.14)}
$$

The two derivations are identical.

6.4 MATLAB Code tdofss.m – Frequency Response Using State Space

6.4.1 Code Description, Plot

The four distinct transfer functions for the default values of m, k and c are plotted using MATLAB in **tdofss.m**, listed below.   The four plots are displayed in Figure 6.1.   The **A**, **B**, **C** and **D** matrices shown in (5.17a) are used as inputs to the program. A MIMO state space model is constructed and the MATLAB function bode.m is used to calculate the magnitude and phase of the resulting frequency responses. As described in the code, the resulting frequency response has dimensions of 6x3x200, where the "6" represents the 6 outputs in the output matrix **C**, the "3" represents the three columns of the input matrix **B** and the "200" represents the 200 frequency points in the frequency vector. The desired magnitude and phase can be extracted from the 6x3x200 matrix by defining the appropriate indices. The default values of c1 and c2 are zero.

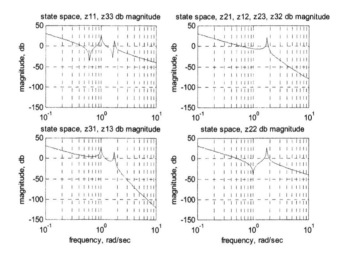

Figure 6.1: Four distinct frequency response amplitudes.

6.4.2 Code Listing

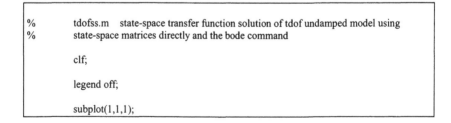

```
%        tdofss.m    state-space transfer function solution of tdof undamped model using
%        state-space matrices directly and the bode command

clf;

legend off;

subplot(1,1,1);
```

```
          clear all;

%         define the values of masses, springs, dampers and Forces

          m1 = 1;
          m2 = 1;
          m3 = 1;

          c1 = input('input value for c1, default 0, ... ');

          if (isempty(c1))
                    c1 = 0;
          else
          end

          c2 = input('input value for c2, default 0, ... ');

          if (isempty(c2))
                    c2 = 0;
          else
          end

          k1 = 1;
          k2 = 1;

          F1 = 1;
          F2 = 1;
          F3 = 1;

%         define the system matrix, a

          a = [    0        1          0              0          0        0
                 -k1/m1   -c1/m1      k1/m1          c1/m1       0        0
                   0        0          0              1          0        0
                 k1/m2    c1/m2    -(k1+k2)/m2    -(c1+c2)/m2   k2/m2    c2/m2
                   0        0          0              0          0        1
                   0        0        k2/m3          c2/m3      -k2/m3   -c2/m3];

%         define the input matrix, b, a 6x3 matrix

          b = [    0        0          0
                 F1/m1      0          0
                   0        0          0
                   0      F2/m2        0
                   0        0          0
                   0        0        F3/m3];

%         define the output matrix, c, the 6x6 identify matrix

          c = eye(6,6);

%         define the direct transmission matrix

          d = 0;

%         solve for the eigenvalues of the system matrix
```

```
            [xm,omega] = eig(a);

%           Define a vector of frequencies to use, radians/sec.  The logspace command uses
%           the log10 value as limits, i.e. -1 is 10^-1 = 0.1 rad/sec, and 1 is
%           10^1 = 10 rad/sec.  The 200 defines 200 frequency points.

            w = logspace(-1,1,200);

%           use the "ss" function to define state space system for three inputs, forces at
%           masses 1, 2 and 3 and for all 6 states, three displacements and three velocities

            sssys = ss(a,b,c,d);

%           use the bode command with left hand magnitude and phase vector arguments
%           to provide values for further analysis/plotting

%           the mag and phs matrices below will be 6x3x200 in size
%           the appropriate magnitude and phase to plot for each transfer function
%           are called by appropriate indexing

%           first index 1-6:  z1 z1dot z2 z2dot z3 z3dot
%           second index 1-3:  F1  F2  F3
%           third index 1-200:  all frequency points, use ":"

            [mag,phs] = bode(sssys,w);

            z11mag = mag(1,1,:);
            z11phs = phs(1,1,:);

            z21mag = mag(3,1,:);
            z21phs = phs(3,1,:);

            z31mag = mag(5,1,:);
            z31phs = phs(5,1,:);

            z22mag = mag(3,2,:);
            z22phs = phs(3,2,:);

%           calculate the magnitude in decibels, db

            z11magdb = 20*log10(z11mag);

            z21magdb = 20*log10(z21mag);

            z31magdb = 20*log10(z31mag);

            z22magdb = 20*log10(z22mag);

%           plot the four transfer functions separately, in a 2x2 subplot form

            subplot(2,2,1)
            semilogx(w,z11magdb(1,:),'k-')
            title('state space, z11, z33 db magnitude')
            ylabel('magnitude, db')
            axis([.1 10 -150 50])
            grid
```

```
      subplot(2,2,2)
      semilogx(w,z21magdb(1,:),'k-')
      title('state space, z21, z12, z23, z32 db magnitude')
      ylabel('magnitude, db')
      axis([.1 10 -150 50])
      grid

      subplot(2,2,3)
      semilogx(w,z31magdb(1,:),'k-')
      title('state space, z31, z13 db magnitude')
      xlabel('frequency, rad/sec')
      ylabel('magnitude, db')
      axis([.1 10 -150 50])
      grid

      subplot(2,2,4)
      semilogx(w,z22magdb(1,:),'k-')
      title('state space, z22 db magnitude')
      xlabel('frequency, rad/sec')
      ylabel('magnitude, db')
      axis([.1 10 -150 50])
      grid

      disp('execution paused to display figure, "enter" to continue'); pause

      subplot(2,2,1)
      semilogx(w,z11phs(1,:),'k-')
      title('state space, z11, z33 phase')
      ylabel('phase, deg')
      %axis([.1 10 -400 -150])
      grid

      subplot(2,2,2)
      semilogx(w,z21phs(1,:),'k-')
      title('state space, z21, z12, z23, z32 phase')
      ylabel('phase, deg')
      %axis([.1 10 -400 -150])
      grid

      subplot(2,2,3)
      semilogx(w,z31phs(1,:),'k-')
      title('state space, z31, z13 phase')
      xlabel('frequency, rad/sec')
      ylabel('phase, deg')
      %axis([.1 10 -400 -150])
      grid

      subplot(2,2,4)
      semilogx(w,z22phs(1,:),'k-')
      title('state space, z22 phase')
      xlabel('frequency, rad/sec')
      ylabel('phase, deg')
      %axis([.1 10 -400 -150])
      grid

      disp('execution paused to display figure, "enter" to continue'); pause
```

6.5 Introduction – Time Domain

Starting with the equations of motion in state space, we will use Laplace transforms to discuss the theoretical solution to the time domain problem. We will define and discuss two methods of calculating the matrix exponential. Then we will use a sdof forced system with position and velocity initial conditions to illustrate the technique. The closed form solution for our tdof example problem with step forces applied to all three masses and with different initial conditions for each mass is too complicated to be shown so we will use only MATLAB for its solution.

6.6 Matrix Laplace Transform – with Initial Conditions

We start with the state equations in general form, (6.1). Taking the matrix Laplace transform of a first order differential equation (DE) with initial conditions (Appendix 2):

$$\mathcal{L}\{\dot{\mathbf{x}}(t)\} = s\mathbf{x}(s) - \mathbf{x}(0)$$
$$\mathcal{L}\{\mathbf{x}(t)\} = \mathbf{x}(s)$$

$$(6.15)$$

Taking the matrix Laplace transform of (6.1) and solving for $\mathbf{x}(s)$:

$$s\mathbf{x}(s) - \mathbf{x}(0) = \mathbf{A}\mathbf{x}(s) + \mathbf{B}u(s)$$
$$(s\mathbf{I} - \mathbf{A})\mathbf{x}(s) = \mathbf{x}(0) + \mathbf{B}u(s) \qquad (6.16\text{a,b,c})$$
$$\mathbf{x}(s) = (s\mathbf{I} - \mathbf{A})^{-1}\mathbf{x}(0) + (s\mathbf{I} - \mathbf{A})^{-1}\mathbf{B}u(s)$$

Solving for the output vector $\mathbf{y}(s)$:

$$\mathbf{y}(s) = \mathbf{C}\mathbf{x}(s)$$
$$= \mathbf{C}(s\mathbf{I} - \mathbf{A})^{-1}\mathbf{x}(0) + \mathbf{C}(s\mathbf{I} - \mathbf{A})^{-1}\mathbf{B}u(s)$$

$$(6.17)$$

The input matrix $\mathbf{B}$ and output matrix $\mathbf{C}$ are familiar from earlier state space presentations. There is a new term in the equation for the Laplace transform of $\mathbf{y}(s)$, the term $(s\mathbf{I} - \mathbf{A})^{-1}$.

There are many methods of calculating the inverse $(s\mathbf{I} - \mathbf{A})^{-1}$ (Chen 1999). If the problem is small, for example 2x2, the inverse can be handled in closed form. Then $\mathbf{y}(s)$ can be back-transformed term by term to get the solution in the time domain, as we shall see in the example in the next section.

For another solution method it is useful to recall the geometric series expansion below, for $|r| < 1$:

$$\frac{1}{1-r} = 1 + r + r^2 + r^3 + \dots \tag{6.18}$$

Expanding $(s\mathbf{I} - \mathbf{A})^{-1}$ with the series expansion analogy above, the inverse results in the infinite series in (6.19).

$$(s\mathbf{I} - \mathbf{A})^{-1} = \frac{1}{s\mathbf{I} - \mathbf{A}} = \frac{\dfrac{1}{s}}{\mathbf{I} - \dfrac{\mathbf{A}}{s}} = \frac{1}{s}\left[\mathbf{I} + \frac{\mathbf{A}}{s} + \frac{\mathbf{A}^2}{s^2} + \frac{\mathbf{A}^3}{s^3} + \dots \right]$$

$$= \frac{\mathbf{I}}{s} + \frac{\mathbf{A}}{s^2} + \frac{\mathbf{A}^2}{s^3} + \frac{\mathbf{A}^3}{s^4} + \dots \tag{6.19}$$

6.7 Inverse Matrix Laplace Transform, Matrix Exponential

Now that we have the inverse in series form, it is easy to back-transform to the time domain, term by term. We introduce two new terms, $\Phi(t)$, the **inverse Laplace transform** of $(s\mathbf{I} - \mathbf{A})^{-1}$ which equals $e^{\mathbf{A}t}$, the **matrix exponential**.

$$\Phi(t) = \mathcal{L}^{-1}\left\{(s\mathbf{I} - \mathbf{A})^{-1}\right\}$$

$$= \mathcal{L}^{-1}\left\{\frac{\mathbf{I}}{s} + \frac{\mathbf{A}}{s^2} + \frac{\mathbf{A}^2}{s^3} + \frac{\mathbf{A}^3}{s^4} + \dots \right\}$$

$$= \mathbf{I} + \mathbf{A}t + \frac{(\mathbf{A}t)^2}{2!} + \frac{(\mathbf{A}t)^3}{3!} + \dots \tag{6.20}$$

$$= e^{\mathbf{A}t}$$

6.8 Back-Transforming to Time Domain

Now that the form of the matrix exponential is known, we can back-transform the entire equation of motion, from (6.16c):

$$\mathcal{L}^{-1}(\mathbf{x}(s)) = \mathcal{L}^{-1}\left[(s\mathbf{I} - \mathbf{A})^{-1}\mathbf{x}(0) + (s\mathbf{I} - \mathbf{A})^{-1}\mathbf{B}u(s)\right] \tag{6.21}$$

The result is:

$$x(t) = e^{At} x(0) + \int_0^t e^{A(t-\tau)} \mathbf{B} u(\tau) \, d\tau \qquad (6.22)$$

The first term in (6.22) is the response due to the initial condition of the state and the second term is the response due to the forcing function. The second term is the **convolution integral**, or **Duhamel integral**, and results from back-transforming the product of two Laplace transforms.

6.9 Single Degree of Freedom System – Calculating Matrix Exponential in Closed Form

Calculating the matrix exponential in closed form for greater than a 2x2 matrix is difficult without the aid of a symbolic algebra program. Even with the program the result can be quite complicated.

A simple, rigid body example will be used to demonstrate how a matrix exponential and transient response are calculated.

We will use the system in Figure 6.2, a mass with position and velocity initial conditions and a step force applied.

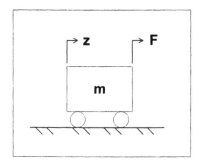

Figure 6.2: sdof system with initial conditions and step force applied.

6.9.1 Equations of Motion, Laplace Transform

Start with the equation of motion:

$$m\ddot{z} = F \qquad (6.23)$$

Defining the states:

$$
\begin{aligned}
x_1 &= z \\
x_2 &= \dot{z}
\end{aligned}
\qquad (6.24)
$$

Defining derivatives and inserting the value for acceleration:

$$\dot{x}_1 = x_2$$
$$\dot{x}_2 = \frac{F}{m} \tag{6.25}$$

The above can be written in matrix form, recognizing that F/m is the acceleration and applying a unity magnitude step:

$$\begin{bmatrix} \dot{x}_1 \\ \dot{x}_2 \end{bmatrix} = \begin{bmatrix} 0 & 1 \\ 0 & 0 \end{bmatrix} \begin{bmatrix} x_1 \\ x_2 \end{bmatrix} + \begin{bmatrix} 0 \\ \left(\dfrac{F}{m}\right) \end{bmatrix} (1) \tag{6.26}$$

Defining the system matrix:

$$A = \begin{bmatrix} 0 & 1 \\ 0 & 0 \end{bmatrix} \tag{6.27}$$

Taking the inverse of the $(sI - A)^{-1}$ term:

$$(sI - A)^{-1} = \left(\begin{bmatrix} s & 0 \\ 0 & s \end{bmatrix} - \begin{bmatrix} 0 & 1 \\ 0 & 0 \end{bmatrix} \right)^{-1} = \begin{bmatrix} s & -1 \\ 0 & s \end{bmatrix}^{-1} = \begin{bmatrix} \dfrac{1}{s} & \dfrac{1}{s^2} \\ 0 & \dfrac{1}{s} \end{bmatrix} \tag{6.28}$$

6.9.2 Defining the Matrix Exponential – Taking Inverse Laplace Transform

Using the table of inverse Laplace transforms from Appendix 2 yields the matrix exponential.

$$e^{At} = \mathcal{L}^{-1} \left\{ \begin{bmatrix} \dfrac{1}{s} & \dfrac{1}{s^2} \\ 0 & \dfrac{1}{s} \end{bmatrix} \right\} = \begin{bmatrix} 1 & t \\ 0 & 1 \end{bmatrix} \tag{6.29}$$

6.9.3 Defining the Matrix Exponential – Using Series Expansion

A Power Series Expansion can also be used to find the matrix exponential for this simple example because higher powers of $\mathbf{A}t$ go to zero quickly:

$$e^{\mathbf{A}t} = \mathbf{I} + \mathbf{A}t + \frac{(\mathbf{A}t)^2}{2!} + \frac{(\mathbf{A}t)^3}{3!} + ...$$

$$= \begin{bmatrix} 1 & 0 \\ 0 & 1 \end{bmatrix} + \begin{bmatrix} 0 & t \\ 0 & 0 \end{bmatrix} + \begin{bmatrix} 0 & 0 \\ 0 & 0 \end{bmatrix} + \text{(all other terms zero)} \qquad (6.30)$$

$$= \begin{bmatrix} 1 & t \\ 0 & 1 \end{bmatrix}$$

This is the same solution as (6.29).

6.9.4 Solving for Time Domain Response

Thus, the general solution for x(t) as a function of time becomes:

$$\mathbf{x}(t) = e^{\mathbf{A}t}\,\mathbf{x}(0) + \int_0^t e^{\mathbf{A}(t-\tau)}\,\mathbf{B}u(t)\,d\tau$$

$$= \begin{bmatrix} 1 & t \\ 0 & 1 \end{bmatrix}\begin{bmatrix} x_1(0) \\ x_2(0) \end{bmatrix} + \int_0^t \begin{bmatrix} 1 & t-\tau \\ 0 & 1 \end{bmatrix}\begin{bmatrix} 0 \\ \dfrac{F}{m} \end{bmatrix}(1)\; d\tau$$

$$= \begin{bmatrix} x_1(0)+t\,x_2(0) \\ x_2(0) \end{bmatrix} + \int_0^t \begin{bmatrix} (t-\tau)\left(\dfrac{F}{m}\right) \\ \left(\dfrac{F}{m}\right) \end{bmatrix} d\tau$$

$$= \begin{bmatrix} x_1(0)+t\,x_2(0) \\ x_2(0) \end{bmatrix} + \left\{ \begin{bmatrix} \left(t\tau-\dfrac{\tau^2}{2}\right)\left(\dfrac{F}{m}\right) \\ \left(\dfrac{F}{m}\right)\tau \end{bmatrix} \right\}\Bigg|_0^t$$

$$
= \begin{bmatrix} x_1(0) + t\,x_2(0) \\ x_2(0) \end{bmatrix} + \begin{bmatrix} \left(t^2 - \dfrac{t^2}{2} \right)\!\left(\dfrac{F}{m} \right) \\ t\!\left(\dfrac{F}{m} \right) \end{bmatrix}
$$

$$
= \begin{bmatrix} x_1(0) + t\,x_2(0) \\ x_2(0) \end{bmatrix} + \begin{bmatrix} \left(\dfrac{t^2}{2} \right)\!\left(\dfrac{F}{m} \right) \\ t\!\left(\dfrac{F}{m} \right) \end{bmatrix}
$$

$$(6.31)$$

This result is the same as the familiar equations for the position and velocity of a mass undergoing a constant acceleration:

$$
\begin{bmatrix} x_1(t) \\ x_2(t) \end{bmatrix} = \begin{bmatrix} \text{initial position} + \text{time} \times (\text{initial velocity}) + \dfrac{(\text{acceleration}) \times (\text{time}^2)}{2} \\ \text{initial velocity} + (\text{acceleration}) \times (\text{time}) \end{bmatrix}
$$

$$(6.32)$$

6.10 MATLAB Code tdof_ss_time_ode45_slnk.m – Time Domain Response of tdof Model

6.10.1 Equations of Motion Review

There are several ways to numerically solve for transient responses using MATLAB. One method uses numerical integration, calling the integration routine from a command line and defining the state equation in a separate MATLAB function.   Another method uses Simulink, a linear/nonlinear graphical block diagram model building tool linked to MATLAB.

We will solve for the transient response of our tdof model using both methods and compare the results with the closed form solution calculated using the modal transient response method in Chapter 9.

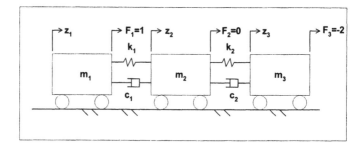

Figure 6.3: tdof model with damping for use in MATLAB/Simulink models.

$$\begin{array}{lll} z_1(0) = x_1(0) = 0 & z_2(0) = x_3(0) = -1 & z_3(0) = x_5(0) = 1 \\ \dot{z}_1(0) = x_2(0) = -1 & \dot{z}_2(0) = x_4(0) = 2 & \dot{z}_3(0) = x_6(0) = -2 \end{array}$$

Table 6.1: Initial conditions for tdof model in Figure 6.3.

Step function forces of amplitudes indicated in Figure 6.3 are applied to masses 1 and 3; mass 2 has no force applied. Initial conditions of position and velocity for each mass are shown in Table 6.1.

The equations of motion in state space are then:

$$\begin{bmatrix} \dot{x}_1 \\ \dot{x}_2 \\ \dot{x}_3 \\ \dot{x}_4 \\ \dot{x}_5 \\ \dot{x}_6 \end{bmatrix} = \begin{bmatrix} 0 & 1 & 0 & 0 & 0 & 0 \\ \dfrac{-k_1}{m_1} & \dfrac{-c_1}{m_1} & \dfrac{k_1}{m_1} & \dfrac{c_1}{m_1} & 0 & 0 \\ 0 & 0 & 0 & 1 & 0 & 0 \\ \dfrac{k_1}{m_2} & \dfrac{c_1}{m_2} & \dfrac{-(k_1+k_2)}{m_2} & \dfrac{-(c_1+c_2)}{m_2} & \dfrac{k_2}{m_2} & \dfrac{c_2}{m_2} \\ 0 & 0 & 0 & 0 & 0 & 1 \\ 0 & 0 & \dfrac{k_2}{m_3} & \dfrac{c_2}{m_3} & \dfrac{-k_2}{m_3} & \dfrac{-c_2}{m_3} \end{bmatrix} \begin{bmatrix} x_1 \\ x_2 \\ x_3 \\ x_4 \\ x_5 \\ x_6 \end{bmatrix} + \begin{bmatrix} 0 \\ \dfrac{1}{m_1} \\ 0 \\ \dfrac{0}{m_2} \\ 0 \\ \dfrac{-2}{m_3} \end{bmatrix} \quad (1)$$

(6.33)

The initial condition vector, x(0) is:

$$\mathbf{x}(0) = \begin{bmatrix} x_1(0) \\ x_2(0) \\ x_3(0) \\ x_4(0) \\ x_5(0) \\ x_6(0) \end{bmatrix} = \begin{bmatrix} z_1(0) \\ \dot{z}_1(0) \\ z_2(0) \\ \dot{z}_2(0) \\ z_3(0) \\ \dot{z}_3(0) \end{bmatrix} = \begin{bmatrix} 0 \\ -1 \\ -1 \\ 2 \\ 1 \\ -2 \end{bmatrix} \qquad (6.34)$$

The output equation for the displacement outputs (no velocities included) with no feedthrough term is:

$$\begin{bmatrix} y_1 \\ y_2 \\ y_3 \end{bmatrix} = \begin{bmatrix} 1 & 0 & 0 & 0 & 0 & 0 \\ 0 & 0 & 1 & 0 & 0 & 0 \\ 0 & 0 & 0 & 0 & 1 & 0 \end{bmatrix} \begin{bmatrix} x_1 \\ x_2 \\ x_3 \\ x_4 \\ x_5 \\ x_6 \end{bmatrix} + (0)(1) \qquad (6.35)$$

These are the system matrices that are used in the MATLAB code below.

6.10.2 Code Description

Two methods will be used to solve for the time domain response.   The MATLAB code **tdof_ss_time_ode45_slnk.m** is used for both methods, prompting the user to define which solution technique is desired.

The first method uses the MATLAB Runge Kutta method ODE45 and calls the function file **tdofssfun.m**, which contains the state equations. The results are then plotted. To use the ODE45 solver, type "tdof_ss_time_ode45_slnk" from the MATLAB prompt and use the default selection.

The second solution uses the Simulink model **tdof_ss_simulink.mdl** and the plotting file **tdof_ss_time_slnk_plot.m**.

To use the Simulink solver:

1) Type "tdof_ss_time_ode45_slnk" and choose the Simulink solver.

2) The program will prompt the reader to type "tdof_ss_simulink" at the MATLAB command prompt. This will bring up the Simulink model on the screen.

3) Click on the "simulation" choice in the model screen and then choose "start." The Simulink model will then run.

4) To see the plotted results, type "tdof_ss_time_slnk_plot."

6.10.3 Code Results – Time Domain Responses

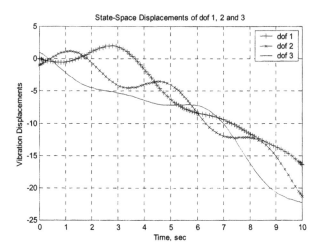

Figure 6.4: ODE45 simulation motion of tdof model.

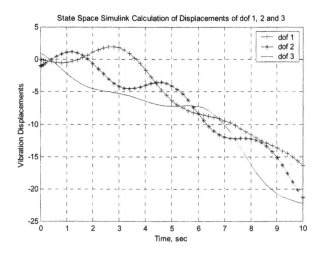

Figure 6.5: Simulink simulation motion of tdof model.

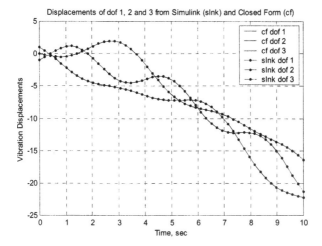

Figure 6.6: Overlay of closed form solution from Chapter 9, Figure 9.4, with Simulink solution.

6.10.4 Code Listing

```
%          tdof_ss_time_ode45_slnk.m        state-space solution of tdof model with
%          initial conditions, step function forcing function and displacement outputs
%          using the ode45 solver or Simulink, user is prompted for damping values

           clear all;

           global a b u          %          this is required to have the parameters available
                                 %          for the function

           which_run = input('enter "1" for Simulink or "enter" for ode45 run ... ');

           if  isempty(which_run)
                      which_run = 0
           end

%          define the values of masses, springs, dampers and Forces

           m1 = 1;
           m2 = 1;
           m3 = 1;

           c1 = input('input value for c1, default 0.0, ... ');

           if  (isempty(c1))
                      c1 = 0.0;
           else
           end

           c2 = input('input value for c2, default 0.0, ... ');
```

```
         if (isempty(c2))
                c2 = 0.0;
         else
         end

         k1 = 1;
         k2 = 1;

         F1 = 1;
         F2 = 0;
         F3 = -2;

%        define the system matrix, a

         a = [ 0       1        0           0          0        0
             -k1/m1   -c1/m1    k1/m1       c1/m1      0        0
              0        0        0           1          0        0
              k1/m2    c1/m2   -(k1+k2)/m2 -(c1+c2)/m2 k2/m2    c2/m2
              0        0        0           0          0        1
              0        0        k2/m3       c2/m3     -k2/m3   -c2/m3];

%        define the input matrix, b

         b = [ 0
              F1/m1
               0
              F2/m2
               0
              F3/m3];

%        define the output matrix for transient response, c, displacements only

         c = [1 0 0 0 0 0
              0 0 1 0 0 0
              0 0 0 0 1 0];

%        define the direct transmission matrix for transient response, d, the same number of
         rows as c and the same number of columns as b

         d = zeros(3,1);

         if which_run == 0    %            transient response using the ode45 command


         u = 1;

         ttotal = input('Input total time for Simulation, default = 10 sec, ... ');

         if (isempty(ttotal))
         ttotal = 10;
         else
         end

         tspan = [0 ttotal];

         x0 = [0 -1 -1 2 1 -2]';              % initial condition vector, note transpose
```

```
      options = [];                      % no options specified for ode45 command

      [t,x] = ode45('tdofssfun',tspan,x0,options);

      y = c*x';              % note transpose, x is calculated as a column vector in time

      plot(t,y(1,:),'k+-',t,y(2,:),'kx-',t,y(3,:),'k-')
      title('State-Space Displacements of dof 1, 2 and 3')
      xlabel('Time, sec')
      ylabel('Vibration Displacements')
      legend('dof 1','dof 2','dof 3')
      grid

      else       % setup Simulink run

%     define the direct transmission matrix for transient response, d, the same number of
      rows as c and the same number of columns as b

%     define time for simulink model

      ttotal = input('Input total time for Simulation, default = 10 sec, ... ');

      if (isempty(ttotal))
      ttotal = 10;
      else
      end

      disp(' ');
      disp(' ');
      disp(' ');
      disp(' ');
      disp(' ');
      disp(' ');
      disp('Run the Simulink model "tdof_ss_simulink.mdl" and then');
      disp('run the plotting file "tdof_ss_time_slnk_plot.m"');

      end
```

6.10.5 MATLAB Function tdofssfun.m –
Called by tdof_ss_time_ode45_slnk.m

```
      function xprime = tdofssfun(t,x)

%     function for calculating the transient response of tdof_ss_time_ode45.m

      global  a b u

      xprime = a*x + b*u;
```

6.10.6 Simulink Model tdofss_simulink.mdl

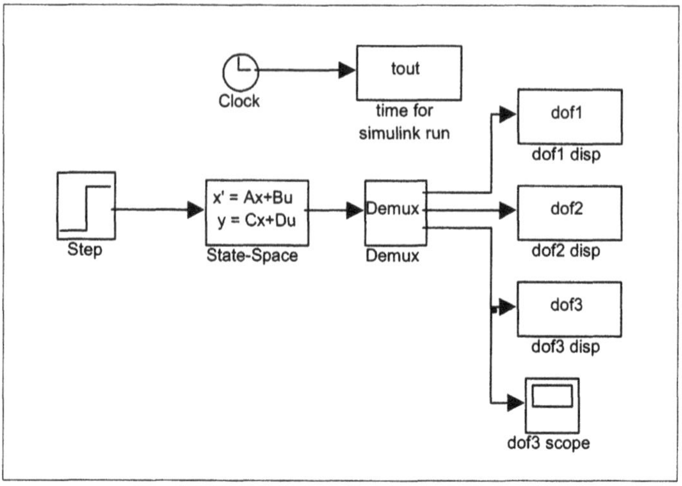

Figure 6.7: Block diagram of Simulink model tdofss_simulink.mdl.

The block diagram was constructed by dragging and dropping blocks from the appropriate Simulink block library and connecting the blocks. The input is the step block. The clock block is used to output time to the tout block for plotting in MATLAB. The model is defined in the state space block, reading in values for the a, b, c and d matrices from the MATLAB workspace, created during execution of **tdof_ss_time_ode45_slnk.m**.   The demux block separates the vector output of the state space block and sends the displacements of the three masses to three blocks for storing for plotting in MATLAB. The scope block brings up a scope screen and shows the position of dof3 versus time as the program executes. This example is so small that the screen displays instantly for the default 10 sec time period, but for a longer time period the scope traces the progress of the simulation.

Problems

Note: All the problems refer to the two dof system shown in Figure P2.2.

P6.1 Set $m_1 = m_2 = m = 1$, $k_1 = k_2 = k = 1$, $c_1 = c_2 = 0$ and define the state space matrices for a step force applied to mass 1 and for output of position of mass 2. Write out by hand the equation for the transfer functions matrix as shown in (6.11). Extra credit: use a symbolic algebra program to take the inverse of the $(s\mathbf{I} - \mathbf{A})$ term and then multiply out the equations to see that they match the results of P2.2.

P6.2 (MATLAB) Modify the code **tdofss.m** for the two dof system and plot the distinct frequency responses.

P6.3 (MATLAB) Modify the code **tdof_ss_time_ode45_slnk.m** for the two dof system with $m_1 = m_2 = m = 1$, $k_1 = k_2 = k = 1$ and $c_1 = c_2 = 0$ for the following step forces and initial conditions:

a) $F_1 = 0$, $F_2 = -3$

b) $z_1 = 0$, $\dot{z}_1 = -2$, $z_2 = -1$, $\dot{z}_2 = 2$

Plot the time domain responses using both MATLAB and Simulink.

CHAPTER 7

MODAL ANALYSIS

7.1 Introduction

In Chapter 2 we systematically defined the equations of motion for a multi dof (mdof) system and transformed to the "s" domain using the Laplace transform. Chapter 3 discussed frequency responses and undamped mode shapes.

Chapter 5 discussed the state space form of equations of motion with arbitrary damping. It also covered the subject of complex modes. Heavily damped structures or structures with explicit damping elements, such as dashpots, result in complex modes and require state space solution techniques using the original coupled equations of motion.

Lightly damped structures are typically analyzed with the "normal mode" method, which is the subject of this chapter. The ability to think about vibrating systems in terms of modal properties is a very powerful technique that serves one well in both performing analysis and in understanding test data. The key to normal mode analysis is to develop tools which allow one to reconstruct the overall response of the system as a superposition of the responses of the different modes of the system. In analysis, the modal method allows one to replace the n-coupled differential equations with n-uncoupled equations, where each uncoupled equation represents the motion of the system for that mode of vibration. If natural frequencies and mode shapes are available for the system, then it is easy to visualize the motion of the system in each mode, which is the first step in being able to understand how to modify the system to change its characteristics.

Summarizing the modal analysis method of analyzing linear mechanical systems and the benefits derived:

1) Solve the undamped eigenvalue problem, which identifies the resonant frequencies and mode shapes (eigenvalues and eigenvectors), useful in themselves for understanding basic motions of the system.

2) Use the eigenvectors to uncouple or diagonalize the original set of coupled equations, allowing the solution of n-uncoupled sdof problems instead of solving a set of n-coupled equations.

3) Calculate the contribution of each mode to the overall response. This also allows one to reduce the size of the problem by eliminating modes that cannot be excited and/or modes that have no outputs at the desired dof's. Also, high frequency modes that have little contribution to the system at lower frequencies can be eliminated or approximately accounted for, further reducing the size of the system to be analyzed.

4) Write the system matrix, **A**, by inspection. Assemble the input and output matrices, **B** and **C**, using appropriate eigenvector terms. Frequency domain and forced transient response problems can be solved at this point. If complete eigenvectors are available, initial condition transient problems can also be solved. For lightly damped systems, proportional damping can be added, while still allowing the equations to be uncoupled.

7.2 Eigenvalue Problem

7.2.1 Equations of Motion

We will start by writing the undamped homogeneous (unforced) equations of motion for the model in Figure 7.1. Then we will define and solve the eigenvalue problem.

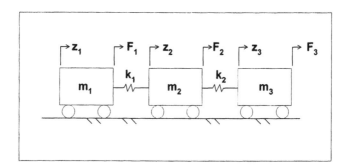

Figure 7.1: Undamped tdof model.

$$m\ddot{z} + kz = 0 \qquad (7.1)$$

From (2.5) with $k_1 = k_2 = k$ and $c1 = c2 = 0$:

$$\begin{bmatrix} m & 0 & 0 \\ 0 & m & 0 \\ 0 & 0 & m \end{bmatrix} \begin{bmatrix} \ddot{z}_1 \\ \ddot{z}_2 \\ \ddot{z}_3 \end{bmatrix} + \begin{bmatrix} k & -k & 0 \\ -k & 2k & -k \\ 0 & -k & k \end{bmatrix} \begin{bmatrix} z_1 \\ z_2 \\ z_3 \end{bmatrix} = \begin{bmatrix} 0 \\ 0 \\ 0 \end{bmatrix} \qquad (7.2)$$

7.2.2 Principal (Normal) Mode Definition

Since the system is conservative (it has no damping), normal modes of vibration will exist. Having normal modes means that at certain frequencies all points in the system will vibrate at the same frequency and in phase, i.e., **all points in the system will reach their minimum and maximum displacements at the same point in time**. Having normal modes can be expressed as (Weaver 1990):

$$z_i = z_{mi} \sin(\omega_i t + \phi_i) = z_{mi} \, \mathrm{Im}(e^{j\omega_i t + \phi_i}) \qquad (7.3)$$

Where:

z_i = vector of displacements for all dof's at the ith frequency

z_{mi} = the ith eigenvector, the mode shape for the ith resonant frequency

ω_i = the ith eigenvalue, i^{th} resonant frequency

ϕ_i = an arbitrary initial phase angle

For our tdof system, for the ith frequency, the equation would appear as:

$$\begin{bmatrix} z_1 \\ z_2 \\ z_3 \end{bmatrix} = \begin{bmatrix} z_{m1i} \\ z_{m2i} \\ z_{m3i} \end{bmatrix} \sin(\omega_i t + \phi_i), \qquad (7.4)$$

where the indices in the z_{mki} term represent the kth dof and the ith mode of the modal matrix z_m.

7.2.3 Eigenvalues / Characteristic Equation

Since the equation of motion

$$m\ddot{z} + kz = 0 \qquad (7.5)$$

and the form of the motion

$$z_i = z_{mi} \sin\left(\omega_i t + \phi_i\right) \tag{7.6}$$

are known, z_i can be differentiated twice and substituted into the equation of motion:

$$\ddot{z}_i = -\omega_i^2 z_{mi} \sin\left(\omega_i t + \phi_i\right) \tag{7.7}$$

$$\mathbf{m}\left[-\omega_i^2 z_{mi} \sin\left(\omega_i t + \phi_i\right)\right] + \mathbf{k}\left[z_{mi} \sin\left(\omega_i t + \phi_i\right)\right] = 0 \tag{7.8}$$

Canceling the sine terms:

$$-\omega_i^2 \mathbf{m} z_{mi} + \mathbf{k} z_{mi} = 0 \tag{7.9}$$

$$\mathbf{k} z_{mi} = \omega_i^2 \mathbf{m} z_{mi} \tag{7.10}$$

Equation (7.10) is the eigenvalue problem in nonstandard form, where the standard form is (Strang 1998):

$$\mathbf{A} \mathbf{z} = \lambda \mathbf{z} \tag{7.11}$$

The solution of the simultaneous equations which make up the standard form eigenvalue problem is a vector $\mathbf{z}$ such that when $\mathbf{z}$ is multiplied by $\mathbf{A}$, the product is a scalar multiple of $\mathbf{z}$ itself.

The nonstandard problem is "nonstandard" because the mass matrix $\mathbf{m}$ falls on the right-hand side. The form of the matrix presents no problem for hand calculations, but for computer calculations it is best transformed to standard form.

Rewriting the nonstandard form eigenvalue problem as a homogeneous equation:

$$\left(\mathbf{k} - \omega_i^2 \mathbf{m}\right) z_{mi} = 0 \tag{7.12}$$

A trivial solution, $z_{mi} = 0$, exists but is of no consequence. The only possibility for a nontrivial solution is if the determinant of the coefficient matrix is zero (Strang 1998). Expanding the matrix entries:

$$\left\{ \begin{bmatrix} k & -k & 0 \\ -k & 2k & -k \\ 0 & -k & k \end{bmatrix} - \omega_i^2 \begin{bmatrix} m & 0 & 0 \\ 0 & m & 0 \\ 0 & 0 & m \end{bmatrix} \right\} \mathbf{z}_{mi} = 0 \qquad (7.13)$$

Performing the matrix subtraction:

$$\begin{bmatrix} k - \omega_i^2 m & -k & 0 \\ -k & 2k - \omega_i^2 m & -k \\ 0 & -k & k - \omega_i^2 m \end{bmatrix} \mathbf{z}_{mi} = 0 \qquad (7.14)$$

Setting the determinant of the coefficient matrix equal to zero:

$$\begin{vmatrix} k - \omega_i^2 m & -k & 0 \\ -k & 2k - \omega_i^2 m & -k \\ 0 & -k & k - \omega_i^2 m \end{vmatrix} = 0 \qquad (7.15)$$

The determinant results in a polynomial in ω_i^2, the characteristic equation, where the roots of the polynomial are the eigenvalues, poles, or resonant frequencies of the system.

$$-m^3\omega^6 + 4km^2\omega^4 - 3k^2 m\omega^2 = 0$$

$$(7.16a,b)$$

$$\omega^2 \left(-m^3\omega^4 + 4km^2\omega^2 - 3k^2 m \right) = 0$$

Two of the roots are at the origin:

$$\omega_1 = 0 \qquad (7.17)$$

Solving for ω^2 as a quadratic in (7.16b) above:

$$\omega^2 = \frac{-4km^2 \pm \left(16k^2 m^4 - 12k^2 m^4 \right)^{\frac{1}{2}}}{-2m^3}$$

$$= \frac{-4km^2 \pm 2km^2}{-2m^3}$$

$$= \frac{-6k}{-2m}, \frac{-2k}{-2m}$$

$$= \frac{3k}{m}, \frac{k}{m} \tag{7.18}$$

$$\omega_2 = \pm\sqrt{\frac{3k}{m}}$$

$$\omega_3 = \pm\sqrt{\frac{k}{m}} \tag{7.19}$$

For each of the three eigenvalue pairs, there exists an eigenvector z_i, which gives the mode shape of the vibration at that frequency.

7.2.4 Eigenvectors

To obtain the eigenvectors of the system, any one of the degrees of freedom, say z_1, is selected as a reference. Then, all but one of the equations of motion is written with that value on the right-hand side:

$$\left(k - \omega_i^2 m\right) z_{mi} = 0 \tag{7.20}$$

$$\begin{bmatrix} \left(k - \omega_i^2 m\right) & -k & 0 \\ -k & \left(2k - \omega_i^2 m\right) & -k \\ 0 & -k & \left(k - \omega_i^2 m\right) \end{bmatrix} \begin{bmatrix} z_{m1i} \\ z_{m2i} \\ z_{m3i} \end{bmatrix} = 0 \tag{7.21}$$

Expanding the first and second equations, dropping the subscripts "i" and "m":

$$\left(k - \omega_i^2 m\right) z_1 - kz_2 = 0$$
$$-kz_1 + \left(2k - \omega_i^2 m\right) z_2 - kz_3 = 0 \tag{7.22a,b}$$

Rewriting with the z_1 term on the right-hand side and solving for the $\left(z_2 / z_1\right)$ ratio from (7.22a):

$$-kz_2 = -\left(k - \omega_i^2 m\right) z_1 \tag{7.23}$$

$$\frac{z_2}{z_1} = \frac{k - \omega_i^2 m}{k} \tag{7.24}$$

Solving for the (z_3 / z_1) ratio from (7.22b):

$$\left(2k - \omega_i^2 m\right) z_2 - k z_3 = k z_1 \tag{7.25}$$

$$\left(2k - \omega_i^2 m\right)\left(\frac{z_2}{z_1}\right) - \frac{k z_3}{z_1} = k \tag{7.26}$$

$$\left(2k - \omega_i^2 m\right)\left(\frac{k - \omega_i^2 m}{k}\right) - \frac{k z_3}{z_1} = k \tag{7.27}$$

$$\frac{z_3}{z_1} = \frac{\left(2k - \omega_i^2 m\right)\left(k - \omega_i^2 m\right)}{k^2} - 1 \tag{7.28}$$

$$\frac{z_3}{z_1} = \frac{m^2 \omega_i^4 - 3 k m \omega_i^2 + k^2}{k^2} \tag{7.29}$$

We now have the general equations for the eigenvector values. If a value is chosen for z_1, say 1.0, then the two ratios above can be solved for corresponding values of z_2 and z_3 for each of the three eigenvalues.

Since at each eigenvalue there are $(n+1)$ unknowns $\left(\omega_i, z_{mi}\right)$ for a system with n equations of motion, the eigenvectors are only known as **ratios** of displacements, not as absolute magnitudes. For the first mode of our tdof system the unknowns are ω_1, z_{m11}, z_{m21} and z_{m31} and we have only three equations of motion.

Substituting values for the three eigenvalues into the general eigenvector ratio equations above, assuming $m_1 = m_2 = m = 1$, $k_1 = k_2 = k = 1$:

For mode 1, $\omega_1^2 = 0$

$$\frac{z_2}{z_1} = \frac{k}{k} = 1 \tag{7.30}$$

$$z_2 = z_1 \tag{7.31}$$

$$\frac{z_3}{z_1} = \frac{(2k)(k)}{k^2} - 1 = 2 - 1 = 1 \tag{7.32}$$

$$z_3 = z_1 \tag{7.33}$$

Arbitrarily assigning $z_1 = 1$:

$$\mathbf{z}_1 = \begin{bmatrix} 1 \\ 1 \\ 1 \end{bmatrix} \tag{7.34}$$

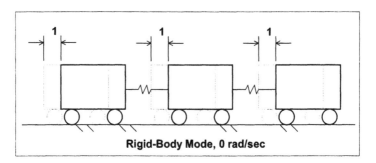

Rigid-Body Mode, 0 rad/sec

Figure 7.2: Mode shape plot for rigid body mode, where all masses move together with no stress in the connecting springs.

For mode 2, $\omega_2^2 = \dfrac{k}{m}$

$$\frac{z_2}{z_1} = \frac{k - \left(\dfrac{k}{m}\right)m}{k} = 0 \tag{7.35}$$

$$z_2 = 0 \tag{7.36}$$

$$\frac{z_3}{z_1} = \frac{\left(2k - \left(\dfrac{k}{m}\right)m\right)\left(k - \left(\dfrac{k}{m}\right)m\right)}{k^2} - 1 = \frac{(2k - k)(0)}{k2} - 1 = -1 \tag{7.37}$$

$$z_3 = -z_1 \tag{7.38}$$

$$\mathbf{z}_2 = \begin{bmatrix} 1 \\ 0 \\ -1 \end{bmatrix} \tag{7.39}$$

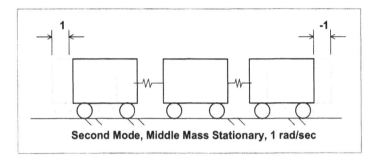

Figure 7.3: Mode shape plot for second mode, middle mass stationary and the two end masses move out of phase with each other with equal amplitude.

For node 3, $\omega_3^2 = \dfrac{3k}{m}$

$$\frac{z_2}{z_1} = \frac{k - \left(\dfrac{3k}{m}\right)m}{k} = \frac{-2k}{k} = -2 \tag{7.40}$$

$$z_2 = -2z_1 \tag{7.41}$$

$$\frac{z_3}{z_1} = \frac{\left(2k - \left(\dfrac{3k}{m}\right)m\right)\left(k - \left(\dfrac{3k}{m}\right)m\right)}{k^2} - 1 = \frac{(-k)(-2k) - 1}{k^2} = \frac{2k^2 - 1}{k^2} = 1 \tag{7.42}$$

$$z_3 = z_1 \tag{7.43}$$

$$\mathbf{z}_3 = \begin{bmatrix} 1 \\ -2 \\ 1 \end{bmatrix} \tag{7.44}$$

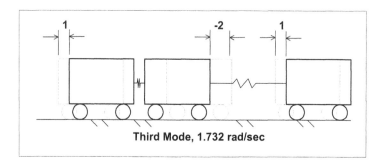

Third Mode, 1.732 rad/sec

Figure 7.4: Mode shape plot for third mode, with two end masses moving in phase with each other and out of phase with the middle mass, which is moving with twice the amplitude of the end masses.

7.2.5 Interpreting Eigenvectors

For the first mode, if all the masses start with either zero or the same initial velocity and with initial displacements of some scalar multiple of $[1 \ 1 \ 1]^T$, where "T" is the transpose, the system will either remain at rest or will continue moving at that velocity with no relative motion between the masses.

For the second and third modes, if the system is released with zero initial velocities but with initial displacements of some scalar multiple of that eigenvector, then the system will vibrate in only that mode with all the masses reaching their minimum and maximum points at the same point in time.

Any other combination of initial displacements will result in a motion which is a combination of the three eigenvectors.

7.2.6 Modal Matrix

Now that the three eigenvectors have been defined, the modal matrix will be introduced. The modal matrix is an (nxn) matrix with columns corresponding to the n system eigenvectors, starting with the first mode in the first column and so on:

mode: 1   2   3

$$\mathbf{z}_m = \begin{bmatrix} z_{m11} & z_{m12} & z_{m13} \\ z_{m21} & z_{m22} & z_{m23} \\ z_{m31} & z_{m32} & z_{m33} \end{bmatrix} \begin{matrix} \leftarrow \text{DOF 1} \\ \leftarrow \text{DOF 2} \\ \leftarrow \text{DOF 3} \end{matrix} \qquad (7.45)$$

$$\uparrow \quad \uparrow \quad \uparrow$$
$$\mathbf{z}_1 \quad \mathbf{z}_2 \quad \mathbf{z}_3$$

For our tdof problem:

$$\mathbf{z}_m = \begin{bmatrix} 1 & 1 & 1 \\ 1 & 0 & -2 \\ 1 & -1 & 1 \end{bmatrix} \qquad (7.46)$$

7.3 Uncoupling the Equations of Motion

At this point the system is well defined in terms of natural frequencies and modes of vibration. If any further information such as transient or frequency response is desired, solving for it would be laborious because the system equations are still coupled. For transient response, the equations would have to be solved simultaneously using a numerical integration scheme unless the problem were simple enough to allow a closed form solution. To calculate the damped frequency response, a complex equation solving routine would have to be used to invert the complex coefficient matrix at each frequency.

In order to facilitate solving for the transient or frequency responses, it is useful to transform the n-coupled second order differential equations to n-uncoupled second order differential equations by transforming from the physical coordinate system to a principal coordinate system. In linear algebra terms, the transformation from physical to principal coordinates is known as a change of basis. There are many options for change of basis, but we will show that when eigenvectors are used for the transformation the principal coordinate system has a physical meaning; each of the uncoupled sdof systems represents the motion of a specific mode of vibration. The n-uncoupled equations in the principal coordinate system can then be solved for the responses in the principal coordinate system using well-known solutions for single degree of freedom systems. The n-responses in the principal coordinate system can then be transformed back to the physical coordinate system to provide the actual response in physical coordinates. This procedure is shown schematically in Figure 7.5.

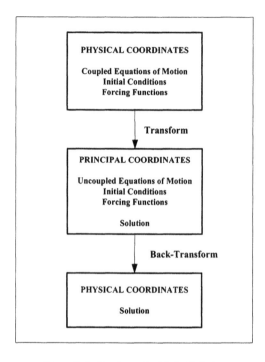

Figure 7.5: Roadmap for Modal Solution

The procedure above is analogous to using Laplace transforms for solving differential equations, where the differential equation is transformed to an algebraic equation, solved algebraically, and back transformed to get the solution of the original problem.

We now need a means of diagonalizing the mass and stiffness matrices, which will yield a set of uncoupled equations.

The condition to guarantee diagonalization is the existence of n-linearly independent eigenvectors, which is always the case if the mass and stiffness matrices are both symmetric or if there are n-different (nonrepeated) eigenvalues (Strang 1998).

Going back to the original homogeneous equation of motion:

$$\mathbf{m\ddot{z}} + \mathbf{kz} = 0 \qquad (7.47)$$

Having normal modes means that at frequency "i":

$$z_i = z_{mi} \sin(\omega_i t + \phi_i) \qquad (7.48)$$

Differentiating twice to get acceleration:

$$\ddot{z}_i = -\omega_i^2 z_{mi} \sin(\omega_i t + \phi_i) \qquad (7.49)$$

Substituting back into the equation of motion:

$$m\{-\omega_i^2 z_{mi} \sin(\omega_i t + \phi_i)\} + k\{z_{mi} \sin(\omega_i t + \phi_i)\} = 0 \qquad (7.50)$$

Canceling sine terms:

$$-\omega_i^2 m z_{mi} + k z_{mi} = 0 \qquad (7.51)$$

Rearranging and writing the above equation for both the "ith" and "jth" modes:

$$k z_{mi} = \omega_i^2 m z_{mi} \qquad (7.52)$$

$$k z_{mj} = \omega_j^2 m z_{mj} \qquad (7.53)$$

z_{mi} and z_{mj} are the "ith" and "jth" eigenvectors, the "ith" and "jth" columns of the modal matrix.

Premultiplying (7.52) by the transpose of z_{mj}, z_{mj}^T:

$$z_{mj}^T k z_{mi} = \omega_i^2 z_{mj}^T m z_{mi} \qquad (7.54)$$

Taking the transpose of (7.53), where the transpose of a product is the product of the individual transposes taken in reverse order, i.e., $[AB]^T = B^T A^T$:

$$z_{mj}^T k^T = \omega_j^2 z_{mj}^T m^T, \qquad (7.55)$$

since m and k are symmetrical, $m^T = m$, and $k^T = k$:

$$z_{mj}^T k = \omega_j^2 z_{mj}^T m \qquad (7.56)$$

Postmultiplying (7.56) by z_{mi}

$$z_{mj}^T k z_{mi} = \omega_j^2 z_{mj}^T m z_{mi} \tag{7.57}$$

Now, subtracting (7.57) from (7.54):

$$
\begin{aligned}
z_{mj}^T k z_{mi} &= \omega_i^2 z_{mj}^T m z_{mi} \\
-(z_{mj}^T k z_{mi} &= \omega_j^2 z_{mj}^T m z_{mi}) \\
\hline
0 &= (\omega_i^2 - \omega_j^2) z_{mj}^T m z_{mi}
\end{aligned}
\tag{7.58}
$$

When $i \neq j$, the term $(\omega_i^2 - \omega_j^2)$ cannot be equal to zero, meaning that the term $z_{mj}^T m z_{mi}$ must be equal to zero.

$$z_{mj}^T m z_{mi} = 0 \tag{7.59}$$

Looking at the sizes of the matrices multiplied:

$$
\begin{aligned}
z_{mj}^T &= 1 \mathrm{xn} \\
m &= \mathrm{nxn} \\
z_{mi} &= \mathrm{nx1}
\end{aligned}
\tag{7.60}
$$

$$(1\mathrm{xn}) \times (\mathrm{nxn}) \times (\mathrm{nx1}) = (1\mathrm{x1}) = \text{scalar} \tag{7.61}$$

Equation (7.59) can be rewritten:

$$z_{mj}^T m z_{mi} = m_{ij} = 0, \tag{7.62}$$

where m_{ij} is an off-diagonal term in the mass matrix of the principal coordinate system.

The two eigenvectors z_{mj} and z_{mi} are said to be orthogonal with respect to m, where orthogonality is defined as the property that causes all the off-diagonal terms in the principal mass matrix to be zero.

Returning to (7.62), for $i = j$, $(\omega_i^2 - \omega_j^2) = 0$. Thus the product $z_{mi}^T m z_{mi}$ can be set equal to **any** arbitrary constant m_{ii}, a diagonal term in the principal mass matrix.

$$z_{mi}^T mz_{mi} = m_{ii} \tag{7.63}$$

This is where various normalization techniques for eigenvectors come into play, discussed in the next section.

The stiffness matrix, k, is normalized in the same manner.

In practice, instead of diagonalizing the mass and stiffness matrices term by term by pre- and postmultiplying by individual eigenvectors, the entire modal matrix is used to diagonalize in one operation using two matrix multiplications:

$$m_n = z_m^T mz_m \tag{7.64}$$

$$k_n = z_m^T kz_m \tag{7.65}$$

7.4 Normalizing Eigenvectors

Because eigenvectors are only known as **ratios** of displacements, not as absolute magnitudes, we can choose how to normalize them. Up to now, when calculating eigenvectors we have arbitrarily set the amplitude of the first dof to 1. We will now discuss two of the most commonly used eigenvector normalization techniques. Different normalizing techniques result in different forms of the resulting uncoupled differential equations.

7.4.1 Normalizing with Respect to Unity

One method is to normalize with respect to unity, making the **largest** element in each eigenvector equal to unity by dividing each column by its largest value. We now add the notation z_n, where the "n" refers to a "**normalized**" modal matrix.

$$z_m = \begin{bmatrix} 1 & 1 & 1 \\ 1 & 0 & -2 \\ 1 & -1 & 1 \end{bmatrix} \quad \Rightarrow \quad z_n = \begin{bmatrix} 1 & 1 & -0.5 \\ 1 & 0 & 1 \\ 1 & -1 & -0.5 \end{bmatrix} \tag{7.66}$$

Using the unity normalized modal matrix to transform the mass matrix in two matrix multiplications:

$$
\mathbf{z}_n^T \mathbf{m} = \begin{bmatrix} 1 & 1 & 1 \\ 1 & 0 & -1 \\ -0.5 & 1 & -0.5 \end{bmatrix} \begin{bmatrix} m & 0 & 0 \\ 0 & m & 0 \\ 0 & 0 & m \end{bmatrix} = \begin{bmatrix} m & m & m \\ m & 0 & -m \\ -.5m & m & -.5m \end{bmatrix} \quad (7.67)
$$

$$
\mathbf{m}_n = \mathbf{z}_n^T \mathbf{m} \mathbf{z}_n = \begin{bmatrix} m & m & m \\ m & 0 & -m \\ -.5m & m & -.5m \end{bmatrix} \begin{bmatrix} 1 & 1 & -.5 \\ 1 & 0 & 1 \\ 1 & -1 & -.5 \end{bmatrix} = \begin{bmatrix} 3m & 0 & 0 \\ 0 & 2m & 0 \\ 0 & 0 & 1.5m \end{bmatrix} \quad (7.68)
$$

Similarly transforming the stiffness matrix:

$$
\mathbf{z}_n^T \mathbf{k} = \begin{bmatrix} 1 & 1 & 1 \\ 1 & 0 & -1 \\ -.5 & 1 & -.5 \end{bmatrix} \begin{bmatrix} k & -k & 0 \\ -k & 2k & -k \\ 0 & -k & k \end{bmatrix} = \begin{bmatrix} 0 & 0 & 0 \\ k & 0 & -k \\ -1.5k & 3k & -1.5k \end{bmatrix} \quad (7.69)
$$

$$
\mathbf{k}_n = \mathbf{z}_n^T \mathbf{k} \mathbf{z}_n = \begin{bmatrix} 0 & 0 & 0 \\ -k & 0 & -k \\ -1.5k & 3k & -1.5k \end{bmatrix} \begin{bmatrix} 1 & 1 & -.5 \\ 1 & 0 & 1 \\ 1 & -1 & -.5 \end{bmatrix} = \begin{bmatrix} 0 & 0 & 0 \\ 0 & 2k & 0 \\ 0 & 0 & 4.5k \end{bmatrix} \quad (7.70)
$$

Note that the original filled stiffness matrix is now diagonal. Also note that if the diagonal elements of the stiffness matrix (7.70) are divided by the corresponding diagonal elements of the mass matrix (7.69), the three terms are the squares of the respective eigenvalues.

7.4.2 Normalizing with Respect to Mass

Another method is to normalize with respect to mass using the equation:

$$
\mathbf{z}_{ni}^T \mathbf{m} \mathbf{z}_{ni} = 1.0 , \quad (7.71)
$$

making each diagonal mass term equal 1.0. This is the method used by default in ANSYS.

Once again, note that modal matrix subscript "ni" in $\mathbf{z}_{ni}$ signifies the normalized i^{th} eigenvector. Each normalized eigenvector is defined as follows:

$$
\mathbf{z}_{ni} = \frac{\mathbf{z}_{mi}}{\left[\mathbf{z}_{mi}^T \mathbf{m} \mathbf{z}_{mi} \right]^{\frac{1}{2}}} = \frac{\mathbf{z}_{mi}}{q_i} \quad (7.72)
$$

Where q_i is defined as:

$$q_i = \left[\sum_{j=i}^{n} z_{mji} \left(\sum_{k=1}^{n} m_{jk} z_{mki} \right) \right]^{\frac{1}{2}} \tag{7.73}$$

For a diagonal mass matrix, q can be simplified since all the m_{jk} terms are zero:

$$q_i = \left[\sum_{k=1}^{n} m_k z_{mki}^2 \right]^{\frac{1}{2}} \tag{7.74}$$

Thus, by operating on $\mathbf{m}$ by $\mathbf{z}_n$, the mass matrix should be transformed into the identity matrix. Starting with $\mathbf{z}_m$ and the "q" values from above:

$$\mathbf{z}_m = \begin{bmatrix} 1 & 1 & 1 \\ 1 & 0 & -2 \\ 1 & -1 & 1 \end{bmatrix} \tag{7.75}$$

$$q_1 = \left[m(1)^2 + m(1)^2 + m(1)^2 \right]^{\frac{1}{2}} = \sqrt{3m}$$

$$q_2 = \left[m(1)^2 + m(0)^2 + m(-1)^2 \right]^{\frac{1}{2}} = \sqrt{2m} \tag{7.76a,b,c}$$

$$q_3 = \left[m(1)^2 + m(-2)^2 + m(1)^2 \right]^{\frac{1}{2}} = \sqrt{6m}$$

The modal matrix normalized with respect to mass becomes:

$$\mathbf{z}_n = \begin{bmatrix} \dfrac{1}{\sqrt{3m}} & \dfrac{1}{\sqrt{2m}} & \dfrac{1}{\sqrt{6m}} \\ \dfrac{1}{\sqrt{3m}} & 0 & \dfrac{-2}{\sqrt{6m}} \\ \dfrac{1}{\sqrt{3m}} & \dfrac{-1}{\sqrt{2m}} & \dfrac{1}{\sqrt{6m}} \end{bmatrix} = \dfrac{1}{\sqrt{m}} \begin{bmatrix} \dfrac{1}{\sqrt{3}} & \dfrac{1}{\sqrt{2}} & \dfrac{1}{\sqrt{6}} \\ \dfrac{1}{\sqrt{3}} & 0 & \dfrac{-2}{\sqrt{6}} \\ \dfrac{1}{\sqrt{3}} & \dfrac{-1}{\sqrt{2}} & \dfrac{1}{\sqrt{6}} \end{bmatrix} \tag{7.77}$$

Using $\mathbf{z}_n$ to transform the mass matrix:

$$\mathbf{z}_n^T \mathbf{m} = \begin{bmatrix} \dfrac{1}{\sqrt{3m}} & \dfrac{1}{\sqrt{3m}} & \dfrac{1}{\sqrt{3m}} \\ \dfrac{1}{\sqrt{2m}} & 0 & \dfrac{-1}{\sqrt{2m}} \\ \dfrac{1}{\sqrt{6m}} & \dfrac{-2}{\sqrt{6m}} & \dfrac{1}{\sqrt{6m}} \end{bmatrix} \begin{bmatrix} m & 0 & 0 \\ 0 & m & 0 \\ 0 & 0 & m \end{bmatrix} = \begin{bmatrix} \dfrac{m}{\sqrt{3m}} & \dfrac{m}{\sqrt{3m}} & \dfrac{m}{\sqrt{3m}} \\ \dfrac{m}{\sqrt{2m}} & 0 & \dfrac{-m}{\sqrt{2m}} \\ \dfrac{m}{\sqrt{6m}} & \dfrac{-2m}{\sqrt{6m}} & \dfrac{m}{\sqrt{6m}} \end{bmatrix} \quad (7.78)$$

$$\mathbf{m}_n = \mathbf{z}_n^T \mathbf{m} \mathbf{z}_n = \begin{bmatrix} \dfrac{m}{\sqrt{3m}} & \dfrac{m}{\sqrt{3m}} & \dfrac{m}{\sqrt{3m}} \\ \dfrac{m}{\sqrt{2m}} & 0 & \dfrac{-m}{\sqrt{2m}} \\ \dfrac{m}{\sqrt{6m}} & \dfrac{-2m}{\sqrt{6m}} & \dfrac{m}{\sqrt{6m}} \end{bmatrix} \begin{bmatrix} \dfrac{1}{\sqrt{3m}} & \dfrac{1}{\sqrt{2m}} & \dfrac{1}{\sqrt{6m}} \\ \dfrac{1}{\sqrt{3m}} & 0 & \dfrac{-2}{\sqrt{6m}} \\ \dfrac{1}{\sqrt{3m}} & \dfrac{-1}{\sqrt{2m}} & \dfrac{1}{\sqrt{6m}} \end{bmatrix} \quad (7.79)$$

$$\mathbf{m}_n = \begin{bmatrix} \left(\dfrac{m}{3m} + \dfrac{m}{3m} + \dfrac{m}{3m} \right) & \left(\dfrac{m}{m\sqrt{3}\sqrt{2}} + 0 - \dfrac{m}{m\sqrt{3}\sqrt{2}} \right) \\ \left(\dfrac{m}{m\sqrt{2}\sqrt{3}} + 0 - \dfrac{m}{m\sqrt{2}\sqrt{3}} \right) & \left(\dfrac{m}{2m} + 0 + \dfrac{m}{2m} \right) \\ \left(\dfrac{m}{m\sqrt{3}\sqrt{6}} - \dfrac{2m}{m\sqrt{3}\sqrt{6}} + \dfrac{m}{m\sqrt{3}\sqrt{6}} \right) & \left(\dfrac{m}{m\sqrt{2}\sqrt{6}} + 0 - \dfrac{m}{m\sqrt{6}\sqrt{2}} \right) \end{bmatrix}$$

$$\begin{bmatrix} \left(\dfrac{m}{m\sqrt{3}\sqrt{6}} - \dfrac{2m}{m\sqrt{3}\sqrt{6}} + \dfrac{m}{m\sqrt{3}\sqrt{6}} \right) \\ \left(\dfrac{m}{m\sqrt{2}\sqrt{6}} + 0 - \dfrac{m}{m\sqrt{2}\sqrt{6}} \right) \\ \left(\dfrac{m}{6m} + \dfrac{4m}{6m} + \dfrac{m}{6m} \right) \end{bmatrix}$$

$$= \begin{bmatrix} 1 & 0 & 0 \\ 0 & 1 & 0 \\ 0 & 0 & 1 \end{bmatrix}$$

$$(7.80)$$

The original mass matrix has been transformed to the identity matrix.

Similarly transforming the stiffness matrix:

$$\mathbf{z}_n^T \mathbf{k} = \frac{1}{\sqrt{m}} \begin{bmatrix} \dfrac{1}{\sqrt{3}} & \dfrac{1}{\sqrt{3}} & \dfrac{1}{\sqrt{3}} \\ \dfrac{1}{\sqrt{2}} & 0 & \dfrac{-1}{\sqrt{2}} \\ \dfrac{1}{\sqrt{6}} & \dfrac{-2}{\sqrt{6}} & \dfrac{1}{\sqrt{6}} \end{bmatrix} k \begin{bmatrix} 1 & -1 & 0 \\ -1 & 2 & -1 \\ 0 & -1 & 1 \end{bmatrix}$$

$$= \frac{k}{\sqrt{m}} \begin{bmatrix} \left(\dfrac{1}{\sqrt{3}} - \dfrac{1}{\sqrt{3}} \right) & \left(\dfrac{-1}{\sqrt{3}} + \dfrac{2}{\sqrt{3}} - \dfrac{1}{\sqrt{3}} \right) & \left(0 - \dfrac{1}{\sqrt{3}} + \dfrac{1}{\sqrt{3}} \right) \\ \left(\dfrac{1}{\sqrt{2}} + 0 + 0 \right) & \left(\dfrac{-1}{\sqrt{2}} + 0 + \dfrac{1}{\sqrt{2}} \right) & \left(0 + 0 - \dfrac{1}{\sqrt{2}} \right) \\ \left(\dfrac{1}{\sqrt{6}} + \dfrac{2}{\sqrt{6}} + 0 \right) & \left(\dfrac{-1}{\sqrt{6}} - \dfrac{4}{\sqrt{6}} - \dfrac{1}{\sqrt{6}} \right) & \left(0 + \dfrac{2}{\sqrt{6}} + \dfrac{1}{\sqrt{6}} \right) \end{bmatrix} \quad (7.81)$$

$$\mathbf{k}_n = \mathbf{z}_n^T \mathbf{k} \mathbf{z}_n = \begin{bmatrix} 0 & 0 & 0 \\ \dfrac{1}{\sqrt{2}} & 0 & \dfrac{-1}{\sqrt{2}} \\ \dfrac{3}{\sqrt{6}} & \dfrac{-6}{\sqrt{6}} & \dfrac{3}{\sqrt{6}} \end{bmatrix} \begin{bmatrix} \dfrac{1}{\sqrt{3}} & \dfrac{1}{\sqrt{2}} & \dfrac{1}{\sqrt{6}} \\ \dfrac{1}{\sqrt{3}} & 0 & \dfrac{-2}{\sqrt{6}} \\ \dfrac{1}{\sqrt{3}} & \dfrac{-1}{\sqrt{2}} & \dfrac{1}{\sqrt{6}} \end{bmatrix} \frac{k}{m} \quad (7.82)$$

$$\mathbf{k}_n = \frac{k}{m} \begin{bmatrix} 0 & & 0 \\ \left(\dfrac{1}{\sqrt{2}\sqrt{3}} - \dfrac{1}{\sqrt{2}\sqrt{3}} \right) & & \left(\dfrac{1}{2} + 0 + \dfrac{1}{2} \right) \\ \left(\dfrac{3}{\sqrt{3}\sqrt{6}} - \dfrac{6}{\sqrt{3}\sqrt{6}} + \dfrac{3}{\sqrt{3}\sqrt{6}} \right) & & \left(\dfrac{3}{\sqrt{2}\sqrt{6}} - 0 - \dfrac{3}{\sqrt{6}\sqrt{2}} \right) \end{bmatrix}$$

$$\qquad (7.83)$$

$$\begin{bmatrix} & 0 & \\ & \left(\dfrac{1}{\sqrt{2}\sqrt{6}} + 0 - \dfrac{1}{\sqrt{2}\sqrt{6}} \right) & \\ & \left(\dfrac{3}{6} + \dfrac{12}{6} + \dfrac{3}{6} \right) & \end{bmatrix}$$

$$\mathbf{k}_n = \begin{bmatrix} 0 & 0 & 0 \\ 0 & 1 & 0 \\ 0 & 0 & 3 \end{bmatrix} \frac{k}{m} \tag{7.84}$$

Note that the normalized stiffness matrix is now diagonal and that the diagonal terms are the squares of the corresponding three eigenvalues. The normalized stiffness matrix is also known as the **spectral matrix** (Weaver 1990).

Because normalizing with respect to mass results in an identity principal mass matrix and squares of the eigenvalues on the diagonal in the principal stiffness matrix, we will use only this normalization in the future. Since we know the form of the principal matrices when normalizing with respect to mass, no multiplying of modal matrices is actually required: **the homogeneous principal equations of motion can be written by inspection knowing only the eigenvalues.**

7.5 Reviewing Equations of Motion in Principal Coordinates – Mass Normalization

7.5.1 Equations of Motion in Physical Coordinate System

$$\begin{bmatrix} m & 0 & 0 \\ 0 & m & 0 \\ 0 & 0 & m \end{bmatrix} \begin{bmatrix} \ddot{z}_1 \\ \ddot{z}_2 \\ \ddot{z}_3 \end{bmatrix} + \begin{bmatrix} k & -k & 0 \\ -k & 2k & -k \\ 0 & -k & k \end{bmatrix} \begin{bmatrix} z_1 \\ z_2 \\ z_3 \end{bmatrix} = \begin{bmatrix} 0 \end{bmatrix} \tag{7.85}$$

Eigenvalues:

$$\omega_1 = 0 \tag{7.86}$$

$$\omega_2 = \pm\sqrt{\frac{3k}{m}}$$

$$\omega_3 = \pm\sqrt{\frac{k}{m}} \tag{7.87a,b}$$

Eigenvectors, normalized with respect to mass:

$$\mathbf{z}_n = \frac{1}{\sqrt{m}} \begin{bmatrix} \dfrac{1}{\sqrt{3}} & \dfrac{1}{\sqrt{2}} & \dfrac{1}{\sqrt{6}} \\[2ex] \dfrac{1}{\sqrt{3}} & 0 & \dfrac{-2}{\sqrt{6}} \\[2ex] \dfrac{1}{\sqrt{3}} & \dfrac{-1}{\sqrt{2}} & \dfrac{1}{\sqrt{6}} \end{bmatrix} \tag{7.88}$$

7.5.2 Equations of Motion in Principal Coordinate System

$$\begin{bmatrix} 1 & 0 & 0 \\ 0 & 1 & 0 \\ 0 & 0 & 1 \end{bmatrix} \begin{bmatrix} \ddot{z}_{p1} \\ \ddot{z}_{p2} \\ \ddot{z}_{p3} \end{bmatrix} + \begin{bmatrix} 0 & 0 & 0 \\ 0 & \dfrac{k}{m} & 0 \\ 0 & 0 & \dfrac{3k}{m} \end{bmatrix} \begin{bmatrix} z_{p1} \\ z_{p2} \\ z_{p3} \end{bmatrix} = \begin{bmatrix} 0 \end{bmatrix} \tag{7.89}$$

7.5.3 Expanding Matrix Equations of Motion in Both Coordinate Systems

Physical Coordinates	Principal Coordinates
$m\ddot{z}_1 + kz_1 - kz_2 = 0$ $m\ddot{z}_2 - kz_1 + 2kz_2 - kz_3 = 0$ $m\ddot{z}_3 - kz_2 + kz_3 = 0$	$\ddot{z}_{p1} = 0$ $\ddot{z}_{p2} + \dfrac{k}{m} z_{p2} = 0$ $\ddot{z}_{p3} + \dfrac{3k}{m} z_{p3} = 0$
These equations are coupled and have to be solved simultaneously.	These homogeneous equations are uncoupled and can be solved independently.

Table 7.1: Summary of equations of motion in physical and principal coordinates.

7.6 Transforming Initial Conditions and Forces

Now that we know how to construct the homogeneous uncoupled equations of motion for the system, we need to know how to transform initial conditions and forces to the principal coordinate system. We can then solve for transient and forced responses in the principal coordinate system using the uncoupled equations.

Starting with the original non-homogeneous equations of motion in physical coordinates:

$$m\ddot{z} + kz = F \qquad (7.90)$$

Premultiplying by z_n^T, the transpose of the modal matrix:

$$z_n^T m\ddot{z} + z_n^T kz = z_n^T F \qquad (7.91)$$

Inserting the identify matrix, $I = z_n z_n^{-1}$:

$$z_n^T m \underbrace{z_n z_n^{-1}}_{I} \ddot{z} + z_n^T k \underbrace{z_n z_n^{-1}}_{I} z = z_n^T F \qquad (7.92)$$

Rewriting and regrouping terms:

$$\underbrace{z_n^T m z_n}_{m_p} \underbrace{z_n^{-1} \ddot{z}}_{\ddot{z}_p} + \underbrace{z_n^T k z_n}_{k_p} \underbrace{z_n^{-1} z}_{z_p} = \underbrace{z_n^T F}_{F_p}, \qquad (7.93)$$

where $z_n^T m z_n$ and $z_n^T k z_n$ were shown to diagonalize the mass and stiffness matrices in the previous section.

Defining terms:

m_p = (nxn) diagonal principal mass matrix

k_p = (nxn) diagonal principal stiffness matrix

$z_n^{-1} \ddot{z} = \ddot{z}_p$ = acceleration vector in principal coordinates

$z_n^{-1} z = z_p$ = displacement vector in principal coordinates

$\mathbf{z}_n^T \mathbf{F} = \mathbf{F}_p$ = force vector in principal coordinates

In the previous section, the definitions for accelerations and displacements in physical and principal coordinates were shown to be:

$$\ddot{\mathbf{z}}_p = \mathbf{z}_n^{-1}\ddot{\mathbf{z}}$$
$$\mathbf{z}_p = \mathbf{z}_n^{-1}\mathbf{z}$$

(7.94)

The same relationships hold for initial conditions of displacement and velocity:

$$\mathbf{z}_{op} = \mathbf{z}_n^{-1}\mathbf{z}_o$$
$$\dot{\mathbf{z}}_{op} = \mathbf{z}_n^{-1}\dot{\mathbf{z}}_o$$

(7.95)

In (7.95), $\mathbf{z}_{op}$ and $\dot{\mathbf{z}}_{op}$ are vectors of initial displacements and velocities, respectively, in the principal coordinate system, and $\mathbf{z}_o$ and $\dot{\mathbf{z}}_o$ are vectors of initial displacements and velocities, respectively, in the physical coordinate system.

Taking the inverse of the modal matrix to convert initial conditions requires that the modal matrix be square, with as many eigenvectors as number of degrees of freedom. We will see in future chapters that there are instances where not all eigenvectors are available. In one case, we may choose to only calculate eigenvalues and eigenvectors up to a certain frequency in order to save calculation time or because the problem only requires knowledge of response in a certain frequency range. In another case, we may build a "reduced" model where only the most significant modes are retained. Fortunately, a large majority of real life problems involve zero initial conditions.

7.7 Summarizing Equations of Motion in Both Coordinate Systems

The two sets of equations, in physical and principal coordinates, are shown in Table 7.2:

Physical Coordinates	Principal Coordinates
$m\ddot{z}_1 + kz_1 - kz_2 = F_1$ $m\ddot{z}_2 - kz_1 + 2kz_2 - kz_3 = F_2$ $m\ddot{z}_3 - kz_2 + kz_3 = F_3$ IC's: $z_1, z_2, z_3, \dot{z}_1, \dot{z}_2, \dot{z}_3$	$\ddot{z}_{p1} = F_{p1}$ $\ddot{z}_{p2} + \dfrac{k}{m} z_{p2} = F_{p2}$ $\ddot{z}_{p3} + \dfrac{3k}{m} z_{p3} = F_{p3}$ IC's: $z_{p1}, z_{p2}, z_{p3}, \dot{z}_{p1}, \dot{z}_{p2}, \dot{z}_{p3}$

Table 7.2: Summary of equations of motion in physical and principal coordinates.

The variables in physical coordinates are the positions and velocities of the masses. The variables in principal coordinates are the displacements and velocities of each mode of vibration.

The equations in principal coordinates can be easily solved, since the equations are uncoupled, yielding the displacements. We now need to back transform the results in the principal coordinate system to the physical coordinate system to get the final answer.

7.8 Back-Transforming from Principal to Physical Coordinates

We showed previously that the relationship between physical and principal coordinates is:

$$z_n^{-1} z = z_p \tag{7.96}$$

Premultiplying by z_n:

$$\underbrace{z_n (z_n^{-1} z)}_{I} = z_n z_p \tag{7.97}$$

$$z = z_n z_p \tag{7.98}$$

Thus, the displacement vector in physical coordinates is obtained by premultiplying the vector of displacements in principal coordinates by the normalized modal matrix z_n.

Similarly for velocity:

$$\dot{\mathbf{z}} = \mathbf{z}_n \dot{\mathbf{z}}_p \qquad (7.99)$$

7.9 Reducing the Model Size When Only Selected Degrees of Freedom are Required

So far we have hinted at the fact that only portions of the eigenvector matrix are needed if selected dof's have forces applied and other (or the same) dof's are needed for output. This section will show how the reduction in dof's occurs. This reduction is one of the key steps to be used later in the book when we cover how to reduce the size of models derived from large finite element simulations.

Reviewing the steps in the modal solution, starting with the equations of motion and initial conditions in physical coordinates:

$$m\ddot{z}_1 + kz_1 - kz_2 = F_1$$
$$m\ddot{z}_2 - kz_1 + 2kz_2 - kz_3 = F_2$$
$$m\ddot{z}_3 - kz_2 + kz_3 = F_3 \qquad (7.100)$$

$$\text{Initial Conditions}: \ z_1, z_2, z_3, \dot{z}_1, \dot{z}_2, \dot{z}_3 = 0$$

Solve for eigenvalues: $\omega_1, \omega_2, \omega_3$

Solve for eigenvectors, normalize with respect to mass and form the modal matrix from columns of eigenvectors:

$$\mathbf{z}_n = \begin{bmatrix} z_{n11} & z_{n12} & z_{n13} \\ z_{n21} & z_{n22} & z_{n23} \\ z_{n31} & z_{n32} & z_{n33} \end{bmatrix} \qquad (7.101)$$

Transform forces from physical to principal coordinates:

$$\mathbf{F}_p = \mathbf{z}_n^T \mathbf{F} \qquad (7.102)$$

Write the equations of motion in principal coordinates:

$$\ddot{z}_{p1} = F_{p1}$$

$$\ddot{z}_{p2} + \omega_2^2 z_{p2} = F_{p2}$$

$$\ddot{z}_{p3} + \omega_3^2 z_{p3} = F_{p3}$$

(7.103a,b,c,d)

$$\text{IC's: } z_{p1}, z_{p2}, z_{p3}, \dot{z}_{p1}, \dot{z}_{p2}, \dot{z}_{p3} = 0$$

Solve the equations in principal coordinates in either time or frequency domain and then back transform to physical coordinates:

$$z = z_n z_p$$

(7.104)

$$\dot{z} = z_n \dot{z}_p$$

Note that the two critical transformations (assuming zero initial conditions) involve premultiplying by the transpose of the modal matrix ($F \rightarrow F_p$) in (7.102) or the modal matrix ($z_p \rightarrow z$) in (7.104).

Let us first examine the force transformation by expanding the equations:

$$F_p = z_n^T F$$

(7.105)

$$z_n^T F = \begin{bmatrix} z_{n11} & z_{n12} & z_{n13} \\ z_{n21} & z_{n22} & z_{n23} \\ z_{n31} & z_{n32} & z_{n33} \end{bmatrix}^T \begin{bmatrix} F_1 \\ F_2 \\ F_3 \end{bmatrix} = \begin{bmatrix} z_{n11} & z_{n21} & z_{n31} \\ z_{n12} & z_{n22} & z_{n32} \\ z_{n13} & z_{n23} & z_{n33} \end{bmatrix} \begin{bmatrix} F_1 \\ F_2 \\ F_3 \end{bmatrix}$$

(7.106)

$$= \begin{bmatrix} z_{n11} F_1 + z_{n21} F_2 + z_{n31} F_3 \\ z_{n12} F_1 + z_{n22} F_2 + z_{n32} F_3 \\ z_{n13} F_1 + z_{n23} F_2 + z_{n33} F_3 \end{bmatrix}$$

Note that the multipliers of F_1 in the first column are the elements of the first row of the modal matrix, the multipliers of F_2 in the second column are the elements of the second row of the modal matrix and the multipliers of F_3 in the third column are the elements of the third row of the modal matrix.

Suppose that force is to be applied at only mass 1, F_1, then only the first row of the modal matrix is required to transform the force in physical coordinates to the force in principal coordinates.

Now let us examine the displacement transformation by expanding the equations:

$$\mathbf{z} = \mathbf{z}_n \mathbf{z}_p \tag{7.107}$$

$$\mathbf{z} = \begin{bmatrix} z_1 \\ z_2 \\ z_3 \end{bmatrix} = \mathbf{z}_n \mathbf{z}_p = \begin{bmatrix} z_{n11} & z_{n12} & z_{n13} \\ z_{n21} & z_{n22} & z_{n23} \\ z_{n31} & z_{n32} & z_{n33} \end{bmatrix} \begin{bmatrix} z_{p1} \\ z_{p2} \\ z_{p3} \end{bmatrix} = \begin{bmatrix} z_{n11}z_{p1} + z_{n12}z_{p2} + z_{n13}z_{p3} \\ z_{n21}z_{p1} + z_{n22}z_{p2} + z_{n23}z_{p3} \\ z_{n31}z_{p1} + z_{n32}z_{p2} + z_{n33}z_{p3} \end{bmatrix}$$

$$\tag{7.108}$$

Note that the coefficients of the principal displacements in the first row above are the elements of the first row of the modal matrix. Similarly, coefficients of the second and third rows are the elements of the second and third rows of the modal matrix.

Suppose that the only physical displacement we are interested in is that of mass 2, z_2, then only the second row of the modal matrix is required to transform the three displacements z_{p1}, z_{p2}, z_{p3} in principal coordinates to z_2. This leads to the following conclusion about reducing the size of the model:

> **Only the rows of the modal matrix that correspond to degrees of freedom to which forces are applied and/or for which displacements are desired are required to complete the model.**

For this tdof model, reducing the size of the problem is not required; however, we will see later that a realistic finite element model, with hundreds of thousands of degrees of freedom, presents an entirely different problem. Having the ability to reduce the problem size is critical in order to use the detailed results of a complicated finite element model to provide accurate results in a lower order MATLAB model.

7.10 Damping in Systems with Principal Modes

7.10.1 Overview

Damping in complex built-up mechanical systems is impossible to predict with the present state of the art. We will discuss in this section the conditions which determine if a damping matrix can be diagonalized, and the criterion to enable the damped equations to be diagonalized. In general, an arbitrary damping matrix cannot be diagonalized by the undamped eigenvectors, as the

mass and stiffness matrices can.   This leads to using what is called
"proportional damping" in most finite element simulations.

If a mechanical system is designed with a specific viscous damping element,
for example a dashpot, that dominates the small amount of inherent structural
damping present, then that element can be added to the system as a viscous
damper. The resulting system is linear, but probably does not exhibit normal
modes as discussed in Section 7.2.2. In general this leads to the inability to
diagonalize and uncouple the equations of motion, requiring a state space
solution of the original, coupled equations of motion.

Viscoelastic damping treatments (damping elastomers) have been used for
years in disk drives, most typically as constrained layer dampers on the thin
sheet metal suspensions which support the read/write head. The effect of this
viscoelastic damping can be approximated at a specific temperature and
frequency as proportional damping by using the "modal strain energy"
technique in association with a finite element structural model (Johnson 1982).

Ignoring specific viscous, coulomb, and viscoelastic damping elements,
damping in typical structures arises from hysteresis losses in the materials as
they are strained, in some cases from viscous losses due to structure/fluid
interaction but more importantly from relative motion at the interfaces and
boundaries where different parts are attached or grounded. Unless a specific
damping element is used in a structural design, most structures have damping
which varies from mode to mode and will be in the range of 0.05% to 2% of
critical damping.

The modes in this chapter are all "real" or "normal" modes as defined earlier.
Once again, having normal modes means that at certain frequencies all points
in the system will vibrate at the same frequency and in phase, i.e., all points in
the system will reach their minimum and maximum displacements at the same
point in time. Chapter 5 discussed "complex" modes, modes in which all
points in the system do not reach their minimum and maximum displacements
at the same point in time.

7.10.2 Conditions Necessary for Existence of Principal Modes in Damped System

With a conservative (no damping) system, normal modes of vibration will
exist. In order to have normal modes in a damped system, the mode shapes
must be the same as for the undamped case, and the various parts of the system
must pass through their minimum and maximum positions at the same instant
in time, expressed as:

$$z_i = z_{mi} \cos(\omega_i t + \phi_i) \quad \text{for the } i^{th} \text{ mode} \qquad (7.109)$$

A sufficient condition for the existence of damped normal modes is that the damping matrix be a linear combination of the mass and stiffness matrices. We know that $\mathbf{m}$ and $\mathbf{k}$ are diagonalized by operating on them with the modal matrix. When $\mathbf{c}$ is a linear combination of $\mathbf{m}$ and $\mathbf{k}$, then the damping matrix $\mathbf{c}$ is also uncoupled (diagonalized) by the same pre- and postmultiplication operations by the modal matrix as with the $\mathbf{m}$ and $\mathbf{k}$ matrices (Weaver 1990, Craig 1981).

The damped equations of motion then become:

$$\mathbf{m\ddot{z}} + \mathbf{c\dot{z}} + \mathbf{kz} = \mathbf{F}, \qquad (7.110)$$

where the damping matrix is a linear combination of $\mathbf{m}$ and $\mathbf{k}$:

$$\mathbf{c} = \mathbf{am} + \mathbf{bk} \qquad (7.111)$$

$$\mathbf{c}_p = \mathbf{z}_n^T \mathbf{c} \mathbf{z}_n , \qquad (7.112)$$

and where $\mathbf{z}_n$ is the normalized (with respect to mass) modal matrix.

Writing out the complete equation:

$$\mathbf{m\ddot{z}} + \mathbf{c\dot{z}} + \mathbf{kz} = \mathbf{F} \qquad (7.113)$$

$$\underbrace{\mathbf{z}_n^T \mathbf{m} \mathbf{z}_n}_{\mathbf{I}} \underbrace{\mathbf{z}_n^{-1} \ddot{\mathbf{z}}}_{\ddot{\mathbf{z}}_p} + \underbrace{\mathbf{z}_n^T \mathbf{c} \mathbf{z}_n}_{\mathbf{c}_p} \underbrace{\mathbf{z}_n^{-1} \dot{\mathbf{z}}}_{\dot{\mathbf{z}}_p} + \underbrace{\mathbf{z}_n^T \mathbf{k} \mathbf{z}_n}_{\mathbf{k}_p} \underbrace{\mathbf{z}_n^{-1} \mathbf{z}}_{\mathbf{z}_p} = \underbrace{\mathbf{z}_n^T \mathbf{F}}_{\mathbf{F}_p} \qquad (7.114)$$

Looking at the $\mathbf{c}$ to $\mathbf{c}_p$ conversion where $\mathbf{c} = \mathbf{am} + \mathbf{bk}$:

$$\mathbf{c} = \mathbf{am} + \mathbf{bk} \qquad (7.115)$$

$$\mathbf{z}_n^T \mathbf{c} \mathbf{z}_n = a \mathbf{z}_n^T \mathbf{m} \mathbf{z}_n + b \mathbf{z}_n^T \mathbf{k} \mathbf{z}_n$$

$$= a\mathbf{I} + b\mathbf{k}_p , \qquad (7.116)$$

where $\mathbf{k}_p$ is a diagonal matrix whose elements are the squares of the eigenvalues.

The equation for the ith mode is:

$$\ddot{z}_{pi} + \left(a + b\omega_i^2\right)\dot{z}_{pi} + \omega_i^2 z_{pi} = F_{pi} \tag{7.117}$$

Rewriting, defining c_p, the $\left(a + b\omega_i^2\right)$ term, using notation:

$$c_{pi} = a + b\omega_i^2 = 2\zeta_i\omega_i \tag{7.118}$$

Where ζ_i is the percentage of critical damping for the ith mode, defined as:

$$\zeta_i = \left(\frac{c_i}{c_{cr}}\right)_i = \frac{c_i}{2\sqrt{k_{pi}m_{pi}}} = \frac{c_i}{2m_{pi}\sqrt{\omega_i^2}} \tag{7.119}$$

Then:

$$\zeta_i = \frac{a + b\omega_i^2}{2\omega_i} \tag{7.120}$$

Rewriting the equation in principal coordinates:

$$\ddot{z}_{pi} + 2\zeta_i\omega_i\dot{z}_{pi} + \omega_i^2 z_{pi} = F_{pi} \tag{7.121}$$

This type of damping is known as proportional damping, where the damping for each mode (they can all be different) is proportional to the critical damping for that mode. Since the damping is also proportional to velocity, it is of a viscous nature. If the same damping value is used for all modes, it will be referred to as "uniform" damping. Damping in which the damping value for each mode can be set individually will be referred to as "non-uniform" damping.

7.10.3 Different Types of Damping

7.10.3.1 Simple Proportional Damping

Viscous damping in each mode is taken to be an arbitrary percentage, ζ, of critical damping:

$$\ddot{z}_{pi} + 2\zeta\omega_i \dot{z}_{pi} + \omega_i^2 z_{pi} = F_{pi}$$

$$\ddot{z}_p + 2\zeta\left[k_p\right]^{\frac{1}{2}} \dot{z}_p + k_p z_p = F_p$$

(7.122)

This is analogous to the familiar notation used for a single degree of freedom system:

$$m\ddot{z} + c\dot{z} + kz = F$$

$$\ddot{z} + \frac{c}{m}\dot{z} + \frac{k}{m}z = \frac{F}{m}$$

(7.123)

Define critical damping $c_{cr} = 2\sqrt{km}$ and define the term multiplying velocity to be:

$$\frac{c}{m} = 2\zeta\omega_n$$

$$= 2\frac{c}{c_{cr}}\sqrt{\frac{k}{m}}$$

$$= \frac{2c}{2\sqrt{km}}\frac{\sqrt{k}}{\sqrt{m}}$$

$$= \frac{c}{m}$$

(7.124)

Rewriting:

$$\ddot{z} + 2\zeta\omega_n\dot{z} + \omega_n^2 z = \frac{F}{m}$$

(7.125)

7.10.3.2 Proportional to Stiffness Matrix – "Relative" Damping

Recognizing that the higher modes of vibration damp out quickly, "relative" damping yields damping in proportion to frequencies in normal modes, basically letting the "a" term for ζ_i go to zero:

$$\zeta_i = \frac{a + b\omega_i^2}{2\omega_i}\bigg|_{a=0} = \frac{b\omega_i}{2}$$

7.126)

If a value of ζ_1, for the first mode, is assumed, a value can be defined for "b":

$$b = \frac{2\zeta_1}{\omega_1},$$
(7.127)

and the value for any other mode i is:

$$\zeta_i = \zeta_1 \frac{\omega_i}{\omega_1}$$
(7.128)

7.10.3.3 Proportional to Mass Matrix – "Absolute" Damping

Absolute damping is based on making "b" equal to zero, in which case the percentage of critical damping is inversely proportional to the natural frequency of each mode. This will give decreasing damping for modes as their frequencies increase.

$$\zeta_i = \left. \frac{a + b\omega_i^2}{2\omega_i} \right|_{b=0} = \frac{a}{2\omega_i}$$
(7.129)

If a value of ζ_1, for the first mode, is assumed, a value can be defined for "a":

$$a = 2\zeta_1\omega_1,$$
(7.130)

and the value for any other mode i is:

$$\zeta_i = \frac{\omega_1\zeta_1}{\omega_i}$$
(7.131)

7.10.4 Defining Damping Matrix When Proportional Damping is Assumed

Figure 7.6: Two degree of freedom for damping example.

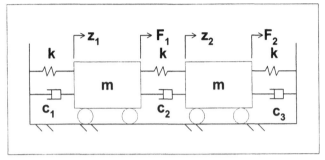

An interesting question to ask is what the elements of the damping matrix should be in the two degree of freedom (2dof) problem shown in Figure 7.6 in order to be able to diagonalize the equations of motion. We will use the eigenvectors from the undamped case to normalize the damping matrix. Then we will solve for the specific values of the individual dampers which will allow the diagonalization. We will see how non-intuitive the values of c_1, c_2 and c_3 are in order to be able to diagonalize. (See Craig [1981] for a general expression to calculate the physical damping matrix when given proportional damping values, the original mass matrix, the diagonalized mass matrix and the eigenvalues and eigenvectors.)

7.10.4.1 Solving for Damping Values

Starting with the undamped eigenvalues and eigenvectors:

$$\mathbf{m} = \begin{bmatrix} m & 0 \\ 0 & m \end{bmatrix} \quad \mathbf{k} = \begin{bmatrix} 2k & -k \\ -k & 2k \end{bmatrix} \quad \mathbf{c} = \begin{bmatrix} c_1 + c_2 & -c_2 \\ -c_2 & c_2 + c_3 \end{bmatrix}$$

$$\mathbf{z}_m = \begin{bmatrix} 1 & 1 \\ 1 & -1 \end{bmatrix} \quad \mathbf{z}_n = \frac{1}{\sqrt{2m}} \begin{bmatrix} 1 & 1 \\ 1 & -1 \end{bmatrix} \quad \mathbf{k}_p = \begin{bmatrix} \dfrac{k}{m} & 0 \\ 0 & \dfrac{3k}{m} \end{bmatrix}$$

$$(7.132)$$

Solve for the diagonalized damping matrix, assuming proportional damping, and knowing that the diagonalized stiffness matrix elements are squares of the eigenvalues:

$$c_p = z_n^T c z_n = 2\zeta \begin{bmatrix} \omega_1 & 0 \\ 0 & \omega_2 \end{bmatrix} = 2\zeta k_p^{\frac{1}{2}} \tag{7.133}$$

Premultiplying by $(z_n^T)^{-1}$ and postmultiplying by $(z_n)^{-1}$:

$$\underbrace{(z_n^T)^{-1} z_n^T}_{I} \ c \ \underbrace{z_n (z_n)^{-1}}_{I} = 2\zeta (z_n^T)^{-1} k_p^{\frac{1}{2}} (z_n)^{-1} \tag{7.134}$$

$$c = 2\zeta (z_n^T)^{-1} k_p^{\frac{1}{2}} (z_n)^{-1} \tag{7.135}$$

Solving for the inverses above, noting that for this 2dof system, $z_n = z_n^T$, and then performing the operations on k_p:

The inverse of a 2 x 2 matrix can be found by:

1.   Interchanging the two diagonal elements.

2.   Changing the signs of the two off-diagonal elements.

3.   Dividing by the determinant of the original matrix.

$$\begin{bmatrix} a & b \\ c & d \end{bmatrix}^{-1} = \frac{\begin{bmatrix} d & -b \\ -c & a \end{bmatrix}}{\begin{vmatrix} a & b \\ c & d \end{vmatrix}}$$

Table 7.2: Inverse of 2x2 matrix.

$$z_n^{-1} = (z_n^T)^{-1} = \frac{\begin{bmatrix} -1 & -1 \\ -1 & 1 \end{bmatrix} \sqrt{2m}}{(-1-1)} = \frac{\sqrt{2m}}{2} \begin{bmatrix} 1 & 1 \\ 1 & -1 \end{bmatrix} \tag{7.136}$$

$$\mathbf{z}_n^{-1}\mathbf{k}_p^{\frac{1}{2}} = \frac{\sqrt{2m}}{2}\begin{bmatrix}1 & 1 \\ 1 & -1\end{bmatrix}\sqrt{\frac{k}{m}}\begin{bmatrix}1 & 0 \\ 0 & \sqrt{3}\end{bmatrix}$$

$$(7.137)$$

$$= \frac{\sqrt{2k}}{2}\begin{bmatrix}1 & \sqrt{3} \\ 1 & -\sqrt{3}\end{bmatrix}$$

$$\mathbf{z}_n^{-1}\mathbf{k}_p^{\frac{1}{2}}\mathbf{z}_n^{-1} = \frac{\sqrt{2k}}{2}\begin{bmatrix}1 & \sqrt{3} \\ 1 & -\sqrt{3}\end{bmatrix}\begin{bmatrix}1 & 1 \\ 1 & -1\end{bmatrix}\frac{\sqrt{2m}}{2}$$

$$(7.138)$$

$$= \frac{2\sqrt{km}}{4}\begin{bmatrix}1+\sqrt{3} & 1-\sqrt{3} \\ 1-\sqrt{3} & 1+\sqrt{3}\end{bmatrix}$$

$$\mathbf{c} = 2\zeta\frac{\sqrt{km}}{2}\begin{bmatrix}1+\sqrt{3} & 1-\sqrt{3} \\ 1-\sqrt{3} & 1+\sqrt{3}\end{bmatrix}$$

$$(7.139)$$

$$= \zeta\sqrt{km}\begin{bmatrix}1+\sqrt{3} & 1-\sqrt{3} \\ 1-\sqrt{3} & 1+\sqrt{3}\end{bmatrix}$$

Now we can solve for the specific values for the three dampers:

$$-c_2 = \zeta\sqrt{km}\left(1-\sqrt{3}\right)$$

$$c_2 = \zeta\sqrt{km}\left(\sqrt{3}-1\right)$$

$$(7.140)$$

$$= \zeta\sqrt{km}\,(.732)$$

$$c_1 + c_2 = c_2 + c_3 = \zeta\sqrt{km}\left(1+\sqrt{3}\right)$$

$$c_1 = c_3 = \zeta\sqrt{km}\left(1+\sqrt{3}\right) - c_2$$

$$= \zeta\sqrt{km}\left[\left(1+\sqrt{3}\right) - \left(\sqrt{3}-1\right)\right] \qquad (7.141)$$

$$= \zeta\sqrt{km}\,(2)$$

$$= 2\zeta\sqrt{km}$$

Summarizing:

$$c_1 = c_3 = 2\zeta\sqrt{km} \qquad (7.142)$$

$$c_2 = \zeta\sqrt{km}\,(.732) \qquad (7.143)$$

Note that the values for the three dampers are not at all intuitive and would have been very difficult if impossible to guess to be able to construct a diagonalizable damping matrix. If defining the diagonalizable damping matrix for this 2x2 problem is difficult, imagine trying to define it for a real life finite element problem with thousands of degrees of freedom. Also, it is highly improbable that the back-calculated damping values in physical coordinates would match the actual damping in the structure.

7.10.4.2 Checking Rayleigh Form of Damping Matrix

We have now defined the values of the c_1, c_2 and c_3, dampers which allow diagonalizing the equations of motion. Another interesting question is whether the Rayleigh form has been satisfied: Is $\mathbf{c}$ a linear combination of $\mathbf{k}$ and $\mathbf{m}$?

$$\mathbf{c} = \zeta\sqrt{km}\begin{bmatrix} 1+\sqrt{3} & 1-\sqrt{3} \\ 1-\sqrt{3} & 1+\sqrt{3} \end{bmatrix} \overset{?}{=} a\left\{m\begin{bmatrix} 1 & 0 \\ 0 & 1 \end{bmatrix}\right\} + b\left\{k\begin{bmatrix} 2 & -1 \\ -1 & 2 \end{bmatrix}\right\} \qquad (7.144)$$

We have two unknowns, a and b, and essentially two equations, since the two diagonal elements are the same and the two off diagonal elements are the same. First, let us look at the two off diagonal terms, equating terms on the two sides above:

$$\zeta\sqrt{km}\left(1-\sqrt{3}\right) = am(0) + bk(-1)$$

(7.145)

$$b = \zeta\sqrt{km}\frac{\left(\sqrt{3}-1\right)}{k} = \zeta\sqrt{\frac{m}{k}}\left(\sqrt{3}-1\right)$$

Now, equating the diagonal terms:

$$\zeta\sqrt{km}\left(1+\sqrt{3}\right) = am + 2bk$$

$$= am + 2\left[\zeta\sqrt{\frac{m}{k}}\left(\sqrt{3}-1\right)\right]k \qquad (7.146)$$

$$= am + 2\zeta\sqrt{mk}\left(\sqrt{3}-1\right)$$

$$am = \zeta\sqrt{km}\left(1+\sqrt{3}\right) - 2\zeta\sqrt{mk}\left(\sqrt{3}-1\right)$$

$$= \zeta\sqrt{km}\left[1+\sqrt{3}-2\sqrt{3}+2\right] \qquad (7.147)$$

$$= \xi\sqrt{km}\left[3-\sqrt{3}\right]$$

$$a = \zeta\sqrt{\frac{k}{m}}\left[3-\sqrt{3}\right] \qquad (7.148)$$

Checking the two values for a and b by substituting back into (7.146).

$$\zeta\sqrt{km}\begin{bmatrix}1+\sqrt{3} & 1-\sqrt{3} \\ 1-\sqrt{3} & 1+\sqrt{3}\end{bmatrix} = \zeta\sqrt{\frac{k}{m}}(m)\begin{bmatrix}1 & 0 \\ 0 & 1\end{bmatrix}\begin{bmatrix}3-\sqrt{3}\end{bmatrix}$$

$$+\zeta\sqrt{\frac{m}{k}}(k)(\sqrt{3}-1)\begin{bmatrix}2 & -1 \\ -1 & 2\end{bmatrix}$$

$$= \zeta\sqrt{km}\left\{\begin{bmatrix}(3-\sqrt{3}) & 0 \\ 0 & (3-\sqrt{3})\end{bmatrix} + \begin{bmatrix}2\sqrt{3}-2 & -\sqrt{3}+1 \\ -\sqrt{3}+1 & 2\sqrt{3}-2\end{bmatrix}\right\}$$

$$= \zeta\sqrt{km}\begin{bmatrix}1+\sqrt{3} & 1-\sqrt{3} \\ 1-\sqrt{3} & 1+\sqrt{3}\end{bmatrix}$$

$$(7.149)$$

So c is a linear combination of k and m and the Rayleigh criterion holds.

Problems

Note: All the problems refer to the two dof system shown in Figure P2.2.

P7.1 Set $m_1 = m_2 = m = 1$, $k_1 = k_2 = k = 1$ and solve for the eigenvalues and eigenvectors of the undamped system. Normalize the eigenvectors to unity, write out the modal matrix and hand plot the mode shapes

P7.2 Normalize the eigenvectors in P7.1 with respect to mass and diagonalize the mass and stiffness matrices. Identify the terms in the normalized mass and stiffness matrices. Write the homogeneous equations of motion in physical and principal coordinates.

P7.3   Convert the following step forcing function and initial conditions in physical coordinates to principal coordinates:

   a) $F_1 = 1$, $F_2 = -3$

   b) $z_1 = 0$, $\dot{z}_1 = -2$, $z_2 = -1$, $\dot{z}_2 = 2$

P7.4   Using the results of P7.2 and P7.3, write the complete equations of motion in physical and principal coordinates assuming proportional damping.

CHAPTER 8

FREQUENCY RESPONSE: MODAL FORM

8.1 Introduction

Now that the theory behind the modal analysis method has been covered, we will solve our tdof problem for its frequency response.

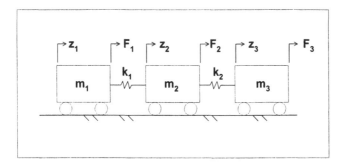

Figure 8.1: tdof undamped model for modal analysis.

We will use eigenvalue/eigenvector results from Chapter 7 to define the equations of motion in principal coordinates and to transform forces to principal coordinates. We will then use Laplace transforms to solve for the transfer functions in principal coordinates and back-transform to physical coordinates, where the individual mode contributions will be evident. We will discuss the relationship between the partial fraction expansion transfer function form and the modal form derived here. We discussed in Section 5.13 how to excite only a single mode of vibration by judicious choice of initial conditions. Here we will describe the forcing function combination required to excite only a single mode.

We will spend considerable time in this chapter on developing a greater understanding of how individual modes of vibration combine to give the overall frequency response. MATLAB code is supplied for the tdof problem to illustrate the point. ANSYS is also used to solve the tdof problem and the ANSYS results are described and compared with the MATLAB results.

8.2 Review from Previous Results

Since the problem we are solving is frequency response, or finding the steady state motion of each mass as a function of frequency and of applied forces, initial conditions are not required.

From previous analyses, (7.85) to (7.88), we know the eigenvalues and eigenvectors normalized with respect to mass, ω_i, z_n :

$$\omega_1 = 0 \qquad \omega_2 = \pm\sqrt{\frac{k}{m}} \qquad \omega_3 = \pm\sqrt{\frac{3k}{m}} \tag{8.1}$$

$$z_n = \frac{1}{\sqrt{m}} \begin{bmatrix} \dfrac{1}{\sqrt{3}} & \dfrac{1}{\sqrt{2}} & \dfrac{1}{\sqrt{6}} \\ \dfrac{1}{\sqrt{3}} & 0 & \dfrac{-2}{\sqrt{6}} \\ \dfrac{1}{\sqrt{3}} & \dfrac{-1}{\sqrt{2}} & \dfrac{1}{\sqrt{6}} \end{bmatrix} \tag{8.2}$$

Knowing that in principal coordinates the mass matrix is the identity matrix and the stiffness matrix is a diagonal matrix with the squares of the respective eigenvalues as terms, we can write the matrices by inspection:

$$m_p = \begin{bmatrix} 1 & 0 & 0 \\ 0 & 1 & 0 \\ 0 & 0 & 1 \end{bmatrix} \qquad k_p = \left(\frac{k}{m}\right) \begin{bmatrix} 0 & 0 & 0 \\ 0 & 1 & 0 \\ 0 & 0 & 3 \end{bmatrix} \tag{8.3}$$

The force vector in principal coordinates is:

$$F_p = z_n^T F = \begin{bmatrix} z_{n11} & z_{n21} & z_{n31} \\ z_{n12} & z_{n22} & z_{n32} \\ z_{n13} & z_{n23} & z_{n33} \end{bmatrix} \begin{bmatrix} F_1 \\ F_2 \\ F_3 \end{bmatrix} \tag{8.4}$$

Expanding:

$$F_{p1} = z_{n11}F_1 + z_{n21}F_2 + z_{n31}F_3$$

$$F_{p2} = z_{n12}F_1 + z_{n22}F_2 + z_{n32}F_3 \qquad (8.5a,b,c)$$

$$F_{p1} = z_{n13}F_1 + z_{n23}F_2 + z_{n33}F_3$$

Performing the actual problem multiplication:

$$F_p = z_n^T F = \frac{1}{\sqrt{m}}\begin{bmatrix} \dfrac{1}{\sqrt{3}} & \dfrac{1}{\sqrt{3}} & \dfrac{1}{\sqrt{3}} \\[2mm] \dfrac{1}{\sqrt{2}} & 0 & \dfrac{-1}{\sqrt{2}} \\[2mm] \dfrac{1}{\sqrt{6}} & \dfrac{-2}{\sqrt{6}} & \dfrac{1}{\sqrt{6}} \end{bmatrix}\begin{bmatrix} F_1 \\ F_2 \\ F_3 \end{bmatrix} = \frac{1}{\sqrt{m}}\begin{bmatrix} \dfrac{F_1}{\sqrt{3}} + \dfrac{F_2}{\sqrt{3}} + \dfrac{F_3}{\sqrt{3}} \\[2mm] \dfrac{F_1}{\sqrt{2}} + 0 - \dfrac{F_3}{\sqrt{2}} \\[2mm] \dfrac{F_1}{\sqrt{6}} - \dfrac{2F_2}{\sqrt{6}} + \dfrac{F_3}{\sqrt{6}} \end{bmatrix} = \begin{bmatrix} F_{p1} \\ F_{p2} \\ F_{p3} \end{bmatrix}$$

$$(8.6)$$

Writing the resulting equations of motion in principal coordinates in matrix form:

$$\begin{bmatrix} 1 & 0 & 0 \\ 0 & 1 & 0 \\ 0 & 0 & 1 \end{bmatrix}\begin{bmatrix} \ddot{z}_{p1} \\ \ddot{z}_{p2} \\ \ddot{z}_{p3} \end{bmatrix} + \left(\frac{k}{m}\right)\begin{bmatrix} 0 & 0 & 0 \\ 0 & 1 & 0 \\ 0 & 0 & 3 \end{bmatrix}\begin{bmatrix} z_{p1} \\ z_{p2} \\ z_{p3} \end{bmatrix} = \frac{1}{\sqrt{m}}\begin{bmatrix} \dfrac{F_1}{\sqrt{3}} + \dfrac{F_2}{\sqrt{3}} + \dfrac{F_3}{\sqrt{3}} \\[2mm] \dfrac{F_1}{\sqrt{2}} + 0 - \dfrac{F_3}{\sqrt{2}} \\[2mm] \dfrac{F_1}{\sqrt{6}} - \dfrac{2F_2}{\sqrt{6}} + \dfrac{F_3}{\sqrt{6}} \end{bmatrix}$$

$$(8.7)$$

Writing out the equations in expanded form:

$$\ddot{z}_{p1} = (F_1 + F_2 + F_3)\frac{1}{\sqrt{3m}} = F_{p1} \qquad (8.8)$$

$$\ddot{z}_{p2} + \frac{k}{m}z_{p2} = \left(F_1 - F_3\right)\frac{1}{\sqrt{2m}} = F_{p2} \qquad\qquad \frac{k}{m} = \omega_2^2 \qquad (8.9)$$

$$\ddot{z}_{p3} + \frac{3k}{m}z_{p3} = \left(F_1 - 2F_2 + F_3\right)\frac{1}{\sqrt{6m}} = F_{p3} \qquad\qquad \frac{3k}{m} = \omega_3^2 \qquad (8.10)$$

8.3 Transfer Functions – Laplace Transforms in Principal Coordinates

We now solve for the transfer functions. Taking the Laplace transform of each equation, ignoring initial conditions and collecting the displacement terms, where $z_{p1}(s)$ is the Laplace transform of z_{p1} (Appendix 2):

$$s^2 z_{p1}(s) = \left(F_1(s) + F_2(s) + F_3(s)\right)\frac{1}{\sqrt{3m}}$$

$$z_{p2}(s)\left(s^2 + \omega_2^2\right) = \left(F_1(s) - F_3(s)\right)\frac{1}{\sqrt{2m}} \qquad (8.11a,b,c)$$

$$z_{p3}(s)\left(s^2 + \omega_3^2\right) = \left(F_1(s) - 2F_2(s) + F_3(s)\right)\frac{1}{\sqrt{6m}}$$

Solving for the three principal displacements and eliminating the "(s)" for simplicity:

$$z_{p1} = \left(F_1 + F_2 + F_3\right)\frac{1}{s^2\sqrt{3m}}$$

$$z_{p2} = \left(F_1 - F_3\right)\frac{1}{\left(s^2 + \omega_2^2\right)\sqrt{2m}} \qquad (8.12a,b,c)$$

$$z_{p3} = \left(F_1 - 2F_2 + F_3\right)\frac{1}{\left(s^2 + \omega_3^2\right)\sqrt{6m}}$$

Taking the forces one at a time, the elements of a transfer function matrix can be defined.

$$\frac{z_{p1}}{F_1} = \frac{1}{s^2\sqrt{3m}}$$

$$\frac{z_{p1}}{F_2} = \frac{1}{s^2\sqrt{3m}} \qquad (8.13a,b,c)$$

$$\frac{z_{p1}}{F_3} = \frac{1}{s^2\sqrt{3m}}$$

$$\frac{z_{p2}}{F_1} = \frac{1}{\left(s^2 + \omega_2^2\right)\sqrt{2m}}$$

$$\frac{z_{p2}}{F_2} = 0 \qquad\qquad\qquad (8.14a,b,c)$$

$$\frac{z_{p2}}{F_3} = \frac{-1}{\left(s^2 + \omega_2^2\right)\sqrt{2m}}$$

$$\frac{z_{p3}}{F_1} = \frac{1}{\left(s^2 + \omega_3^2\right)\sqrt{6m}}$$

$$\frac{z_{p3}}{F_2} = \frac{-2}{\left(s^2 + \omega_3^2\right)\sqrt{6m}} \qquad (8.15a,b,c)$$

$$\frac{z_{p3}}{F_3} = \frac{1}{\left(s^2 + \omega_3^2\right)\sqrt{6m}}$$

Writing out the principal coordinate transfer functions for each external force, F_1, F_2, and F_3:

$$\frac{z_p}{F_1} = \begin{bmatrix} \dfrac{z_{p1}}{F_1} \\[2mm] \dfrac{z_{p2}}{F_1} \\[2mm] \dfrac{z_{p3}}{F_1} \end{bmatrix} = \begin{bmatrix} \dfrac{1}{s^2\sqrt{3m}} \\[3mm] \dfrac{1}{\left(s^2 + \omega_2^2\right)\sqrt{2m}} \\[3mm] \dfrac{1}{\left(s^2 + \omega_3^2\right)\sqrt{6m}} \end{bmatrix} = \begin{bmatrix} z_{p11} \\[2mm] z_{p21} \\[2mm] z_{p31} \end{bmatrix} \qquad (8.16)$$

$$\frac{z_p}{F_2} = \begin{bmatrix} \dfrac{z_{p1}}{F_2} \\[2mm] \dfrac{z_{p2}}{F_2} \\[2mm] \dfrac{z_{p3}}{F_2} \end{bmatrix} = \begin{bmatrix} \dfrac{1}{s^2\sqrt{3m}} \\[3mm] 0 \\[3mm] \dfrac{-2}{\left(s^2 + \omega_3^2\right)\sqrt{6m}} \end{bmatrix} = \begin{bmatrix} z_{p12} \\[2mm] z_{p22} \\[2mm] z_{p32} \end{bmatrix} \qquad (8.17)$$

$$\frac{z_p}{F_3} = \begin{bmatrix} \dfrac{z_{p1}}{F_3} \\[6pt] \dfrac{z_{p2}}{F_3} \\[6pt] \dfrac{z_{p3}}{F_3} \end{bmatrix} = \begin{bmatrix} \dfrac{1}{s^2\sqrt{3m}} \\[10pt] \dfrac{-1}{\left(s^2+\omega_2^2\right)\sqrt{2m}} \\[10pt] \dfrac{1}{\left(s^2+\omega_3^2\right)\sqrt{6m}} \end{bmatrix} = \begin{bmatrix} z_{p13} \\[6pt] z_{p23} \\[6pt] z_{p33} \end{bmatrix} \qquad (8.18)$$

8.4 Back-Transforming Mode Contributions to Transfer Functions in Physical Coordinates

Now the transfer functions in principal coordinates can be back-transformed to physical coordinates. This allows one to see the contributions of each mode, where z_{ij} is the physical displacement at dof "i" due to a force at dof "j."

$$\mathbf{z} = \mathbf{z}_n \mathbf{z}_p = \begin{bmatrix} z_{n11} & z_{n12} & z_{n13} \\ z_{n21} & z_{n22} & z_{n23} \\ z_{n31} & z_{n32} & z_{n33} \end{bmatrix} \begin{bmatrix} z_{p11} & z_{p12} & z_{p13} \\ z_{p21} & z_{p22} & z_{p23} \\ z_{p31} & z_{p32} & z_{p33} \end{bmatrix} = \begin{bmatrix} z_{11} & z_{12} & z_{13} \\ z_{21} & z_{22} & z_{23} \\ z_{31} & z_{32} & z_{33} \end{bmatrix} \qquad (8.19)$$

The equations below show how the results from each of the principal equations (modes) combine to give the overall response. The overall transfer function is seen to be a combination of the three modes of vibration and is referred to as the "modal form."

$$\frac{z_1}{F_1} = \underbrace{z_{n11}z_{p11}}_{\text{1st mode}} + \underbrace{z_{n12}z_{p21}}_{\text{2nd mode}} + \underbrace{z_{n13}z_{p31}}_{\text{3rd mode}} \qquad \text{contributions to total } \frac{z_1}{F_1} \text{ transfer function.}$$

$$\frac{z_2}{F_1} = \underbrace{z_{n21}z_{p11}}_{\text{1st mode}} + \underbrace{z_{n22}z_{p21}}_{\text{2nd mode}} + \underbrace{z_{n23}z_{p31}}_{\text{3rd mode}} \qquad \text{contributions to total } \frac{z_2}{F_1} \text{ transfer function.}$$

$$\frac{z_3}{F_1} = \underbrace{z_{n31}z_{p11}}_{\text{1st mode}} + \underbrace{z_{n32}z_{p21}}_{\text{2nd mode}} + \underbrace{z_{n33}z_{p31}}_{\text{3rd mode}} \qquad \text{contributions to total } \frac{z_3}{F_1} \text{ transfer function.}$$

$$\frac{z_1}{F_2} = \underbrace{z_{n11}z_{p12}}_{\text{1st mode}} + \underbrace{z_{n12}z_{p22}}_{\text{2nd mode}} + \underbrace{z_{n13}z_{p32}}_{\text{3rd mode}} \qquad \text{contributions to total } \frac{z_1}{F_2} \text{ transfer function.}$$

$$\frac{z_2}{F_2} = \underbrace{z_{n21}z_{p12}}_{\text{1st mode}} + \underbrace{z_{n22}z_{p22}}_{\text{2nd mode}} + \underbrace{z_{n23}z_{p32}}_{\text{3rd mode}} \qquad \text{contributions to total } \frac{z_2}{F_2} \text{ transfer function.}$$

$$\frac{z_3}{F_2} = \underbrace{z_{n31}z_{p12}}_{\text{1st mode}} + \underbrace{z_{n32}z_{p22}}_{\text{2nd mode}} + \underbrace{z_{n33}z_{p32}}_{\text{3rd mode}} \qquad \text{contributions to total } \frac{z_3}{F_2} \text{ transfer function.}$$

$$\frac{z_1}{F_3} = \underbrace{z_{n11}z_{p13}}_{\text{1st mode}} + \underbrace{z_{n12}z_{p23}}_{\text{2nd mode}} + \underbrace{z_{n13}z_{p33}}_{\text{3rd mode}} \qquad \text{contributions to total } \frac{z_1}{F_3} \text{ transfer function.}$$

$$\frac{z_2}{F_3} = \underbrace{z_{n21}z_{p13}}_{\text{1st mode}} + \underbrace{z_{n22}z_{p23}}_{\text{2nd mode}} + \underbrace{z_{n23}z_{p33}}_{\text{3rd mode}} \qquad \text{contributions to total } \frac{z_2}{F_3} \text{ transfer function.}$$

$$\frac{z_3}{F_3} = \underbrace{z_{n31}z_{p13}}_{\text{1st mode}} + \underbrace{z_{n32}z_{p23}}_{\text{2nd mode}} + \underbrace{z_{n33}z_{p33}}_{\text{3rd mode}} \qquad \text{contributions to total } \frac{z_3}{F_3} \text{ transfer function.}$$

We saw earlier that because of symmetry there are only four distinctly different transfer functions of the total of nine:

$$\frac{z_1}{F_1}, \frac{z_2}{F_1}, \frac{z_3}{F_1} \text{ and } \frac{z_2}{F_2}$$

Expanding the four transfer functions:

$$\frac{z_1}{F_1} = \frac{1}{\sqrt{m}} \left[\frac{z_{p11}}{\sqrt{3}} + \frac{z_{p21}}{\sqrt{2}} + \frac{z_{p31}}{\sqrt{6}} \right]$$

$$= \frac{1}{\sqrt{m}} \left[\frac{1}{s^2 \sqrt{3m}\sqrt{3}} + \frac{1}{\left(s^2 + \omega_2^2\right)\sqrt{2m}\sqrt{2}} + \frac{1}{\left(s^2 + \omega_3^2\right)\sqrt{6m}\sqrt{6}} \right]$$

$$= \frac{1}{s^2 (3m)} + \frac{1}{\left(s^2 + \omega_2^2\right)2m} + \frac{1}{\left(s^2 + \omega_3^2\right)6m}$$

$$= \frac{\left(\frac{1}{3m}\right)}{s^2} + \frac{\left(\frac{1}{2m}\right)}{s^2 + \omega_2^2} + \frac{\left(\frac{1}{6m}\right)}{s^2 + \omega_3^2} \tag{8.20}$$

$$\frac{z_2}{F_1} = \frac{1}{\sqrt{m}} \left[\frac{1}{s^2 \sqrt{3m}\sqrt{3}} + 0 - \frac{2}{\left(s^2 + \omega_3^2\right)\sqrt{6m}\sqrt{6}} \right]$$

$$= \frac{\left(\frac{1}{3m}\right)}{s^2} + 0 - \frac{\left(\frac{2}{6m}\right)}{s^2 + \omega_3^2} \tag{8.21}$$

$$\frac{z_3}{F_1} = \frac{1}{\sqrt{m}} \left[\frac{1}{s^2 \sqrt{3m}\sqrt{3}} - \frac{1}{\left(s^2 + \omega_2^2\right)\sqrt{2m}\sqrt{2}} + \frac{1}{\left(s^2 + \omega_3^2\right)\sqrt{6m}\sqrt{6}} \right]$$

$$= \frac{\left(\frac{1}{3m}\right)}{s^2} - \frac{\left(\frac{1}{2m}\right)}{s^2 + \omega_2^2} + \frac{\left(\frac{1}{6m}\right)}{s^2 + \omega_3^2} \tag{8.22}$$

$$\frac{z_2}{F_2} = \frac{1}{\sqrt{m}} \left[\frac{1}{s^2 \sqrt{3m}\sqrt{3}} + 0 + \frac{4}{\left(s^2 + \omega_3^2\right)\sqrt{6m}\sqrt{6}} \right]$$

$$= \frac{\left(\frac{1}{3m}\right)}{s^2} + 0 + \frac{\left(\frac{4}{6m}\right)}{s^2 + \omega_3^2} \tag{8.23}$$

Taking $m = k = 1$ yields: $\omega_1^2 = 0, \ \omega_2^2 = \frac{k}{m} = 1, \ \omega_3^2 = \frac{3k}{m} = 3,$ and substituting above:

$$\frac{z_1}{F_1} = \frac{\frac{1}{3}}{s^2} + \frac{\frac{1}{2}}{s^2 + 1} + \frac{\frac{1}{6}}{s^2 + 3} \tag{8.24}$$

$$\frac{z_2}{F_1} = \frac{\frac{1}{3}}{s^2} - \frac{\frac{1}{3}}{s^2 + 3} \tag{8.25}$$

$$\frac{Z_3}{F_1} = \frac{\frac{1}{3}}{s^2} - \frac{\frac{1}{2}}{s^2+1} + \frac{\frac{1}{6}}{s^2+3} \tag{8.26}$$

$$\frac{Z_2}{F_2} = \frac{\frac{1}{3}}{s^2} + \frac{\frac{2}{3}}{s^2+3} \tag{8.27}$$

8.5 Partial Fraction Expansion and the Modal Form

Another way of finding the modal form (not as insightful, but does not require solving the eigenvalue problem) is to take the original transfer functions derived in the Chapter 2 and perform a partial fraction expansion. Partial fraction expansion gives the same results as the modal form in Section 8.4. The four unique transfer functions from (2.62) to (2.65) are repeated below:

$$\frac{Z_1}{F_1} = \frac{m^2s^4 + 3mks^2 + k^2}{s^2\left(m^3s^4 + 4m^2ks^2 + 3mk^2\right)} \tag{8.28}$$

$$\frac{Z_2}{F_1} = \frac{k}{s^2(m^2s^2 + 3km)} \tag{8.29}$$

$$\frac{Z_3}{F_1} = \frac{k^2}{s^2\left(m^3s^4 + 4m^2ks^2 + 3mk^2\right)} \tag{8.30}$$

$$\frac{Z_2}{F_2} = \frac{m^2s^4 + 2mks^2 + k^2}{s^2\left(m^3s^4 + 4m^2ks^2 + 3mk^2\right)} \tag{8.31}$$

In order to perform a partial fraction expansion, we need the roots of the characteristic equation, found earlier to be:

$$\omega_1^2 = 0 \qquad \omega_2^2 = \frac{k}{m} \qquad \omega_3^2 = \frac{3k}{m} \tag{8.32}$$

Taking the z_1/F_1 transfer function and expanding in partial fraction form, setting $m = k = 1$;

$$\frac{Z_1}{F_1} = \frac{m^2s^4 + 3mks^2 + k^2}{s^2\left(m^3s^4 + 4m^2ks^2 + 3mk^2\right)} = \frac{s^4 + 3s^2 + 1}{s^2(s^4 + 4s^2 + 3)} = \frac{s^4 + 3s^2 + 1}{s^2(s^2 + \omega_2^2)(s^2 + \omega_3^2)}$$

$$\frac{s^4 + 3s^2 + 1}{s^2(s^2 + \omega_2^2)(s^2 + \omega_3^2)} = \frac{A}{s^2 + \omega_1^2} + \frac{B}{s^2 + \omega_2^2} + \frac{C}{s^2 + \omega_3^2}$$

$$(8.33a,b)$$

The terms A, B and C, known as "residues," are evaluated using the "cover-up" method, where each coefficient is evaluated by "covering up" its term in the transfer function and evaluating the remaining expression at $s^2 = -\omega_i^2$.

Evaluating A:

$$A = \frac{s^4 + 3s^2 + 1}{(s^2 + \omega_2^2)(s^2 + \omega_3^2)} \quad \text{evaluated at } s^2 = -\omega_1^2 = 0$$

$$(8.34)$$

$$= \frac{1}{\omega_2^2 \omega_3^2} = \frac{1}{(1)(3)} = \frac{1}{3}$$

Evaluating B:

$$B = \frac{s^4 + 3s^2 + 1}{s^2(s^2 + \omega_3^2)} \quad \text{evaluated at } s^2 = -\omega_2^2 = -1$$

$$(8.35)$$

$$= \frac{\omega_2^4 - 3\omega_2^2 + 1}{-\omega_2^2(-\omega_2^2 + \omega_3^2)} = \frac{1 - 3 + 1}{-1(-1 + 3)} = \frac{-1}{-2} = \frac{1}{2}$$

Evaluating C:

$$C = \frac{s^4 + 3s^2 + 1}{s^2(s^2 + \omega_2^2)(s^2 + \omega_3^2)} \qquad \text{evaluated at } s^2 = -\omega_3^2 = -3$$

$$= \frac{\omega_3^4 - 3\omega_3^2 + 1}{-\omega_3^2(-\omega_3^2 + \omega_2^2)} = \frac{3^2 - 3(-3) + 1}{-3(-3+1)} = \frac{1}{6}$$

(8.36)

Combining terms:

$$\frac{z_1}{F_1} = \frac{s^4 + 3s^2 + 1}{s^2(s^2 + \omega_2^2)(s^2 + \omega_3^2)} = \frac{\begin{bmatrix} 1 \\ 3 \end{bmatrix}}{s^2 + \omega_1^2} + \frac{\begin{bmatrix} 1 \\ 2 \end{bmatrix}}{s^2 + \omega_2^2} + \frac{\begin{bmatrix} 1 \\ 6 \end{bmatrix}}{s^2 + \omega_3^2} \qquad (8.37)$$

This expression is the same as the term for z_1/F_1 in (8.20). Converting the other three transfer functions to partial fraction form also reveals their modal form.

8.6 Forcing Function Combinations to Excite Single Mode

It is instructive at this point to see what types of forcing function combinations will excite each of the three modes separately. From the definition of normal modes, we know that if the system is started from initial displacement conditions that match one of the normal modes, the system will respond at only that mode. An analogous situation exists for combinations of forcing functions. Repeating the transformed equations of motion in principal coordinates from (8.12a,b,c) with $m = 1$.

$$z_{p1} = (F_1 + F_2 + F_3)\frac{1}{s^2\sqrt{3}} = F_{p1}$$

$$z_{p2} = (F_1 - F_3)\frac{1}{(s^2 + \omega_2^2)\sqrt{2}} = F_{p2} \qquad (8.38a,b,c)$$

$$z_{p3} = (F_1 - 2F_2 + F_3)\frac{1}{(s^2 + \omega_3^2)\sqrt{6}} = F_{p3}$$

To excite only the first mode, we can start with initial displacements being any multiple of the first eigenvector, which has equal displacements for all masses.

Now let us see if applying equal forces to all three masses with zero initial conditions excites only the first mode. Set $F_1 = F_2 = F_3 \neq 0$, which should excite only the first, rigid body mode:

$$z_{p1} = \left(F_1 + F_2 + F_3\right)\frac{1}{s^2\sqrt{3}} = 3F_1\,\frac{1}{s^2\sqrt{3}}$$

$$z_{p2} = \left(F_1 - F_3\right)\frac{1}{\left(s^2 + \omega_2^2\right)\sqrt{2}} = 0 \qquad (8.39a,b,c)$$

$$z_{p3} = \left(F_1 - 2F_2 + F_3\right)\frac{1}{\left(s^2 + \omega_3^2\right)\sqrt{6}} = 0$$

We can see above that the motion for the second and third modes is zero. It is the information contained in the eigenvector, which, when multiplied by the force vector in physical coordinates, determines the force vector in principal coordinates.

To excite the second mode only, applying zero force at mass 1 and equal and opposite sign forces at masses 1 and 2 should work: $F_1 = -F_3$, $F_2 = 0$:

$$z_{p1} = \left(F_1 + F_2 + F_3\right)\frac{1}{s^2\sqrt{3}} = 0$$

$$z_{p2} = \left(F_1 - F_3\right)\frac{1}{\left(s^2 + \omega_2^2\right)\sqrt{2}} = 2F_1\,\frac{1}{\left(s^2 + \omega_2^2\right)\sqrt{2}} \qquad (8.40a,b,c)$$

$$z_{p3} = \left(F_1 - 2F_2 + F_3\right)\frac{1}{\left(s^2 + \omega_3^2\right)\sqrt{6}} = 0$$

In this case the combination of the eigenvectors and forcing function signs cancel out the first and third modes, leaving only the second mode.

To excite the third mode only, applying the same force to masses 1 and 3 and twice the force with opposite sign to mass 2 should work: $F_1 = F_3$, $F_2 = -2F_1$:

$$z_{p1} = \left(F_1 + F_2 + F_3\right)\frac{1}{s^2\sqrt{3}} = 0$$

$$z_{p2} = \left(F_1 - F_3\right)\frac{1}{\left(s^2 + \omega_2^2\right)\sqrt{2}} = 0 \qquad (8.41a,b,c)$$

$$z_{p3} = \left(F_1 - 2F_2 + F_3\right)\frac{1}{\left(s^2 + \omega_3^2\right)\sqrt{6}} = 6F_1\,\frac{1}{\left(s^2 + \omega_3^2\right)\sqrt{6}}$$

In this case the combination of the eigenvectors and forcing function signs cancel out the first and second modes, leaving only the third mode.

8.7 How Modes Combine to Create Transfer Functions

We have shown that both the normal mode method and partial fraction expansion of transfer functions yield additive combinations of sdof systems for the overall response. The purpose of this section is to develop a general equation for any transfer function, again showing that the system frequency response is an additive combination of sdof systems. **Each sdof system has a gain determined by the appropriate eigenvector entries and a resonant frequency given by the appropriate eigenvalue.**

The three equations of motion in principal coordinates are:

$$\ddot{z}_{p1} + \omega_1^2 z_{p1} = F_{p1}$$

$$\ddot{z}_{p2} + \omega_2^2 z_{p2} = F_{p2} \qquad (8.42a,b,c)$$

$$\ddot{z}_{p3} + \omega_3^2 z_{p3} = F_{p3}$$

$$\omega_1^2 = 0 \qquad \omega_2^2 = \frac{k}{m} \qquad \omega_3^2 = \frac{3k}{m} \qquad (8.43a,b,c)$$

Where the forces in principal coordinates are given by:

$$F_p = z_n^T F = \begin{bmatrix} z_{n11} & z_{n21} & z_{n31} \\ z_{n12} & z_{n22} & z_{n32} \\ z_{n13} & z_{n23} & z_{n33} \end{bmatrix}\begin{bmatrix} F_1 \\ F_2 \\ F_3 \end{bmatrix}$$

$$(8.44)$$

$$= \begin{bmatrix} z_{n11}F_1 + z_{n21}F_2 + z_{n31}F_3 \\ z_{n12}F_1 + z_{n22}F_2 + z_{n32}F_3 \\ z_{n13}F_1 + z_{n23}F_2 + z_{n33}F_3 \end{bmatrix}$$

Taking the Laplace transform of the differential equations (8.42a,b,c) and dividing by the coefficients of each principal displacement:

$$
\mathbf{z}_p = \begin{bmatrix} z_{p1} \\ z_{p2} \\ z_{p3} \end{bmatrix} = \begin{bmatrix} \dfrac{F_{p1}}{s^2 + \omega_1^2} \\[3mm] \dfrac{F_{p2}}{s^2 + \omega_2^2} \\[3mm] \dfrac{F_{p3}}{s^2 + \omega_3^2} \end{bmatrix} = \begin{bmatrix} \dfrac{z_{n11}F_1 + z_{n21}F_2 + z_{n31}F_3}{s^2 + \omega_1^2} \\[3mm] \dfrac{z_{n12}F_1 + z_{n22}F_2 + z_{32}F_3}{s^2 + \omega_2^2} \\[3mm] \dfrac{z_{n13}F_1 + z_{n23}F_2 + z_{n33}F_3}{s^2 + \omega_3^2} \end{bmatrix} \qquad (8.45)
$$

The equation above shows how the individual eigenvector matrix elements contribute to the displacements in principal coordinates.

Since we are only interested in SISO transfer functions that arise from a force applied to a single dof, we will look at the F_1, F_2, F_3 cases individually.

For force F_1:

$$
\mathbf{z}_p = \begin{bmatrix} z_{p1} \\ z_{p2} \\ z_{p3} \end{bmatrix} = \begin{bmatrix} \dfrac{F_{p1}}{s^2 + \omega_1^2} \\[3mm] \dfrac{F_{p2}}{s^2 + \omega_2^2} \\[3mm] \dfrac{F_{p3}}{s^2 + \omega_3^2} \end{bmatrix} = \begin{bmatrix} \dfrac{z_{n11}F_1}{s^2 + \omega_1^2} \\[3mm] \dfrac{z_{n12}F_1}{s^2 + \omega_2^2} \\[3mm] \dfrac{z_{n13}F_1}{s^2 + \omega_3^2} \end{bmatrix} \qquad (8.46)
$$

For force F_2:

$$
\mathbf{z}_p = \begin{bmatrix} z_{p1} \\ z_{p2} \\ z_{p3} \end{bmatrix} = \begin{bmatrix} \dfrac{F_{p1}}{s^2 + \omega_1^2} \\[3mm] \dfrac{F_{p2}}{s^2 + \omega_2^2} \\[3mm] \dfrac{F_{p3}}{s^2 + \omega_3^2} \end{bmatrix} = \begin{bmatrix} \dfrac{z_{n21}F_2}{s^2 + \omega_1^2} \\[3mm] \dfrac{z_{n22}F_2}{s^2 + \omega_2^2} \\[3mm] \dfrac{z_{n23}F_2}{s^2 + \omega_3^2} \end{bmatrix} \qquad (8.47)
$$

For force F_3

$$\mathbf{z}_p = \begin{bmatrix} z_{p1} \\ z_{p2} \\ z_{p3} \end{bmatrix} = \begin{bmatrix} \dfrac{F_{p1}}{s^2 + \omega_1^2} \\[6pt] \dfrac{F_{p2}}{s^2 + \omega_2^2} \\[6pt] \dfrac{F_{p3}}{s^2 + \omega_3^2} \end{bmatrix} = \begin{bmatrix} \dfrac{z_{n31} F_3}{s^2 + \omega_1^2} \\[6pt] \dfrac{z_{32} F_3}{s^2 + \omega_2^2} \\[6pt] \dfrac{z_{n33} F_3}{s^2 + \omega_3^2} \end{bmatrix} \qquad (8.48)$$

The equations for displacements in physical coordinates are found by premultiplying the above three equations by $\mathbf{z}_n$ (7.107).

For F_1:

$$\mathbf{z} = \begin{bmatrix} z_1 \\ z_2 \\ z_3 \end{bmatrix} = \begin{bmatrix} \dfrac{z_{n11} z_{n11} F_1}{s^2 + \omega_1^2} + \dfrac{z_{n12} z_{n12} F_1}{s^2 + \omega_2^2} + \dfrac{z_{n13} z_{n13} F_1}{s^2 + \omega_3^2} \\[8pt] \dfrac{z_{n21} z_{n11} F_1}{s^2 + \omega_1^2} + \dfrac{z_{n22} z_{n12} F_1}{s^2 + \omega_2^2} + \dfrac{z_{n23} z_{n13} F_1}{s^2 + \omega_3^2} \\[8pt] \dfrac{z_{n31} z_{n11} F_1}{s^2 + \omega_1^2} + \dfrac{z_{n32} z_{n12} F_1}{s^2 + \omega_2^2} + \dfrac{z_{n33} z_{n13} F_1}{s^2 + \omega_3^2} \end{bmatrix} \qquad (8.49)$$

Dividing by F_1:

$$\frac{\mathbf{z}}{F_1} = \begin{bmatrix} \dfrac{z_1}{F_1} \\[6pt] \dfrac{z_2}{F_1} \\[6pt] \dfrac{z_3}{F_1} \end{bmatrix} = \begin{bmatrix} \dfrac{z_{n11} z_{n11}}{s^2 + \omega_1^2} + \dfrac{z_{n12} z_{n12}}{s^2 + \omega_2^2} + \dfrac{z_{n13} z_{n13}}{s^2 + \omega_3^2} \\[8pt] \dfrac{z_{n21} z_{n11}}{s^2 + \omega_1^2} + \dfrac{z_{n22} z_{n12}}{s^2 + \omega_2^2} + \dfrac{z_{n23} z_{n13}}{s^2 + \omega_3^2} \\[8pt] \dfrac{z_{n31} z_{n11}}{s^2 + \omega_1^2} + \dfrac{z_{n32} z_{n12}}{s^2 + \omega_2^2} + \dfrac{z_{n33} z_{n13}}{s^2 + \omega_3^2} \end{bmatrix} \qquad (8.50)$$

Similarly for F_1 and F_2 :

$$\frac{\mathbf{z}}{F_2} = \begin{bmatrix} \dfrac{z_1}{F_2} \\[2mm] \dfrac{z_2}{F_2} \\[2mm] \dfrac{z_3}{F_2} \end{bmatrix} = \begin{bmatrix} \dfrac{z_{n11}z_{n21}}{s^2+\omega_1^2} + \dfrac{z_{n12}z_{n22}}{s^2+\omega_2^2} + \dfrac{z_{n13}z_{n23}}{s^2+\omega_3^2} \\[4mm] \dfrac{z_{n21}z_{n21}}{s^2+\omega_1^2} + \dfrac{z_{n22}z_{n22}}{s^2+\omega_2^2} + \dfrac{z_{n23}z_{n23}}{s^2+\omega_3^2} \\[4mm] \dfrac{z_{n31}z_{n21}}{s^2+\omega_1^2} + \dfrac{z_{n32}z_{n22}}{s^2+\omega_2^2} + \dfrac{z_{n33}z_{n23}}{s^2+\omega_3^2} \end{bmatrix} \qquad (8.51)$$

$$\frac{\mathbf{z}}{F_3} = \begin{bmatrix} \dfrac{z_1}{F_3} \\[2mm] \dfrac{z_2}{F_3} \\[2mm] \dfrac{z_3}{F_3} \end{bmatrix} = \begin{bmatrix} \dfrac{z_{n11}z_{n31}}{s^2+\omega_1^2} + \dfrac{z_{n12}z_{n32}}{s^2+\omega_2^2} + \dfrac{z_{n13}z_{n33}}{s^2+\omega_3^2} \\[4mm] \dfrac{z_{n21}z_{n31}}{s^2+\omega_1^2} + \dfrac{z_{n22}z_{n32}}{s^2+\omega_2^2} + \dfrac{z_{n23}z_{n33}}{s^2+\omega_3^2} \\[4mm] \dfrac{z_{n31}z_{n31}}{s^2+\omega_1^2} + \dfrac{z_{n32}z_{n32}}{s^2+\omega_2^2} + \dfrac{z_{n33}z_{n33}}{s^2+\omega_3^2} \end{bmatrix} \qquad (8.52)$$

The nine transfer functions above may be generalized by the following equation. For the transfer function with the force applied at dof "k," the displacement taken at dof "j" and for mode "i":

$$\frac{z_j}{F_k} = \frac{z_{nji}z_{nk1}}{s^2+\omega_1^2} + \frac{z_{nj2}z_{nk2}}{s^2+\omega_2^2} + \frac{z_{nj3}z_{nk3}}{s^2+\omega_3^2} \qquad (8.53)$$

Rewriting in summation form, and generalizing from our tdof system to a general system where "m" is the total number of modes for the system for an undamped (8.54a) and damped (8.54b) system:

$$\frac{z_j}{F_k} = \sum_{i=1}^{m} \frac{z_{nji}z_{nki}}{s^2+\omega_i^2}$$

$$(8.54a,b)$$

$$\frac{z_j}{F_k} = \sum_{i=1}^{m} \frac{z_{nji}z_{nki}}{s^2+2\zeta_i\omega_i s+\omega_i^2}$$

Equations (8.54a,b) shows that in general every transfer function is made up of additive combinations of single degree of freedom systems, with each system having its dc gain (transfer function evaluated with s = j0) determined by the appropriate eigenvector entry product divided by the

square of the eigenvalue, $z_{nji}z_{nki}/\omega_i^2$, and with resonant frequency
defined by the appropriate eigenvalue, ω_i.

For our tdof system, substituting for the ω_i values:

$$\frac{z_j}{F_k} = \frac{z_{nj1}z_{nk1}}{s^2} + \frac{z_{nj2}z_{nk2}}{s^2+k/m} + \frac{z_{nj3}z_{nk3}}{s^2+3k/m} \qquad (8.55)$$

This equation makes the graphical combining of modal contributions to the
final transfer function more clear. The contribution of each mode is a simple
harmonic oscillator at frequency ω_i with dc gain $z_{nji}z_{nki}/\omega_i^2$, where "i" is the
mode number.

8.8 Plotting Individual Mode Contributions

Taking z_1/F_1 for example, the separate contributions of each mode to the total
response can be plotted as follows. First we calculate the DC response of the
non-rigid body mode:

Rigid body response: at $\omega = 0.1$ rad/sec $z_{111} = \dfrac{1}{0.03} = 33.33 = 30.457$ db,

$\qquad\qquad\qquad$ slope $= -2$

Now we calculate the dc gain of the non-rigid body modes:

Second mode response: at DC , $z_{112} = \dfrac{1}{2} = 0.5 = -6$db, slope $= 0$

$\qquad\qquad\qquad\qquad$ resonance at $\omega_2^2 = 1$, slope at $\infty = -2$

Third mode response: at DC , $z_{113} = \dfrac{1}{18} = 0.0555 = -25.1$db, slope $= 0$

$\qquad\qquad\qquad\qquad$ resonance at $\omega_3^2 = 3$, slope at $\infty = -2$, where the
$\qquad\qquad\qquad\qquad$ "ijk" notation in z_{ijk} indicates: dof "i," due to force
$\qquad\qquad\qquad\qquad$ "j," for mode k.

Thus, the total response is defined by the additive combination of three single
degree of freedom responses, each with its own spring-dominated low

frequency section, damping dominated resonant section and mass dominated high frequency section.

The MATLAB code **tdof_modal_xfer.m** is used to calculate and plot the individual mode contributions to the overall frequency response of all four unique transfer functions for the tdof model. The program plots the frequency responses using several different magnitude scalings. We will discuss below the results for the z_1 / F_1 frequency response, using plots from the MATLAB code to illustrate. The notation "z113" below signifies the transfer function z1/F1 for mode 3, and so forth.

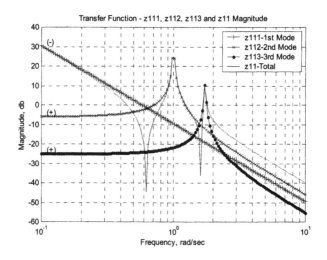

Figure 8.2: z11 transfer function frequency response plot with individual mode contributions overlaid.

Figure 8.2 shows the overall z_1 / F_1 (z11) transfer function and the individual modal contributions which add to create it. Because the magnitude scale in Figure 8.2 is in log or "db" units, the individual mode contributions cannot be added graphically. To add graphically requires a linear magnitude axis. We cannot use the log magnitude or db scale for adding directly because adding with log or db coordinates is equivalent to the multiplication of responses, not addition.

There is zero damping in this model, so the amplitudes at the two poles in Figure 8.2 should go to infinity. The peak amplitudes do not go to infinity because they are limited by the resolution of the frequency scale chosen for the plot. The two zeros should go to zero, but once again they do not because of the frequency resolution chosen.

Figures 8.3 and 8.4 show the same frequency responses plotted on a linear magnitude scale.

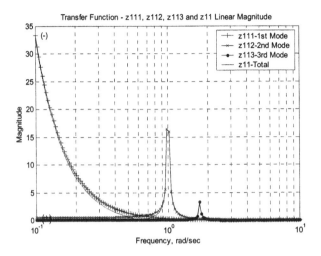

Figure 8.3: z11 frequency response and modal contributions plotted with linear magnitude scale.

Figure 8.4 uses an expanded magnitude axis to more clearly show how the three individual mode sdof responses combine graphically to create the overall frequency response. It also contains notation that shows how the signs change through the resonance. In Chapter 3 we learned how to sketch frequency response plots by hand, knowing the high and low frequency asymptotes and the locations of the poles and zeros. Similarly, we can combine modes by hand if we know the signs (phases) of the individual modes that are being combined. In our tdof example, it just so happens that the signs of the low frequency portions of the second and third modes (1.0 and 1.7 rps) were both positive. In general, it is not the case that all low frequency signs will be positive (see the z31 example below). The discussion below will show how to define the sign (phase) of the low frequency portion of each mode by knowing the signs of the eigenvector entries for the input and output degrees of freedom.

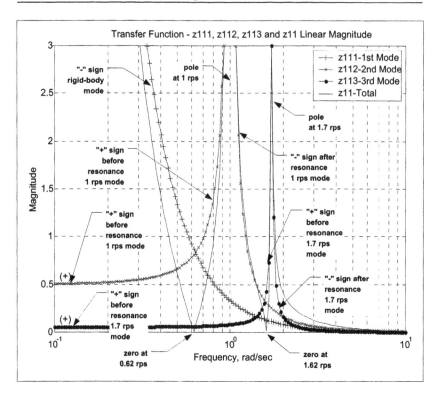

Figure 8.4: z11 frequency response with expanded magnitude scale to see contributors to the zeros.

Since the phase at frequencies much lower than the resonant frequency is zero for a spring mass sdof system (2.19b), and since each mode in principal coordinates is a sdof system, **the phase for each mode contribution to the overall response at low frequency is given by the sign of the eigenvector for the dof whose displacement is desired times the sign of the dof where the force is applied.** For the three modes and the transfer function z11, where we are interested in measuring the displacement of mass 1 and in the force being applied to mass 1, the signs for the three modes at low frequencies are found as follows. The normalized modal matrix is repeated to see the signs of the entries:

	Mode 1	Mode 2	Mode 3
dof 1	−0.5774	−0.7071	0.4082
dof2	−0.5774	0.0000	−0.8165
dof3	−0.5774	0.7071	0.4082

Table 8.1: Normalized modal matrix.

<u>Sign of mode 1 low frequency asymptote for z11 frequency response:</u>

dof 1, mode 1: −0.5774   (−)

dof 1, mode 1: −0.5774 4 (−)

Low frequency sign (phase) = (−) times (−) = (+), but since resonance is rigid body at zero rad/sec, all frequencies of interest to us are "after resonance" and thus the sign is (−) because the phase is −180°.

<u>Sign of mode 2 low frequency asymptote for z11 frequency response:</u>

dof 1, mode 2: −0.7071   (−)

dof 1, mode 2: −0.7071   (−)

Low frequency sign (phase) = (−) times (−) = (+) , 0°

<u>Sign of mode 3 low frequency asymptote for z11 frequency response:</u>

dof 1, mode 3: +0.4082   (−)

dof 1, mode 3: +0.4082   (−)

Low frequency sign (phase) = (+) times (+) = (+), 0°

As mentioned earlier, the signs of the low frequency portions of the second and third modes were both (+). The signs of the eigenvector entries above show why this was the case.

The "sign" of the rigid body mode is always "−" because the phase is always −180°. The signs of the 1 rad/sec (rps) and 1.732 rps modes are both "+" at low frequencies because their phases are 0°. After the resonance, their signs change to "−" as phase goes to −180°. Exactly at resonance the phase of each is −90°, however, away from resonance the phases are either 0° or −180° because the problem has no damping.

Thus, if the low frequency asymptote sign (phase) is known for each mode, the SISO **frequency response zeros can be identified as frequencies where the appropriate modes add to zero algebraically**, as can be seen graphically on Figure 8.4.

The z11 zeros at 0.62 and 1.62 rps arise when the contributions of the three modes combine algebraically to zero.

For other transfer functions, for example z31, the low frequency signs would be different, as can be seen below:

Sign of mode 1 low frequency asymptote for z31 frequency response:

> dof 3, mode 1: −0.5774   (−)

> dof 1, mode 1: −0.5774   (−)

> Low frequency sign (phase) = (−) times (−) = (+), but is after resonance so sign is (−)

Sign of mode 2 low frequency asymptote for z31 frequency response:

> dof 3, mode 2: +0.7071   (−)

> dof 1, mode 2: −0.7071   (−)

> Low frequency sign (phase) = (+) times (−) = (−), −180°

Sign of mode 3 low frequency asymptote for z31 frequency response:

> dof 3, mode 3: +0.4082   (−)

> dof 1, mode 3: +0.4082   (−)

Low frequency sign (phase) = (+) times (+) = (+), 0°

Now that the low frequency phases of the individual modes are defined, we can follow the combining of modes to get the overall response, indicated by the "+" signs.

Because we are dealing with a linear magnitude axis above, we can graphically add or subtract the contribution of each to get the overall response.

To get the overall response we combine the amplitudes of each mode depending on its sign. For example, at 0.4 rad/sec frequency, we combine the amplitude of the rigid body mode with a negative sign with the two oscillatory modes, each of which has a positive sign:

Rigid body response: at $\omega = 0.4$ rad/sec, $\omega_1 = 0$:

$$z_{111} = \frac{1}{3s^2} = \frac{1}{3(j\omega)^2} = \frac{-1}{3\omega^2} = \frac{-1}{3(0.4)^2} = -2.083 \qquad (8.56)$$

Second mode response: at $\omega = 0.4$ rad/sec, $\omega_2 = 1$:

$$z_{112} = \frac{1}{2(s^2 + \omega_2^2)} = \frac{1}{2\left[(j\omega)^2 + \omega_2^2\right]} = \frac{1}{2\left[-\omega^2 + \omega_2^2\right]} = \frac{1}{2\left[-(0.4)^2 + 1\right]} = 0.595$$

$$(8.57)$$

Third mode response: at $\omega = 0.4$ rad/sec, $\omega_3 = 1.732$:

$$z_{113} = \frac{1}{6(s^2 + \omega_3^2)} = \frac{1}{6\left[(j\omega)^2 + \omega_3^2\right]} = \frac{1}{6\left[-\omega^2 + \omega_3^2\right]} = \frac{1}{6\left[-(0.4)^2 + 3\right]} = 0.0586$$

$$(8.58)$$

Adding (with proper signs) the three contributions at 0.4 rad/sec gives the amplitude and phase of the overall response at 0.4 rad/sec:

$$-2.083 + 0.595 + 0.0586 = -1.4294 \qquad (8.59)$$

The amplitude is 1.4294 and the phase is $-180°$, as indicated by the negative sign. Because the model has no damping, each mode has 0° phase before resonance and immediately after resonance switches phase to $-180°$. Exactly at resonance the amplitudes are theoretically infinite.

Let us now track what happens at the frequency of the first zero, which we showed in (2.85) to be 0.618 rad/sec. We will carry out the same calculations as above for a frequency of 0.618 rad/sec:

Rigid body response: at $\omega = 0.618$ rad/sec, $\omega_1 = 0$:

$$z_{111} = \frac{1}{3s^2} = \frac{1}{3(j\omega)^2} = \frac{-1}{3\omega^2} = \frac{-1}{3(0.618)^2} = -0.8727 \tag{8.60}$$

Second mode response: at $\omega = 0.618$ rad/sec, $\omega_2 = 1$:

$$z_{112} = \frac{1}{2(s^2 + \omega_2^2)} = \frac{1}{2\left[(j\omega)^2 + \omega_2^2\right]} = \frac{1}{2\left[-(0.618)^2 + 1\right]} = 0.8089 \tag{8.61}$$

Third mode response: at $\omega = 0.618$ rad/sec, $\omega_3 = 1.732$:

$$z_{113} = \frac{1}{6(s^2 + \omega_3^2)} = \frac{1}{6\left[(j\omega)^2 + \omega_3^2\right]} = \frac{1}{6\left[-(0.618)^2 + 3\right]} = 0.0636 \tag{8.62}$$

Adding (with proper signs) the three contributions gives the amplitude and phase of the overall response at 0.618 rad/sec:

$$-0.8727 + 0.8089 + 0.0636 = -0.0002 \cong 0 \tag{8.63}$$

The amplitude is -0.0002. With greater accuracy in the values used for the eigenvalues and the frequency of the zero, the solution would have been exactly zero.

In Chapter 2, we showed that the zeros for SISO transfer functions arose from the roots of the numerator. The modal analysis method shows another explanation of how zeros of transfer functions arise: **when modes combine with appropriate signs (phases) it is possible at some frequencies to have no motion.**

We will calculate the response at one more frequency to show how the phase changes for a mode when the frequency is higher than the resonant frequency.

We will choose a frequency of 1.3 rad/sec, which is higher than the second mode but lower than the third mode. We should see that the sign of the contribution for the second mode changes sign from positive to negative. Signs for the first and third mode should remain unchanged.

Rigid body response: at $\omega = 1.3$ rad/sec, $\omega_1 = 0$:

$$z_{111} = \frac{1}{3s^2} = \frac{1}{3(j\omega)^2} = \frac{-1}{3\omega^2} = \frac{-1}{3(1.3)^2} = -0.1972 \qquad (8.64)$$

Second mode response: at $\omega = 1.3$ rad/sec, $\omega_2 = 1$:

$$z_{112} = \frac{1}{2(s^2 + \omega_2^2)} = \frac{1}{2\left[(j\omega)^2 + \omega_2^2\right]} = \frac{1}{2\left[-(1.3)^2 + 1\right]} = -0.7246$$

$$(8.65)$$

Third mode response: at $\omega = 1.3$ rad/sec, $\omega_3 = 1.732$:

$$z_{113} = \frac{1}{6(s^2 + \omega_3^2)} = \frac{1}{6\left[(j\omega)^2 + \omega_3^2\right]} = \frac{1}{6\left[-(1.3)^2 + 3\right]} = 0.1272$$

$$(8.66)$$

Adding (with proper signs) the three contributions at 1.3 rad/sec gives the amplitude and phase of the overall response at 1.3 rad/sec:

$$-0.1972 - 0.7246 + 0.1272 = -0.7946 \qquad (8.67)$$

The amplitude is 0.7946 and the phase is $-180°$. Note that the sign of the second   mode contribution changed from positive to negative when the resonant frequency was passed.

The same calculations can be repeated for any desired frequency. Also, knowing the high and low frequency asymptotes, their signs and resonant frequencies, one can plot the overall frequency response roughly by hand, similar to what was done in Section 3.3. Here, unlike the previous hand plotting, we have not calculated any zeros; they occur by additive combinations of individual modes.

8.9 MATLAB Code tdof_modal_xfer.m – Plotting Frequency Responses, Modal Contributions

8.9.1 Code Overview

Figures 8.2 to 8.4 were plotted using this code. The code uses (8.24 to 8.27) to evaluate the four transfer functions z11, z21, z31 and z22. Each of the transfer functions also has its modal contributions calculated and plotted as overlays.   The frequency response plots are all plotted with log and db magnitude scales as well as a linear scale which is expanded in the fourth plot of the series. Because of the amount of code used for the plotting, only the code for the z11 transfer function will be listed.   All the other transfer functions are calculated in a similar fashion.

8.9.2 Code Listing, Partial

```
%        tdof_modal_xfer.m    plotting modal transfer functions of three dof model

         clf;

         legend off;

         subplot(1,1,1);

         clear all;

%        Define a vector of frequencies to use, radians/sec.  The logspace command uses
%        the log10 value as limits, i.e. -1 is 10^-1 = 0.1 rad/sec, and 1 is
%        10^1 = 10 rad/sec.  The 200 defines 200 frequency points.

         w = logspace(-1,1,150);

%        calculate the rigid-body motions for low and high frequency portions
%        of all the transfer functions

%        z11, output 1 due to force 1 transfer functions

         z111num = 1/3;

         z111den = [1 0 0];

         z112num = 1/2;

         z112den = [1 0 1];

         z113num = 1/6;

         z113den = [1 0 3];

         [z111mag,z111phs] = bode(z111num,z111den,w);
```

```
        [z112mag,z112phs] = bode(z112num,z112den,w);

        [z113mag,z113phs] = bode(z113num,z113den,w);

        if  abs(z111phs(1)) >= 10

                z111text = '(-)';

        else

                z111text = '(+)';

        end

        if  abs(z112phs(1)) >= 10

                z112text = '(-)';

        else

                z112text = '(+)';

        end

        if  abs(z113phs(1)) >= 10

                z113text = '(-)';

        else

                z113text = '(+)';

        end

        z111magdb = 20*log10(z111mag);

        z112magdb = 20*log10(z112mag);

        z113magdb = 20*log10(z113mag);

%       calculate the complete transfer function

        z11 = ((1/3)./((j*w).^2) + ((1/2)./((j*w).^2 + 1)) + ((1/6)./((j*w).^2 + 3)));

        z11mag = abs(z11);

        z11magdb = 20*log10(z11mag);

        z11phs = 180*angle(z11)/pi ;

%       truncate peaks for microsoft word plotting of expanded linear scale

        z11plotmag = z11mag;

        z11plotmag = z11mag;
```

```
                z112plotmag = z112mag;

                z113plotmag = z113mag;


                for  cnt = 1:length(z11mag)

                        if  z11plotmag(cnt) >= 3.0

                                z11plotmag(cnt) = 3.0;

                        end

                        if  z111plotmag(cnt) >= 3.0

                                z111plotmag(cnt) = 3.0;

                        end

                        if  z112plotmag(cnt) >= 3.0

                                z112plotmag(cnt) = 3.0;

                        end

                        if  z113plotmag(cnt) >= 3.0

                                z113plotmag(cnt) = 3.0;

                        end

                end


%       plot the three modal contribution transfer functions and the total using
%       log magnitude versus frequency

        loglog(w,z111mag,'k+-',w,z112mag,'kx-',w,z113mag,'k.-',w,z11mag,'k-')
        title('Transfer Functions - z111, z112, z113 and z11 magnitude')
        legend('z111-1st Mode','z112-2nd Mode','z113-3rd Mode','z11-Total')
        text(.11,1.2*z111mag(1),z111text)
        text(.11,1.2*z112mag(1),z112text)
        text(.11,1.2*z113mag(1),z113text)
        xlabel('Frequency, rad/sec')
        ylabel('Magnitude')
        grid

        disp('execution paused to display figure, "enter" to continue'); pause

%       plot the four transfer functions using db

        semilogx(w,z111magdb,'k+-',w,z112magdb,'kx-',w,z113magdb,'k.-',w,z11magdb,'k-')
        title('Transfer Function - z111, z112, z113 and z11 Magnitude')
        legend('z111-1st Mode','z112-2nd Mode','z113-3rd Mode','z11-Total')
```

```
            text(.11,2+z111magdb(1),z111text)
            text(.11,2+z112magdb(1),z112text)
            text(.11,2+z113magdb(1),z113text)
            xlabel('Frequency, rad/sec')
            ylabel('Magnitude, db')
            grid

            disp('execution paused to display figure, "enter" to continue'); pause

%           plot the four transfer functions using a linear magnitude scale so that
%           the amplitudes can be added directly

            semilogx(w,z111mag,'k+-',w,z112mag,'kx-',w,z113mag,'k.-',w,z11mag,'k-')
            title('Transfer Function - z111, z112, z113 and z11 Linear Magnitude')
            legend('z111-1st Mode','z112-2nd Mode','z113-3rd Mode','z11-Total')
            text(.11,1.0*z111mag(1),z111text)
            text(.11,1.1*z112mag(1),z112text)
            text(.11,1.1*z113mag(1),z113text)
            xlabel('Frequency, rad/sec')
            ylabel('Magnitude')
            grid

            disp('execution paused to display figure, "enter" to continue'); pause

            semilogx(w,z111plotmag,'k+-',w,z112plotmag,'kx-',w,z113plotmag,' ...
                             k.-',w,z11plotmag,'k-')
            title('Transfer Function - z111, z112, z113 and z11 Linear Magnitude')
            legend('z111-1st Mode','z112-2nd Mode','z113-3rd Mode','z11-Total')
            text(.11,1.0*z111mag(1),z111text)
            text(.11,1.1*z112mag(1),z112text)
            text(.11,1.1*z113mag(1),z113text)
            xlabel('Frequency, rad/sec')
            ylabel('Magnitude')
            axis([.1 10 0 3]);
            grid

            disp('execution paused to display figure, "enter" to continue'); pause

%           plot phase

            semilogx(w,z111phs,'k+-',w,z112phs,'kx-',w,z113phs,'k.-',w,z11phs,'k-')
            title('Transfer Function - z111, z112, z113 and z11 Phase')
            legend('z111-1st Mode','z112-2nd Mode','z113-3rd Mode','z11-Total')
            xlabel('Frequency, rad/sec')
            ylabel('Phase, Deg')
            grid

            disp('execution paused to display figure, "enter" to continue'); pause
```

8.10 tdof Eigenvalue Problem Using ANSYS

An ANSYS solution to the tdof problem is now shown in order to start becoming familiar with how ANSYS presents its eigenvalue/eigenvector results.

8.10.1 ANSYS Code threedof.inp Description

The ANSYS code **threedof.inp** below is used to build the model, calculate eigenvalues and eigenvectors, output the frequency listing and eigenvectors, plot the mode shapes, calculate and plot all three transfer functions for a forcing function applied to mass 1: z_1 / F_1, z_2 / F_1, and z_3 / F_1.   The hand calculated values for masses and stiffnesses are used, m1 = m2 = m3 = 1.0, k1 = k2 = 1.0.

To run the code, from the "begin" level in ANSYS, type "/input,threedof,inp," and the program will run unattended. The various outputs are available as follows:

threedof.frq	frequency list, ascii file
threedof.eig	eigenvector list, ascii file
threedof.grp2	mode shape plots
threedof.grp1	frequency response plots

Use the ANSYS Display program to load and display the two plot files.

8.10.2 ANSYS Code Listing

```
/title, threedof.inp, three dof vibration class model, Ansys Version 5.5

/prep7                              ! enter model preparation section

! element type definitions

et,1,21                             ! element type for mass
et,2,14                             ! element type for spring

! real value definitions

r,1,1,1,1                           ! mass of 1kg
r,2,1                               ! spring stiffness of 1mn/mm, or 1n/m

! define plotting characteristics

/view,1,-1,0,0                      ! z-y plane
/angle,1,0                          ! not iso
```

```
/pnum,real,1                    ! color by real
/num,1                          ! numbers off
/type,1,0                       ! hidden plot
/pbc,all,1                      ! show all boundary conditions

csys,0                          ! define global coordinate system

! nodes

n,1,0,0,-1                      ! left hand mass at x = -1.0 mm
n,2,0,0,0                       ! middle mass at x = 0 mm
n,3,0,0,1                       ! right hand mass at x = +1.0 mm

! define masses

type,1
real,1
e,1
e,2
e,3

! define springs

type,2
real,2
e,1,2
e,2,3

! define constraints, ux and uy zero, leaving only uz motion

nsel,s,node,,1,3
d,all,ux,0
d,all,uy,0

allsel
eplo

! ************** eigenvalue run *****************

fini                ! fini just in case not in begin

/solu               ! enters the solution processor, needs to be here to do editing below

allsel              ! default selects all items of specified entity type, typically nodes, elements

! define masters for frequency response (transfer function) run

m,1,uz
m,2,uz
m,3,uz

antype,modal,new
modopt,reduc,3                  ! method – Block Lanczost
expass,off                      ! key = off, no expansion pass, key = on, do expansion
mxpand,3,,,no                   ! number of modes to expand
```

```
total,3,1                        ! total masters, all translational dof

allsel

solve                 ! starts the solution of one load step of a solution sequence, modal

fini

! ***************** output frequencies *******************

/post1

/output,threedof,frq              ! write out frequency list to ascii file .frq

set,list

/output,term                  ! returns output to terminal

! ***************** output eigenvectors ********************

! define nodes for output

allsel

/output,threedof,eig              ! write out eigenvectors to ascii file .eig

*do,i,1,3
         set,,i
         prdisp
*enddo

/output,term
************** plot modes *****************

/show,threedof,grp2,0            ! raster plot, 1 is vector plot, write out to graph file .grp2

allsel

*do,i,1,3
         set,1,i
   pldi,1
*enddo

/show,term

! **************** calculate and plot transfer functions *****************

fini

/assign,rst,junk,rst              ! reassigns a file name to an ANSYS identifier

/solu

dmprat,0                   ! sets a constant damping ratio for all modes, zeta = 0
```

```
allsel
eplo                          ! show forces applied

f,1,fz,1                      ! 1 mn force applied to node 1, left-hand mass

/title, threedof.inp, tdof, force at mass 1

antype,harmic                 ! harmonic (frequency response) analysis

hropt,msup,3                  ! mode superposition method, nummodes modes used

harfrq,0.0159,1.59            ! frequency range, hz, for solution, -1 to 10 rad/sec

hrout,off,off                 ! amplitude/phase, cluster off

kbc,1

nsubst,200                    ! 200 frequency points

outres,nsol,all,              ! controls solution set written to database, nodal dof solution, all
                              ! frequencies, component name for selected set of nodes

solve

fini

/post26

file,,rfrq           ! frequency response results

xvar,0               ! display versus frequency

lines,10000          ! specifies the length of a printed page for frequency response listing

nsol,2,1,u,z,z1      ! specifies nodal data to be stored in results file
                     ! u - displacement, z direction
                     ! note that nsol,1 is frequency vector

nsol,3,2,u,z,z2

nsol,4,3,u,z,z3

! plot magnitude

plcplx,0
/grid,1
/axlab,x,frequency, hz
/axlab,y,amplitude, mm
/gropt,logx,1                          ! log plot for frequency
/gropt,logy,1                          ! log plot for amplitude

/show,threedof,grp1                    ! file name for storing
plvar,2,3,4
/show,term
```

```
! plot phase

plcplx,1
/grid,1
/axlab,x,freq
/axlab,y,phase, deg          ! label for y axis
/gropt,logx,1                ! log plot for frequency
/gropt,logy,0                ! linear plot for phase

/show,threedof,grp1
plvar,2,3,4
/show,term

! save ascii data to file

prcplx,1                     ! stores phase angle in asci file .dat

/output,threedof,dat
prvar,2,3,4
/output,term

fini
```

8.10.3 ANSYS Results

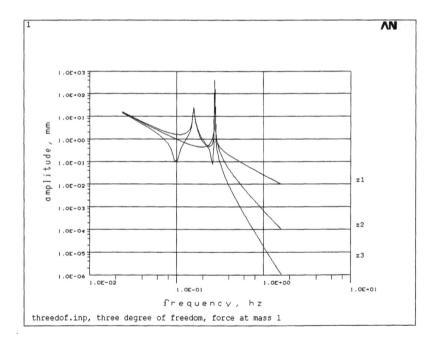

Figure 8.5: ANSYS frequency responses for force at mass 1.

The resulting ANSYS transfer function plot is shown in Figure 8.5, with the frequency axis in Hz, not rad/sec.

The ANSYS frequency listing from **threedof.frq** is shown below, in hz units:

```
***** INDEX OF DATA SETS ON RESULTS FILE *****


SET  TIME/FREQ  LOAD STEP  SUBSTEP  CUMULATIVE

 1    0.47280E-06      1          1          1

 2     0.15915         1          2          2

 3     0.27566         1          3          3
```

Note that the rigid body mode is calculated to be 0.4726e-6, close to 0 hz. The second and third modes are calculated to be 0.15915 and 0.27566 hz, or 0.999969 and 1.732 rad/sec, respectively. This is the same as our hand-calculated results.

The ANSYS eigenvector listing from threedof.eig is below:

```
*DO LOOP ON PARAMETER= I     FROM 1.0000   TO 3.0000   BY 1.0000


USE LOAD STEP   1 SUBSTEP   1 FOR LOAD CASE 0

SET COMMAND GOT LOAD STEP=   1 SUBSTEP=   1 CUMULATIVE ITERATION=
1
  TIME/FREQUENCY= 0.47280E-06
TITLE= threedof.inp, three dof vibration class model, Ansys Version 5.5

PRINT DOF  NODAL SOLUTION PER NODE

***** POST1 NODAL DEGREE OF FREEDOM LISTING *****

LOAD STEP=   1 SUBSTEP=    1
  FREQ=  0.47280E-06  LOAD CASE=  0

THE FOLLOWING DEGREE OF FREEDOM RESULTS ARE IN GLOBAL COORDINATES

  NODE    UX      UY      UZ     ROTX    ROTY     ROTZ
```

```
1     0.0000    0.0000    0.57735
2     0.0000    0.0000    0.57735
3     0.0000    0.0000    0.57735

MAXIMUM ABSOLUTE VALUES
NODE    0        0        1        0        0        0
VALUE  0.0000   0.0000   0.57735  0.0000   0.0000   0.0000

*ENDDO  INDEX= I

***** POST1 NODAL DEGREE OF FREEDOM LISTING *****

LOAD STEP=  1 SUBSTEP=   2
FREQ=  0.15915    LOAD CASE=  0

THE FOLLOWING DEGREE OF FREEDOM RESULTS ARE IN GLOBAL COORDINATES

NODE   UX      UY      UZ       ROTX     ROTY    ROTZ
  1   0.0000  0.0000  -0.70711
  2   0.0000  0.0000  0.75552E-14
  3   0.0000  0.0000  0.70711

MAXIMUM ABSOLUTE VALUES
NODE    0        0        3        0        0        0
VALUE  0.0000   0.0000   0.70711  0.0000   0.0000   0.0000

***** POST1 NODAL DEGREE OF FREEDOM LISTING *****

LOAD STEP=  1 SUBSTEP=   3
FREQ=  0.27566    LOAD CASE=  0

THE FOLLOWING DEGREE OF FREEDOM RESULTS ARE IN GLOBAL COORDINATES

NODE   UX      UY      UZ       ROTX     ROTY    ROTZ
  1   0.0000  0.0000  -0.40825
  2   0.0000  0.0000  0.81650
  3   0.0000  0.0000  -0.40825

MAXIMUM ABSOLUTE VALUES
NODE    0        0        2        0        0        0
VALUE  0.0000   0.0000   0.81650  0.0000   0.0000   0.0000
```

The ANSYS calculated eigenvectors, the three "UZ" listings highlighted in bold type above, arranged in the modal matrix:

$$\text{ANSYS } z_n = \begin{bmatrix} 0.57735 & -0.707 & -0.40825 \\ 0.57735 & 0 & 0.81649 \\ 0.57735 & 0.707 & -0.40825 \end{bmatrix} \quad (8.68)$$

The hand-calculated modal matrix is below, only differing from the ANSYS calculated values in the arbitrary "−1" multiplier for the second and third modes:

$$z_n = \frac{1}{\sqrt{m}} \begin{bmatrix} \dfrac{1}{\sqrt{3}} & \dfrac{1}{\sqrt{2}} & \dfrac{1}{\sqrt{6}} \\ \dfrac{1}{\sqrt{3}} & 0 & \dfrac{-2}{\sqrt{6}} \\ \dfrac{1}{\sqrt{3}} & \dfrac{-1}{\sqrt{2}} & \dfrac{1}{\sqrt{6}} \end{bmatrix} = \begin{bmatrix} 0.57735 & 0.707 & 0.40825 \\ 0.57735 & 0 & -0.81649 \\ 0.57735 & -0.707 & 0.40825 \end{bmatrix}$$

$$(8.69)$$

Problems

Note: All the problems refer to the two dof system shown in Figure P2.2.

P8.1 Using the eigenvalues and eigenvectors normalized with respect to mass from P7.2 and forces F_1 and F_2 applied to mass 1 and mass 2, respectively, write the equations in motion and physical and principal coordinates in matrix form. Identify the components of the forcing function vector in principal coordinates – which eigenvector components and which force, F_1 or F_2, are involved.

P8.2 Solve for the four transfer functions for the system of P8.1 and write them in transfer function matrix form. Separate each transfer function in principal coordinates to show z_p / F_1 and z_p / F_2 as in (8.16).

P8.3 Back transform the transfer functions in principal coordinates to physical coordinates. Identify the contributions to the transfer function from mode 1 and from mode 2 for all transfer functions.

P8.4   Take the transfer function results of P2.2 with $m_1 = m_2 = m = 1$, $k_1 = k_2 = k = 1$ and zero damping and perform a partial fraction expansion on each transfer function. Show that the results are identical to P8.3, the modal form.

P8.5 What is the relationship between F_1 and F_2 in order to excite mode 1 only? To excite only mode 2?

P8.6 Plot by hand the individual mode contributions to the z_2 / F_1 frequency response for zero damping. Note the sign of the dc gain portion of each contribution and add the two contributions appropriately to obtain the overall frequency response. Extra Credit: Plot all three unique frequency responses, showing the individual mode contributions.

P8.7 (MATLAB) Modify **tdof_modal_xfer.m** for the undamped two dof system with $m_1 = m_2 = m = 1$, $k_1 = k_2 = k = 1$ and plot the overlaid frequency responses.

P8.8 (ANSYS) Modify the **threedof.inp** code for the two dof system. Run the code and plot the frequency responses for both masses for $F_1 = 1, F_2 = 0$. Print out the eigenvalue and eigenvector results. Pick out the appropriate entries of the eigenvector output and write out the modal matrix that ANSYS calculates. Compare it with the modal matrix from P7.1 and identify any differences.

CHAPTER 9

TRANSIENT RESPONSE: MODAL FORM

9.1 Introduction

The transient response example shown in Figure 9.1 will be solved by hand, using the modal analysis method derivation from Chapter 7. As in the frequency response analysis in the previous chapter, we will again start with the eigenvalues and eigenvectors from Chapter 7. We will use them to transform initial conditions and forces to principal coordinates and write the equations of motion in principal coordinates. Laplace transforms will be used to solve for the motions in principal coordinates and we will then back transform to physical coordinates. Once again, the individual mode contributions to the overall transient response of each of the masses will be evident. The closed form solution is then coded in MATLAB and the results plotted, highlighting the individual mode contributions.

9.2 Review of Previous Results

The applied step forces are as shown in Figure 9.1 and the initial conditions of position and velocity for each of the three masses are shown in Table 9.1.

From previous results, (7.86) to (7.88), we know the eigenvalues and eigenvectors normalized with respect to mass, z_n :

$$\omega_1 = 0, \quad \omega_2 = \sqrt{\frac{k}{m}}, \quad \omega_3 = \sqrt{\frac{3k}{m}} \tag{9.1}$$

$$z_n = \frac{1}{\sqrt{m}} \begin{bmatrix} \dfrac{1}{\sqrt{3}} & \dfrac{1}{\sqrt{2}} & \dfrac{1}{\sqrt{6}} \\ \dfrac{1}{\sqrt{3}} & 0 & \dfrac{-2}{\sqrt{6}} \\ \dfrac{1}{\sqrt{3}} & \dfrac{-1}{\sqrt{2}} & \dfrac{1}{\sqrt{6}} \end{bmatrix} \tag{9.2}$$

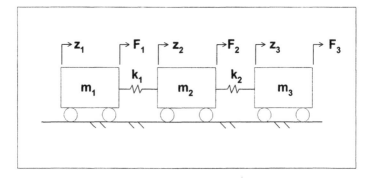

Figure 9.1: Step forces applied to tdof system.

Mass 1	Mass 2	Mass 3
$z_{o1} = 0$	$z_{o2} = -1$	$z_{o3} = 1$
$\dot{z}_{o1} = -1$	$\dot{z}_{o2} = 2$	$\dot{z}_{o3} = -2$

Table 9.1: Initial conditions applied to tdof system.

By inspection, the mass and stiffness matrices in principal coordinates can be written as:

$$\mathbf{m}_p = \begin{bmatrix} 1 & 0 & 0 \\ 0 & 1 & 0 \\ 0 & 0 & 1 \end{bmatrix}, \quad \mathbf{k}_p = \left(\frac{k}{m}\right) \begin{bmatrix} 0 & 0 & 0 \\ 0 & 1 & 0 \\ 0 & 0 & 3 \end{bmatrix} \tag{9.3}$$

9.3 Transforming Initial Conditions and Forces

9.3.1 Transforming Initial Conditions

The initial condition vectors are transformed to principal coordinates by:

$$\begin{aligned} \dot{\mathbf{z}}_{po} &= \mathbf{z}_n^{-1}\dot{\mathbf{z}}_o \\ \mathbf{z}_{po} &= \mathbf{z}_n^{-1}\mathbf{z}_o \end{aligned} \tag{9.4}$$

The inverse of $\mathbf{z}_n$, found using a symbolic algebra program:

$$\mathbf{z}_n^{-1} = \sqrt{m} \begin{bmatrix} \dfrac{\sqrt{3}}{3} & \dfrac{\sqrt{3}}{3} & \dfrac{\sqrt{3}}{3} \\ \dfrac{\sqrt{2}}{2} & 0 & \dfrac{-\sqrt{2}}{2} \\ \dfrac{\sqrt{6}}{6} & \dfrac{-\sqrt{6}}{3} & \dfrac{\sqrt{6}}{6} \end{bmatrix} \tag{9.5}$$

$$\mathbf{z}_{po} = \mathbf{z}_n^{-1} \mathbf{z}_o = \sqrt{m} \begin{bmatrix} \dfrac{\sqrt{3}}{3} & \dfrac{\sqrt{3}}{3} & \dfrac{\sqrt{3}}{3} \\ \dfrac{\sqrt{2}}{2} & 0 & \dfrac{-\sqrt{2}}{2} \\ \dfrac{\sqrt{6}}{6} & \dfrac{-\sqrt{6}}{3} & \dfrac{\sqrt{6}}{6} \end{bmatrix} \begin{bmatrix} 0 \\ -1 \\ 1 \end{bmatrix} = \sqrt{m} \begin{bmatrix} 0 \\ \dfrac{-\sqrt{2}}{2} \\ \dfrac{\sqrt{6}}{2} \end{bmatrix} \tag{9.6}$$

$$\dot{\mathbf{z}}_{po} = \mathbf{z}_n^{-1} \dot{\mathbf{z}}_o = \sqrt{m} \begin{bmatrix} \dfrac{\sqrt{3}}{3} & \dfrac{\sqrt{3}}{3} & \dfrac{\sqrt{3}}{3} \\ \dfrac{\sqrt{2}}{2} & 0 & \dfrac{-\sqrt{2}}{2} \\ \dfrac{\sqrt{6}}{6} & \dfrac{-\sqrt{6}}{3} & \dfrac{\sqrt{6}}{6} \end{bmatrix} \begin{bmatrix} -1 \\ 2 \\ -2 \end{bmatrix} = \sqrt{m} \begin{bmatrix} \dfrac{-\sqrt{3}}{3} \\ \dfrac{\sqrt{2}}{2} \\ \dfrac{-7\sqrt{6}}{6} \end{bmatrix} \tag{9.7}$$

9.3.2 Transforming Forces

The force vector in principal coordinates is:

$$\mathbf{F}_p = \mathbf{z}_n^T \mathbf{F} = \dfrac{1}{\sqrt{m}} \begin{bmatrix} \dfrac{1}{\sqrt{3}} & \dfrac{1}{\sqrt{3}} & \dfrac{1}{\sqrt{3}} \\ \dfrac{1}{\sqrt{2}} & 0 & \dfrac{-1}{\sqrt{2}} \\ \dfrac{1}{\sqrt{6}} & \dfrac{-2}{\sqrt{6}} & \dfrac{1}{\sqrt{6}} \end{bmatrix} \begin{bmatrix} 1 \\ 0 \\ -2 \end{bmatrix} = \dfrac{1}{\sqrt{m}} \begin{bmatrix} \dfrac{-\sqrt{3}}{3} \\ \dfrac{3\sqrt{2}}{2} \\ \dfrac{-\sqrt{6}}{6} \end{bmatrix} \tag{9.8}$$

9.4 Complete Equations of Motion in Principal Coordinates

Now the equations in principal coordinates can be written in matrix form:

$$
\begin{bmatrix} 1 & 0 & 0 \\ 0 & 1 & 0 \\ 0 & 0 & 1 \end{bmatrix} \begin{bmatrix} \ddot{z}_{p1} \\ \ddot{z}_{p2} \\ \ddot{z}_{p3} \end{bmatrix} + \left(\frac{k}{m} \right) \begin{bmatrix} 0 & 0 & 0 \\ 0 & 1 & 0 \\ 0 & 0 & 3 \end{bmatrix} \begin{bmatrix} z_{p1} \\ z_{p2} \\ z_{p3} \end{bmatrix} = \begin{bmatrix} \dfrac{-\sqrt{3}}{3} \\ \dfrac{3\sqrt{2}}{2} \\ \dfrac{-\sqrt{6}}{6} \end{bmatrix} \frac{1}{\sqrt{m}} \tag{9.9}
$$

With initial conditions:

$$
\mathbf{z}_{po} = \sqrt{m} \begin{bmatrix} 0 \\ \dfrac{-\sqrt{2}}{2} \\ \dfrac{\sqrt{6}}{2} \end{bmatrix}, \quad \dot{\mathbf{z}}_{po} = \sqrt{m} \begin{bmatrix} \dfrac{-\sqrt{3}}{3} \\ \dfrac{\sqrt{2}}{2} \\ \dfrac{-7\sqrt{6}}{6} \end{bmatrix} \tag{9.10}
$$

Summarizing the equations in tabular form:

Equations of Motion, Principal Coordinates	Displacement Initial Conditions: Principal Coordinates	Velocity Initial Conditions: Principal Coordinates
$\ddot{z}_{p1} = \dfrac{-\sqrt{3}}{3\sqrt{m}}$	$z_{p1o} = 0$	$\dot{z}_{p1o} = \dfrac{-\sqrt{3m}}{3}$
$\ddot{z}_{p2} + \left(\dfrac{k}{m} \right) z_{p2} = \dfrac{3\sqrt{2}}{2\sqrt{m}}$	$z_{p2o} = \dfrac{-\sqrt{2m}}{2}$	$\dot{z}_{p2o} = \dfrac{\sqrt{2m}}{2}$
$\ddot{z}_{p3} + \left(\dfrac{3k}{m} \right) z_{p3} = \dfrac{-\sqrt{6}}{6\sqrt{m}}$	$z_{p3o} = \dfrac{\sqrt{6m}}{2}$	$\dot{z}_{p3o} = \dfrac{-7\sqrt{6m}}{6}$

Table 9.2: Equations of motion and initial conditions in principal coordinates.

9.5 Solving Equations of Motion Using Laplace Transform

We will now take the Laplace transform of each equation and solve for the transient response resulting from a combination of the forcing function and the initial conditions.

Note that taking the Laplace transform of first and second order differential equations (DE) with initial conditions is (Appendix 2):

First Order DE: $\mathcal{L}\{\dot{x}(t)\} = sX(s) - x(0)$ (9.11)

Second Order DE: $\mathcal{L}\{\ddot{x}(t)\} = s^2X(s) - sx(0) - \dot{x}(0)$ (9.12)

Solving for z_{pl} using Laplace transforms:

$$\ddot{z}_{pl} = \frac{-\sqrt{3}}{3\sqrt{m}} \qquad (9.13)$$

$$s^2 z_{pl}(s) - s z_{pl}(0) - \dot{z}_{pl}(0) = \frac{-\sqrt{3}}{s3\sqrt{m}} \qquad (9.14)$$

$$s^2 z_{pl}(s) - s(0) - \left(\frac{-\sqrt{3m}}{3}\right) = \frac{-\sqrt{3}}{s3\sqrt{m}} \qquad (9.15)$$

$$s^2 z_{pl}(s) = \frac{-\sqrt{3}}{s3\sqrt{m}} - \frac{\sqrt{3m}}{3} \qquad (9.16)$$

$$z_{pl}(s) = \frac{-\sqrt{3}}{s^3 3\sqrt{m}} - \frac{\sqrt{3m}}{3s^2}$$

$$= \frac{-1}{s^3 \sqrt{3m}} - \frac{\sqrt{3m}}{3s^2}$$

$$= \frac{-1}{s^3 \sqrt{3m}} - \frac{\sqrt{3m}}{3s^2} \qquad (9.17)$$

Back-transforming to time domain, noting that:

$$t^n \rightarrow \frac{n!}{s^{n+1}} \quad \text{or} \quad t^2 = \frac{2!}{s^{(2+1)}} \tag{9.18}$$

$$z_{p1}(t) = \frac{-t^2}{2\sqrt{3m}} \qquad \text{Forced Response}$$

$$+0 \qquad \text{Initial Displacement} \tag{9.19}$$

$$-\frac{\sqrt{3m}}{3}t \qquad (\text{Initial Velocity}) \times (\text{Time})$$

Substituting $m = 1$, $k = 1$:

$$z_{p1} = \frac{-t^2}{2\sqrt{3}} - \frac{\sqrt{3}\,t}{3} \tag{9.20}$$

Solving for z_{p2} using Laplace transforms:

$$\ddot{z}_{p2} + \left(\frac{k}{m}\right)z_{p2} = \frac{3\sqrt{2}}{2\sqrt{m}} \tag{9.21}$$

$$s^2 z_{p2}(s) - s z_{p2}(0) - \dot{z}_{p2}(0) + \left(\frac{k}{m}\right)z_{p2}(s) = \frac{3\sqrt{2}}{2\sqrt{ms}} \tag{9.22}$$

$$s^2 z_{p2}(s) - s\left(\frac{-\sqrt{2m}}{2}\right) - \frac{\sqrt{2m}}{2} + \left(\frac{k}{m}\right)z_{p2}(s) = \frac{3\sqrt{2}}{2\sqrt{ms}} \tag{9.23}$$

$$z_{p2}(s)\left[s^2 + \left(\frac{k}{m}\right)\right] = \frac{3\sqrt{2}}{s2\sqrt{m}} - \frac{s\sqrt{2m}}{2} + \frac{\sqrt{2m}}{2} = \frac{3\sqrt{2}}{s2\sqrt{m}} + \frac{\sqrt{2m}}{2}(-s+1) \tag{9.24}$$

$$z_{p2}(s) = \frac{\dfrac{3\sqrt{2}}{2\sqrt{m}}}{s\left(s^2 + \dfrac{k}{m}\right)} - \frac{s\left(\dfrac{\sqrt{2m}}{2}\right)}{s^2 + \dfrac{k}{m}} + \frac{\dfrac{\sqrt{2m}}{2}}{s^2 + \dfrac{k}{m}}, \qquad \omega_{3,4}^2 = \frac{k}{m} \tag{9.25}$$

Back-transforming to the time domain:

$$z_{p2}(t) = \frac{3\sqrt{2}}{2\sqrt{m}}\left[\frac{1}{\omega_2^2}(1-\cos\omega_2 t)\right] - \frac{\sqrt{2m}}{2}\underbrace{\left(\frac{\omega_2}{\omega_2}\right)\sin\left(\omega_2 t + 90°\right)}_{\cos(\omega_2 t)}$$

$$+\frac{\sqrt{2m}}{2}\frac{1}{\omega_2}\sin\left(\omega_2 t\right)$$

(9.26)

Substituting $m = k = 1$, $\omega_2 = 1$:

$$z_{p2}(t) = \frac{3\sqrt{2}}{2} - \frac{3\sqrt{2}}{2}\cos(t) - \frac{\sqrt{2}}{2}\cos(t) + \frac{\sqrt{2}}{2}\sin(t)$$

(9.27)

Solving for z_{p3} using Laplace transforms:

$$\ddot{z}_{p3} + \omega_3^2 z_{p3} = \frac{-\sqrt{6}}{6\sqrt{m}}$$

(9.28)

$$s^2 z_{p3}(s) - sz_{p3}(0) - \dot{z}_{p3}(0) + \omega_3^2 z_{p3}(s) = \frac{-\sqrt{6}}{6s\sqrt{m}}$$

(9.29)

$$s^2 z_{p3}(s) + \omega_3^2 z_{p3}(s) - s\left(\frac{\sqrt{6m}}{2}\right) - \left(\frac{-7\sqrt{6m}}{\sqrt{6}}\right) = \frac{\sqrt{6}}{6s\sqrt{m}}$$

(9.30)

$$z_{p3}(s)\left(s^2 + \omega_3^2\right) = \frac{-\sqrt{6}}{6s\sqrt{m}} + \frac{s\sqrt{6m}}{2} - \frac{7\sqrt{6m}}{\sqrt{6}}$$

(9.31)

$$z_{p3}(s) = \frac{-\sqrt{6}}{6\sqrt{m}}\left[\frac{1}{s\left(s^2 + \omega_3^2\right)}\right] + \frac{\sqrt{6m}}{2}\left[\frac{s}{\left(s^2 + \omega_3^2\right)}\right] - \frac{7\sqrt{6m}}{\sqrt{6}\left(s^2 + \omega_3^2\right)}$$

(9.32)

Back-transforming to the time domain:

$$z_{p3}(t) = \frac{-\sqrt{6}}{6\sqrt{m}}\left[\frac{1}{\omega_3^2}(1 - \cos\omega_3 t)\right]$$

$$+\frac{\sqrt{6m}}{2}\underbrace{\left(\frac{\omega_3}{\omega_3}\right)\sin\left(\omega_3 t + 90°\right)}_{\cos(\omega_3 t)} - \frac{7\sqrt{6m}}{\sqrt{6}}\frac{1}{\omega_3}\sin\omega_3 t$$

(9.33)

Substituting $m = k = 1$, $\omega_3^2 = \dfrac{3k}{m}$:

$$z_{p3}(t) = \frac{-\sqrt{6}}{18} + \frac{\sqrt{6}}{18}\cos(\sqrt{3}t) + \frac{\sqrt{6}}{2}\cos(\sqrt{3}t) - \frac{7}{\sqrt{6}}\frac{\sqrt{6}}{\sqrt{3}}\sin(\sqrt{3}t)$$

$$= \frac{\sqrt{6}}{6}\left[\frac{-1}{3} + \frac{1}{3}\cos(\sqrt{3}t) + 3\cos(\sqrt{3}t) - \frac{7}{\sqrt{3}}\sin(\sqrt{3}t)\right] \tag{9.34}$$

Note: $\dfrac{-\sqrt{26}}{\sqrt{6}} - \dfrac{-\sqrt{2}\sqrt{3}\sqrt{2}\sqrt{3}\sqrt{2}}{\sqrt{3}\sqrt{2}} = -2\sqrt{3}$ and $\dfrac{\sqrt{6}}{\sqrt{3}} = \dfrac{\sqrt{2}\sqrt{3}}{\sqrt{3}} = \sqrt{2}$

Now that the displacements in principal coordinates are available, they can be plotted to see the motions of each individual mode of vibration.

Displacements in principal coordinates can be back-transformed to physical coordinates:

$$\mathbf{z} = \mathbf{z}_n\mathbf{z}_p \tag{9.35}$$

$$\mathbf{z}_p = \begin{bmatrix} z_{p1} \\ z_{p2} \\ z_{p3} \end{bmatrix} = \begin{bmatrix} \dfrac{-t^2}{2\sqrt{3}} - \dfrac{\sqrt{3}t}{3} \\[2mm] \dfrac{3\sqrt{2}}{2} - \dfrac{3\sqrt{2}}{2}\cos t - \dfrac{\sqrt{2}}{2}\cos t + \dfrac{\sqrt{2}}{2}\sin t \\[2mm] -\dfrac{\sqrt{6}}{18} + \dfrac{\sqrt{6}}{18}\cos\sqrt{3}t + \dfrac{\sqrt{6}}{2}\cos\sqrt{3}t - \dfrac{7}{\sqrt{3}}\sin\sqrt{3}t \end{bmatrix} \tag{9.36}$$

$$\mathbf{z} = \mathbf{z}_n\mathbf{z}_p = \begin{bmatrix} \dfrac{1}{\sqrt{3}} & \dfrac{1}{\sqrt{2}} & \dfrac{1}{\sqrt{6}} \\[2mm] \dfrac{1}{\sqrt{3}} & 0 & \dfrac{-2}{\sqrt{6}} \\[2mm] \dfrac{1}{\sqrt{3}} & \dfrac{-1}{\sqrt{2}} & \dfrac{1}{\sqrt{6}} \end{bmatrix} \begin{bmatrix} \dfrac{-t^2}{2\sqrt{3}} - \dfrac{\sqrt{3}t}{3} \\[2mm] \dfrac{3\sqrt{2}}{2} - \dfrac{3\sqrt{2}}{2}\cos t - \dfrac{\sqrt{2}}{2}\cos t + \dfrac{\sqrt{2}}{2}\sin t \\[2mm] -\dfrac{\sqrt{6}}{18} + \dfrac{\sqrt{6}}{18}\cos\sqrt{3}t + \dfrac{\sqrt{6}}{2}\cos\sqrt{3}t - \dfrac{7}{\sqrt{3}}\sin\sqrt{3}t \end{bmatrix}$$

$$\tag{9.37}$$

Rewriting the equations to highlight the contributions to the total motion in physical coordinates of each mode:

$$z = \begin{bmatrix} z_1 \\ z_2 \\ z_3 \end{bmatrix} = \begin{bmatrix} z_{n11} & z_{n12} & z_{n13} \\ z_{n21} & z_{n22} & z_{n23} \\ z_{n31} & z_{n32} & z_{n33} \end{bmatrix} \begin{bmatrix} z_{p1} \\ z_{p2} \\ z_{p3} \end{bmatrix} \qquad (9.38)$$

$z_1 = \underbrace{z_{n11}z_{p1}}_{\text{1st mode}} + \underbrace{z_{n12}z_{p2}}_{\text{2nd mode}} + \underbrace{z_{n13}z_{p3}}_{\text{3rd mode}}$ Mode contributions to total z_1 motion

$z_2 = \underbrace{z_{n21}z_{p1}}_{\text{1st mode}} + \underbrace{z_{n22}z_{p2}}_{\text{2nd mode}} + \underbrace{z_{n23}z_{p3}}_{\text{3rd mode}}$ Mode contributions to total z_2 motion

$z_3 = \underbrace{z_{n31}z_{p1}}_{\text{1st mode}} + \underbrace{z_{n32}z_{p2}}_{\text{2nd mode}} + \underbrace{z_{n33}z_{p3}}_{\text{3rd mode}}$ Mode contributions to total z_3 motion

$$(9.39a,b,c)$$

Because the first mode motion for each degree of freedom is rigid body, and its displacement eventually goes to infinity, it masks the vibration motion of the second and third modes for long time period simulations. If the first mode (rigid body) motion is subtracted from the total motion of z_1, z_2, and z_3, the motion due to the vibration can be seen, as shown in Figure 9.8.

9.6 MATLAB code tdof_modal_time.m – Time Domain Displacements in Physical/Principal Coordinates

9.6.1 Code Description

The MATLAB code **tdof_modal_time.m** is used to plot the displacements versus time in principal coordinates using (9.19), (9.27) and (9.34) with $m = k = 1$.   Displacements in physical coordinates are obtained by premultiplying principal displacements by the modal matrix.

9.6.2 Code Results

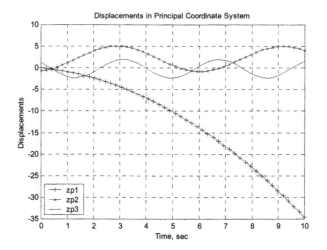

Figure 9.2: Displacements in principal coordinates, motion of the three modes of vibration.

The initial conditions in principal coordinates were 0, -0.707 and 1.225 for z_{p1}, z_{p2} and z_{p3}, respectively, which match the results shown in Figure 9.3.

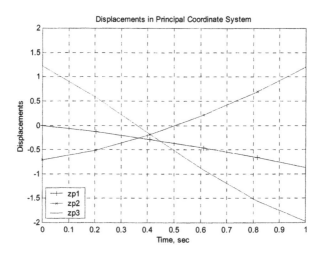

Figure 9.3: Displacements in principal coordinates, expanded vertical scale to check initial conditions.

Plotting the displacements in physical coordinates, where the initial displacement conditions in physical coordinates were 0, −1 and 1 for z_1, z_2 and z_3, respectively.

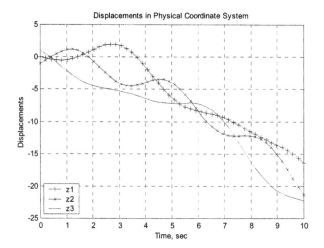

Figure 9.4: Displacement in physical coordinates.

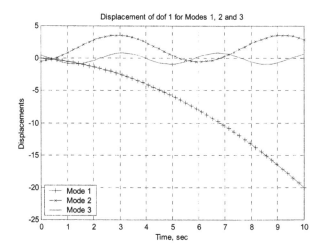

Figure 9.5: Displacements of mass 1 for all three modes of vibration.

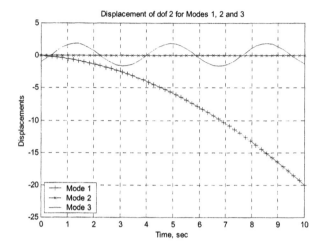

Figure 9.6: Displacements of mass 2 for all three modes of vibration.

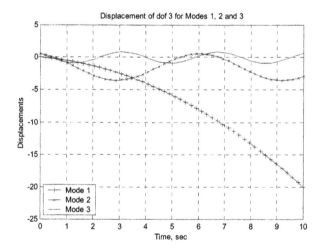

Figure 9.7: Displacements of mass 3 for all three modes of vibration.

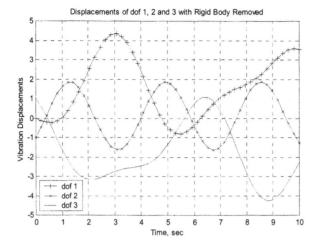

Figure 9.8: Displacements in physical coordinates, with the rigid body motion removed to show more clearly the oscillatory motion of the three masses.

9.6.3 Code Listing

```
%          tdof_modal_time.m    hand solution of modal equations

           clf;

           clear all;

%          define time vector for plotting responses

           t = linspace(0,10,50);

%          solve for and plot the modal displacements

           zp1 = (-t.^2/(2*sqrt(3))) - sqrt(3)*t/3;

           zp2 = 3*sqrt(2)/2 - (3*sqrt(2)/2)*cos(t) - (sqrt(2)/2)*cos(t) + (sqrt(2)/2)*sin(t);

           zp3 = (sqrt(6)/6)*((-1/3) + (1/3)*cos(sqrt(3)*t) + 3*cos(sqrt(3)*t) - ...
                          (7/sqrt(3))*sin(sqrt(3)*t));

           plot(t,zp1,'k+-',t,zp2,'kx-',t,zp3,'k-')
           title('Displacements in Principal Coordinate System')
           xlabel('Time, sec')
           ylabel('Displacements')
           legend('zp1','zp2','zp3',3)
           grid

           disp('execution paused to display figure, "enter" to continue'); pause
```

```
            axis([0 1 -2 2])

            disp('execution paused to display figure, "enter" to continue'); pause

%           define the normalized modal matrix, m = 1

            zn = [1/sqrt(3)  1/sqrt(2)  1/sqrt(6)
                  1/sqrt(3)    0         -2/sqrt(6)
                  1/sqrt(3) -1/sqrt(2)  1/sqrt(6)];

%           define the principal displacement matrix, column vectors of principal displacements
%           at each time step

            zp = [zp1; zp2; zp3];

%           multiply zn times zp to get z

            z = zn*zp;

            z1 = z(1,:);
            z2 = z(2,:);
            z3 = z(3,:);

            plot(t,z1,'k+-',t,z2,'kx-',t,z3,'k-')
            title('Displacements in Physical Coordinate System')
            xlabel('Time, sec')
            ylabel('Displacements')
            legend('z1','z2','z3',3)
            grid

            disp('execution paused to display figure, "enter" to continue'); pause

%           define the motion of each each dof for each mode, zij below refers to the
%           motion of dof "i" due to mode "j"

            z11 = zn(1,1)*zp1;

            z12 = zn(1,2)*zp2;

            z13 = zn(1,3)*zp3;

            z21 = zn(2,1)*zp1;

            z22 = zn(2,2)*zp2;

            z23 = zn(2,3)*zp3;

            z31 = zn(3,1)*zp1;

            z32 = zn(3,2)*zp2;

            z33 = zn(3,3)*zp3;

            plot(t,z11,'k+-',t,z12,'kx-',t,z13,'k-')
            title('Displacement of dof 1 for Modes 1, 2 and 3')
```

```
          xlabel('Time, sec')
          ylabel('Displacements')
          legend('Mode 1','Mode 2','Mode 3',3)
          grid

          disp('execution paused to display figure, "enter" to continue'); pause

          plot(t,z21,'k+-',t,z22,'kx-',t,z23,'k-')
          title('Displacement of dof 2 for Modes 1, 2 and 3')
          xlabel('Time, sec')
          ylabel('Displacements')
          legend('Mode 1','Mode 2','Mode 3',3)
          grid

          disp('execution paused to display figure, "enter" to continue'); pause

          plot(t,z31,'k+-',t,z32,'kx-',t,z33,'k-')
          title('Displacement of dof 3 for Modes 1, 2 and 3')
          xlabel('Time, sec')
          ylabel('Displacements')
          legend('Mode 1','Mode 2','Mode 3',3)
          grid

          disp('execution paused to display figure, "enter" to continue'); pause

%         define the motion of each each dof with the rigid body motion for that
%         mode subtracted

          z1vib = z1 - z11;

          z2vib = z2 - z21;

          z3vib = z3 - z31;

          plot(t,z1vib,'k+-',t,z2vib,'kx-',t,z3vib,'k-')
          title('Displacements of dof 1, 2 and 3 with Rigid Body Removed')
          xlabel('Time, sec')
          ylabel('Vibration Displacements')
          legend('dof 1','dof 2','dof 3',3)
          grid

          disp('execution paused to display figure, "enter" to continue'); pause

          tplot = t;

          plot(tplot,z1,'k+-',t,z2,'kx-',t,z3,'k-')
          title('Displacements of dof 1, 2 and 3')
          xlabel('Time, sec')
          ylabel('Vibration Displacements')
          legend('dof 1','dof 2','dof 3',3)
          grid

          disp('execution paused to display figure, "enter" to continue'); pause

          save tdof_modal_time_z1z2z3  tplot z1 z2 z3
```

Problems

Note: All the problems refer to the two dof system shown in Figure P2.2.

P9.1 Using the equations, initial conditions and forcing functions from P7.4, solve for the closed form time domain response in principal coordinates using Laplace transforms. Back transform to physical coordinates and identify the components of the response associated with each mode.

P9.2 (MATLAB) Modify the **tdof_modal_time.m** code for the two dof system and solve for the time domain responses in both principal and physical coordinates using the equations, initial conditions and forcing functions from P7.4.

CHAPTER 10

MODAL ANALYSIS: STATE SPACE FORM

10.1 Introduction

In Chapters 5, 6 and 7 we developed the state space (first order differential equation) form of the equations of motion and used them to solve for the eigenvalues and eigenvectors (with real **or** complex modes) and frequency and transient responses. The state space methodology presented so far was independent of the amount of damping in the system, hence the possibility of complex modes.

In Chapters 8 and 9 we developed the modal analysis method using the second order differential equation form. If the amount of damping in the system is low, we can make the approximation that normal modes exist and solve for the undamped (real) modes of the system. Proportional damping can then be added to the equations of motion in principal coordinates while keeping the equations uncoupled.

In the next three chapters we will combine the state space techniques in Chapters 5, 6 and 7 with the modal analysis techniques in Chapters 8 and 9. In real world situations, finite element models are used to describe dynamic systems. The finite element program is used to solve for eigenvalues and eigenvectors, which are then used to create a state space model in MATLAB. However, one may have the need to solve for eigenvalues and eigenvectors in state space form for a model that is not created using finite elements. For this reason, the chapter will start out with a closed form solution to the tdof eigenvalue problem in state space form. The eigenvalues and eigenvectors which result from the state space eigenvalue problem will contain the same information as in the second order eigenvalue problem, but will be in a different form. The differences will be highlighted and discussed.

We then will use the eigenvalues to form the uncoupled homogeneous equations of motion in the state space principal coordinate system by inspection. Forcing function and initial conditions will then be converted to principal coordinates using the normalized modal matrix, creating the final state equations of motion in the principal coordinate system. As in the second order form, proportional damping can be added to the modal formulation and the solution in principal coordinates back-transformed to physical coordinates for the final result. We will use a method of formulating the input matrix **B** such that the transformation of forces to principal coordinates and conversion to state space form can happen in one step instead of two. A similar formulation will be developed for the output matrix **C**, where we will define the output vector and convert back to physical coordinates in one step. The method described here can be used for both transient and frequency response solutions.

One might ask why we are going to all the trouble of doing a state space version of modal analysis. Chapter 5 showed that given the state space equations of motion of a system, we can use MATLAB to solve for both frequency and time domain responses without knowing anything about eigenvalues and eigenvectors. The reason we are going to this trouble is that most mechanical simulations are performed using finite element techniques, where the equations of motion are too numerous to be able to be used directly in MATLAB or in a servo system simulation. Since modal analysis results, the eigenvalues and eigenvectors, are available from an ANSYS eigenvalue solution, it would be nice if we knew how to use these results by developing them into a MATLAB state space model. We could then use the power of MATLAB to perform any further analysis.

The techniques described above can be further extended by taking the results set from a large finite element problem and defining a small state space model that accurately describes the pertinent dynamics of the system (Chapters 15 to 19). The small MATLAB state space model can then be used in lieu of the frequency and transient analysis capabilities in the finite element program. The MATLAB state space model can also be combined with a servo system model, allowing complete servo-mechanical system simulations.

10.2 Eigenvalue Problem

We start with the undamped homogeneous equations of motion in state space form:

$$\dot{x} = Ax \tag{10.1}$$

In Chapter 7 we defined a normal mode as:

$$x_i = x_{mi} \sin\left(\omega_i t + \phi_i\right) = x_{mi} \, \text{Im}(e^{j\omega_i t + \phi_i}) \tag{10.2}$$

For our three degree of freedom (z_1 to z_3), six state (x_1 to x_6) system, for the ith eigenvalue and eigenvector, the equation would appear as:

$$\begin{bmatrix} z_{1i} \\ \dot{z}_{1i} \\ z_{2i} \\ \dot{z}_{2i} \\ z_{3i} \\ \dot{z}_{3i} \end{bmatrix} = \begin{bmatrix} x_{1i} \\ x_{2i} \\ x_{3i} \\ x_{4i} \\ x_{5i} \\ x_{6i} \end{bmatrix} = \mathbf{x}_{mi} \sin\left(\omega_i t + \phi_i\right) = \begin{bmatrix} x_{m1i} \\ x_{m2i} \\ x_{m3i} \\ x_{m4i} \\ x_{m5i} \\ x_{m6i} \end{bmatrix} \sin\left(\omega_i t + \phi_i\right) \qquad (10.3)$$

Differentiating the modal displacement equation above to get the modal velocity equation:

$$\begin{aligned} \frac{d}{dt}\left[\mathbf{x}_{mi} \sin\left(\omega_i t + \phi_i\right)\right] &= \mathbf{x}_{mi} \frac{d}{dt} \operatorname{Im}\left[e^{j(\omega_i t + \phi_i)}\right] \\ &= \mathbf{x}_{mi} \operatorname{Im}\left[j\omega_i e^{j(\omega_i t + \phi_i)}\right] \\ &= \mathbf{x}_{mi} \operatorname{Im}\left[j\omega_i \left(\cos(\omega_i t + \phi_i) + j\sin(\omega_i t + \phi_i)\right)\right] \qquad (10.4) \\ &= \mathbf{x}_{mi} \operatorname{Im}\left[j\omega_i \cos(\omega_i t + \phi_i) - \omega_i \sin(\omega_i t + \phi_i)\right] \\ &= \mathbf{x}_{mi} \omega_i \cos(\omega_i t + \phi_i) \end{aligned}$$

Substituting the derivatives into the state equation we arrive at the eigenvalue problem:

$$\begin{aligned} \dot{\mathbf{x}} &= \mathbf{A}\mathbf{x} \\ j\omega_i \mathbf{x}_{mi} \sin\left(\omega_i t + \phi_i\right) &= \mathbf{A}\mathbf{x}_{mi} \sin\left(\omega_i t + \phi_i\right) \\ j\omega_i \mathbf{x}_{mi} &= \mathbf{A}\mathbf{x}_{mi} \\ (j\omega_i \mathbf{I} - \mathbf{A})\mathbf{x}_{mi} &= 0 \end{aligned} \qquad (10.5)$$

10.3 Eigenvalue Problem – Laplace Transform

We can also use Laplace transforms to define the eigenvalue problem. Taking the matrix Laplace transform of the homogeneous state equation and solving for $\mathbf{x}(s)$:

$$\begin{aligned} s\mathbf{I}\mathbf{x}(s) &= \mathbf{A}\mathbf{x}(s) \\ (s\mathbf{I} - \mathbf{A})\mathbf{x}(s) &= 0 \end{aligned} \qquad (10.6a,b)$$

This is another form of the eigenvalue problem, again where the determinant of the term $(s\mathbf{I} - \mathbf{A})$ has to equal zero to have anything other than a trivial solution.

$$\left|(s\mathbf{I} - \mathbf{A})\right| = 0 \tag{10.7}$$

Letting $m_1 = m_2 = m_3 = m$, $k_1 = k_2 = k$, $c_1 = c_2 = 0$ and rewriting the matrix equations of motion to match the original undamped problem used in (6.8).

$$(s\mathbf{I} - \mathbf{A}) = \begin{bmatrix} s & -1 & 0 & 0 & 0 & 0 \\ \dfrac{k}{m} & s & \dfrac{-k}{m} & 0 & 0 & 0 \\ 0 & 0 & s & -1 & 0 & 0 \\ \dfrac{-k}{m} & 0 & \dfrac{2k}{m} & s & \dfrac{-k}{m} & 0 \\ 0 & 0 & 0 & 0 & s & -1 \\ 0 & 0 & \dfrac{-k}{m} & 0 & \dfrac{k}{m} & s \end{bmatrix} \tag{10.8}$$

In Section 6.3 we used this form of the equation to find the state space transfer function matrix, where we needed the inverse of $(s\mathbf{I} - \mathbf{A})$. Here we need the determinant of $(s\mathbf{I} - \mathbf{A})$. Using a symbolic algebra program results in the following characteristic equation:

$$s^2(m^3s^4 + 4m^2ks^2 + 3mk^2) = 0$$
$$m^3s^6 + 4m^2ks^4 + 3mk^2s^2 = 0 \tag{10.9a,b}$$

This is the same equation we found in (2.58) for the characteristic equation, whose roots were found to be the poles (eigenvalues). Repeating from Chapter 2, (2.67), (2.70) and (2.71):

$$s_{1,2} = 0$$

$$s_{3,4} = \pm j\sqrt{\dfrac{k}{m}} = \pm j1$$

$$s_{5,6} = \pm j\sqrt{\dfrac{3k}{m}} = \pm j1.732 \tag{10.10a,b,c}$$

In Chapter 5, the state space chapter, we showed that for arbitrary damping the eigenvalues would be complex numbers with both real and imaginary components, where the real part was indicative of there being damping in the system as the poles were offset to the left of the imaginary axis (Figure 5.3).

We defined the damped eigenvalues as $(\lambda_{n1,n2} = \sigma_{n1} \pm j\omega_{n1})$ (5.48). Note for the undamped eigenvalues above, the σ values are zero, with all poles lying on the imaginary axis.

10.4 Eigenvalue Problem – Eigenvectors

Let us now solve for the eigenvectors in state space form, going back to the original equations of motion for the ith mode, similar to (10.5):

$$
\begin{bmatrix}
j\omega_i & -1 & 0 & 0 & 0 & 0 \\
\dfrac{k}{m} & j\omega_i & \dfrac{-k}{m} & 0 & 0 & 0 \\
0 & 0 & j\omega_i & -1 & 0 & 0 \\
\dfrac{-k}{m} & 0 & \dfrac{2k}{m} & j\omega_i & \dfrac{-k}{m} & 0 \\
0 & 0 & 0 & 0 & j\omega_i & -1 \\
0 & 0 & \dfrac{-k}{m} & 0 & \dfrac{k}{m} & j\omega_i
\end{bmatrix}
\begin{bmatrix}
x_{m1i} \\
x_{m2i} \\
x_{m3i} \\
x_{m4i} \\
x_{m5i} \\
x_{m6i}
\end{bmatrix}
= 0 \qquad (10.11)
$$

Expanding the equations:

$$j\omega_i x_{m1i} - x_{m2i} = 0$$

$$\frac{k}{m} x_{m1i} + j\omega_i x_{m2i} - \frac{k}{m} x_{m3i} = 0$$

$$j\omega_i x_{m3i} - x_{m4i} = 0$$

$$-\frac{k}{m} x_{m1i} + \frac{2k}{m} x_{m3i} + j\omega_i x_{m4i} - \frac{k}{m} x_{m5i} = 0 \qquad (10.12a\text{-}f)$$

$$j\omega_i x_{m5i} - x_{m6i} = 0$$

$$-\frac{k}{m} x_{m3i} + \frac{k}{m} x_{m5i} + j\omega_i x_{m6i} = 0$$

Dropping the "m" and "i" terms from the eigenvectors:

$$j\omega_i x_1 - x_2 = 0$$

$$\frac{k}{m}x_1 + j\omega_i x_2 - \frac{k}{m}x_3 = 0$$

$$j\omega_i x_3 - x_4 = 0$$

$$-\frac{k}{m}x_1 + \frac{2k}{m}x_3 + j\omega_i x_4 - \frac{k}{m}x_5 = 0 \qquad (10.13\text{a-f})$$

$$j\omega_i x_5 - x_6 = 0$$

$$-\frac{k}{m}x_3 + \frac{k}{m}x_5 + j\omega_i x_6 = 0$$

Selecting the first state, x_1, as a reference and solving for x_2 through x_6 in terms of x_1.

Solving for x_2 from (10.13a):

$$j\omega_i x_1 - x_2 = 0$$

$$x_2 = j\omega_i x_1 \qquad (10.14)$$

$$\frac{x_2}{x_1} = j\omega_i$$

Solving for x_3 from (10.13b):

$$\frac{k}{m}x_1 + j\omega_i x_2 - \frac{k}{m}x_3 = 0$$

$$j\omega_i (j\omega_i x_1) - \frac{k}{m}x_3 = -\frac{k}{m}x_1$$

$$-\frac{k}{m}x_3 = -\frac{k}{m}x_1 + \omega_i^2 x_1 \qquad (10.15)$$

$$x_3 = \left(\frac{k - \omega_i^2 m}{k}\right) x_1$$

$$\frac{x_3}{x_1} = \frac{k - \omega_i^2 m}{k}$$

Solving for x_4 from (10.13c):

$$j\omega_i x_3 - x_4 = 0$$

$$j\omega_i \left(\frac{k - \omega_i^2 m}{k} \right) x_1 - x_4 = 0$$

$$x_4 = j\omega_i \left(\frac{k - \omega_i^2 m}{k} \right) x_1 = j\omega_i x_3 \qquad (10.16)$$

$$\frac{x_4}{x_1} = j\omega_i \left(\frac{k - \omega_i^2 m}{k} \right) = j\omega_i \frac{x_3}{x_1}$$

Solving for x_5 from (10.13d):

$$-\frac{k}{m} x_1 + \frac{2k}{m} x_3 + j\omega_i x_4 - \frac{k}{m} x_5 = 0$$

$$-\frac{k}{m} x_1 + \frac{2k}{m} \left(\frac{k - \omega_i^2 m}{k} \right) x_1$$

$$+ j\omega_i \left[j\omega_i \left(\frac{k - \omega_i^2 m}{k} \right) \right] x_1 - \frac{k}{m} x_5 = 0$$

$$x_5 = \left(\frac{m^2 \omega_i^4 - 3mk\omega_i^2 + k^2}{k^2} \right) x_1$$

$$\frac{x_5}{x_1} = \frac{m^2 \omega_i^4 - 3mk\omega_i^2 + k^2}{k^2}$$

$$(10.17)$$

Solving for x_6 from (10.13e):

$$j\omega_i x_5 - x_6 = 0$$

$$x_6 = j\omega_i \left(\frac{m^2 \omega_i^4 - 3mk\omega_i^2 + k^2}{k^2} \right) x_1 = j\omega_i x_5 \qquad (10.18)$$

$$\frac{x_6}{x_1} = j\omega_i \left(\frac{m^2 \omega_i^4 - 3mk\omega_i^2 + k^2}{k^2} \right) = j\omega_i \frac{x_5}{x_1}$$

Note that the results for the displacement eigenvector components in (10.15) and (10.17) match the two displacement eigenvectors calculated in (7.24) and (7.29), respectively. Also note that all three velocity eigenvector components are equal to $j\omega_i$ times their respective displacement eigenvector components.

Unlike the complex eigenvectors found in Chapter 5 for the damped model, these undamped eigenvector displacement states are all real; they have no complex terms.

10.5 Modal Matrix

We will see that when we transform to principal coordinates, create the state equations in principal coordinates and back transform results to physical coordinates we only require a 3x3 displacement modal matrix. This is because we can transform positions and velocities separately. The modal matrix (7.46) and normalized modal matrix (7.77) are repeated below, again for $m = k = 1$:

$$\mathbf{z}_m = \begin{bmatrix} 1 & 1 & 1 \\ 1 & 0 & -2 \\ 1 & -1 & 1 \end{bmatrix} \tag{10.19}$$

$$\mathbf{z}_n = \frac{1}{\sqrt{m}} \begin{bmatrix} \dfrac{1}{\sqrt{3}} & \dfrac{1}{\sqrt{2}} & \dfrac{1}{\sqrt{6}} \\ \dfrac{1}{\sqrt{3}} & 0 & \dfrac{-2}{\sqrt{6}} \\ \dfrac{1}{\sqrt{3}} & \dfrac{-1}{\sqrt{2}} & \dfrac{1}{\sqrt{6}} \end{bmatrix} = \begin{bmatrix} 0.5774 & -0.707 & 0.4082 \\ 0.5774 & 0 & -0.8165 \\ 0.5774 & 0.707 & 0.4082 \end{bmatrix} \tag{10.20}$$

10.6   MATLAB Code tdofss_eig.m: Solving for Eigenvalues and Eigenvectors

10.6.1 Code Description

The MATLAB code **tdofss_eig.m** solves for the eigenvalues and eigenvectors in the state space form of the system. The code will be listed in sections with commented results and explanations following each section.

10.6.2 Eigenvalue Calculation

```
%   tdofss_eig.m eigenvalue problem solution for tdof undamped model

    clear all;

%   define the values of masses, springs, dampers and forces

    m1 = 1;
    m2 = 1;
```

```
      m3 = 1;

      c1 = 0;
      c2 = 0;

      k1 = 1;
      k2 = 1;

%    define the system matrix, a

      a = [ 0          1          0          0          0          0
           -k1/m1    -c1/m1     k1/m1      c1/m1      0          0
            0          0          0          1          0          0
            k1/m2     c1/m2    -(k1+k2)/m2 -(c1+c2)/m2 k2/m2      c2/m2
            0          0          0          0          0          1
            0          0          k2/m3      c2/m3     -k2/m3     -c2/m3];

%    solve for the eigenvalues of the system matrix

      [xm,omega] = eig(a)
```

The resulting eigenvalues, in units of rad/sec, are below. Note that MATLAB uses "i" for imaginary numbers instead of "j" which is used in the text.

```
omega =
  Columns 1 through 4
     0 + 1.7321i      0              0              0
     0                0 - 1.7321i    0              0
     0                0              0              0
     0                0              0              0 + 1.0000i
     0                0              0              0
     0                0              0              0
  Columns 5 through 6
     0                0
     0                0
     0                0
     0                0
     0 - 1.0000i      0
     0                0
```

The eigenvalues, what MATLAB calls "generalized eigenvalues," are the diagonal elements of the omega matrix. The six values that MATLAB calculates are: $1.7321i$, $-1.732i$, 0, $1.0000i$, $-1.0000i$, 0, in that order. These are the same values we found using our closed form calculations. Also, the values are all imaginary, as we would expect with a system with no damping and as we found above from our $|(s\mathbf{I} - \mathbf{A})| = 0$ derivation.

10.6.3 Eigenvector Calculation

The resulting eigenvectors, directly from MATLAB output are:

```
xm =
 Columns 1 through 4
  0.2041          0.2041          0.5774        0 + 0.5000i
  0 + 0.3536i     0 - 0.3536i     0             -0.5000
  -0.4082         -0.4082         0.5774        0 + 0.0000i
  0 - 0.7071i     0 + 0.7071i     0             0.0000
  0.2041          0.2041          0.5774        0 - 0.5000i
  0 + 0.3536i     0 - 0.3536i     0             0.5000
 Columns 5 through 6
  0 - 0.5000i     -0.5774
  -0.5000         0.0000
  0 - 0.0000i     -0.5774
  0.0000          0.0000
  0 + 0.5000I     -0.5774
  0.5000          0.0000
```

Note that unlike the eigenvectors calculated in the Modal Analysis section, which had three rows, these eigenvectors each have six rows, each row corresponding to its respective state. Repeating the state definitions from (5.4) to (5.9):

$$x_1 = z_1 \quad \text{Position of Mass 1}$$

$$x_2 = \dot{z}_1 \quad \text{Velocity of Mass 1}$$

$$x_3 = z_2 \quad \text{Position of Mass 2}$$

$$x_4 = \dot{z}_2 \quad \text{Velocity of Mass 2}$$

$$x_5 = z_3 \quad \text{Position of Mass 3}$$

$$x_6 = \dot{z}_3 \quad \text{Velocity of Mass 3}$$

Thus, the first, third and fifth rows represent the positions of the three masses for each mode, and the second, fourth and sixth rows represent the velocities of the three masses for each mode. Separating into position and velocity components:

```
xm(position) =
 0.2041      0.2041      0.5774      0 + 0.5000i     0 - 0.5000i     -0.5774
 -0.4082     -0.4082     0.5774      0 + 0.0000i     0 - 0.0000i     -0.5774
 0.2041      0.2041      0.5774      0 - 0.5000i     0 + 0.5000i     -0.5774

xm(velocity) =
 0 + 0.3536i     0 - 0.3536i     0     -0.5000     -0.5000     0.0000
 0 - 0.7071i     0 + 0.7071i     0     0.0000      0.0000      0.0000
 0 + 0.3536i     0 - 0.3536i     0     0.5000      0.5000      0.0000
```

What is the relationship between the position and velocity terms in each of the eigenvectors? Once again, knowing that at each undamped frequency a normal mode exists and that the position and velocity can be defined as:

$$\mathbf{z}_i = \mathbf{z}_{ni} e^{j\omega_i t}$$

$$\dot{\mathbf{z}}_i = j\omega_i \mathbf{z}_{ni} e^{j\omega_i t}$$

(10.21a,b)

Taking the amplitudes of the position and velocity:

$$\left| \dot{\mathbf{z}}_n \right| = \omega \left| \mathbf{z}_n \right|$$

(10.22)

The amplitude of the velocity eigenvector terms should be equal to the eigenvalue times its respective position eigenvector term. The fact that the velocity entries are complex numbers by virtue of multiplying the "real" position eigenvector entries by the eigenvalue does not make the eigenvectors "complex," but refers to the fact that in the undamped case velocity is 90° out of phase with position.

Checking the first eigenvector by multiplying the position term (state 1) by the eigenvalue to get the velocity term (state 2): (highlighted in bold type above)

$$0.2041 * 1.7321j = .3535j$$

(10.23)

Note that for the third and sixth eigenvectors, which have zero eigenvalues, the velocity entries are zero because the position entry is multiplied by zero.

10.6.4 MATLAB Eigenvectors – Real and Imaginary Values

It is interesting to see how MATLAB handles real and imaginary values in its eigenvectors.

```
xm =
  0.2041        0.2041       0.5774    0 + 0.5000i    0 - 0.5000i    -0.5774
  0 + 0.3536i   0 - 0.3536i  0          -0.5000        -0.5000        0.0000
  -0.4082       -0.4082      0.5774    0 + 0.0000i    0 - 0.0000i    -0.5774
  0 - 0.7071i   0 + 0.7071i  0          0.0000         0.0000         0.0000
  0.2041        0.2041       0.5774    0 - 0.5000i    0 + 0.5000i    -0.5774
  0 + 0.3536i   0 - 0.3536i  0          0.5000         0.5000         0.0000
```

We know that the position and velocity entries are related by "j" times the eigenvalue, but why are some position eigenvector entries real and some imaginary? For example, the position eigenvector entries for all except the mode at 1 rad/sec (the fourth and fifth columns), are real, while the fourth and fifth column position entries are imaginary. From the original normal modes analysis, we know that only the **ratios** of eigenvector entries are important, and that the eigenvectors can be normalized in several fashions. Therefore, each eigenvector can be multiplied by an arbitrary constant. The fourth and fifth eigenvectors can be multiplied by " i " to make their position entries real for consistency with the hand-calculated results.

10.6.5 Sorting Eigenvalues / Eigenvectors

Typically some housekeeping is done on the eigenvalues and eigenvectors before continuing, sorting the eigenvalues from small to large (done by default in ANSYS), rearranging the eigenvectors accordingly and checking for eigenvectors with imaginary position entries and converting them to real by multiplying by " i ." Also, the signs of the real portion of state 1 are set positive to ensure that sets of eigenvectors are complex conjugates of each other for consistency.

Continuing the listing of **tdofss_eig.m**, showing the sorting code:

```
%   take the diagonal elements of the generalized eigenvalue matrix omega

    omegad = diag(omega);

%   in real problems, we would next convert to hz from radians/sec

    omegahz = omegad/(2*pi);

%   now reorder the eigenvalues and eigenvectors from low to high frequency,
%   keeping track of how the eigenvalues are ordered to reorder the
%   eigenvectors to match, using indexhz

    [omegaorder,indexhz] = sort(abs(imag(omegad)))

    for  cnt = 1:length(omegad)

        omegao(cnt,1) = omegad(indexhz(cnt));% reorder eigenvalues

        xmo(:,cnt) = xm(:,indexhz(cnt));    % reorder eigenvector columns

    end

    omegao

    xmo

%   check for any eigenvectors with imaginary position elements by checking
%   the first three position entries for each eigenvector (first, third and
%   and fifth rows) and convert to real
```

```
for  cnt = 1:length(omegad)

if  (real(xmo(1,cnt)) & real(xmo(3,cnt)) & real(xmo(5,cnt))) == 0

    xmo(:,cnt) = i*(xmo(:,cnt));    % convert whole column if imaginary

else

end

end

xmo
```

```
%   check for any eigenvectors with negative position elements for the first
%   displacement, if so change to positive to that eigenvectors for the same mode
%   are complex conjugates

for  cnt = 1:length(omegad)

    if  real(xmo(1,cnt)) < 0

    xmo(:,cnt) = -1*(xmo(:,cnt));   % convert whole column if negative

    else

    end

end

xmo
```

Printing the results of the MATLAB reordering:

```
omegaorder =
     0
     0
  1.0000        These are the re-ordered eigenvalues, from low to high.
  1.0000
  1.7321
  1.7321
indexhz =
   3
   6
   4          This is the ordering of the original eigenvalues.
   5
   1
   2
omegao =
     0
     0
   0 + 1.0000i
   0 - 1.0000i
   0 + 1.7321i
   0 - 1.7321i
```

Here are the reordered eigenvectors.

```
xmo =
  Columns 1 through 4
    0.5774        -0.5774       0 + 0.5000i      0 - 0.5000i
    0              0.0000      -0.5000          -0.5000
    0.5774        -0.5774       0 + 0.0000i      0 - 0.0000i
    0              0.0000       0.0000           0.0000
    0.5774        -0.5774       0 - 0.5000i      0 + 0.5000i
    0              0.0000       0.5000           0.5000
  Columns 5 through 6
    0.2041         0.2041
    0 + 0.3536i    0 - 0.3536i
   -0.4082        -0.4082
    0 - 0.7071i    0 + 0.7071i
    0.2041         0.2041
    0 + 0.3536i    0 - 0.3536i
```

Here the converting of imaginary position values to real is performed, note that the third and fourth eigenvectors are converted.

```
xmo =
Columns 1 through 4
   0.5774        -0.5774        -0.5000        0.5000
   0              0.0000        0 - 0.5000i     0 - 0.5000i
   0.5774        -0.5774        0.0000         0.0000
   0              0.0000        0 - 0.0000i     0 - 0.0000i
   0.5774        -0.5774        0.5000         -0.5000
   0              0.0000        0 + 0.5000i     0 + 0.5000i
Columns 5 through 6
   0.2041        0.2041
   0 + 0.3536i    0 - 0.3536i
   -0.4082       -0.4082
   0 - 0.7071i    0 + 0.7071i
   0.2041        0.2041
   0 + 0.3536i    0 - 0.3536i
```

In this step the first row elements are checked to see that they are positive; if not, the column is multiplied by –1.

```
xmo =

Columns 1 through 4

   0.5774        0.5774         0.5000         0.5000
   0             -0.0000        0 + 0.5000i     0 - 0.5000i
   0.5774        0.5774         0.0000         0.0000
   0             -0.0000        0 + 0.0000i     0 - 0.0000i
   0.5774        0.5774         -0.5000        -0.5000
   0             -0.0000        0 - 0.5000i     0 + 0.5000i

Columns 5 through 6

   0.2041        0.2041
   0 + 0.3536i    0 - 0.3536i
   -0.4082       -0.4082
   0 - 0.7071i    0 + 0.7071i
   0.2041        0.2041
   0 + 0.3536i    0 - 0.3536i
```

10.6.6 Normalizing Eigenvectors

Now that the eigenvalues and eigenvectors are available, we can normalize the eigenvectors with respect to mass.   Then we will check the resulting diagonalization by multiplying the original mass and stiffness matrices by the normalized eigenvectors to see if the mass matrix becomes the identity matrix and the stiffness matrix becomes a diagonal matrix with squares of the eigenvalues on the diagonal (spectral matrix).

Since we need to deal only with the displacement entries of the 6x6 modal matrix in order to transform the 3x3 mass and stiffness matrices, the x_m matrix below is a 3x3 matrix with only displacement entries.

Reviewing, the mass matrix is diagonalized by pre- and postmultiplying by the normalized eigenvector matrix:

$$x_n^T m x_n = I ,$$

(10.24)

yielding the identity matrix. The stiffness matrix is also diagonalized by pre- and postmultiplying by the normalized eigenvector matrix:

$$x_n^T k x_n = k_p ,$$

(10.25)

yielding the stiffness matrix in principal coordinates, the spectral matrix, a diagonal matrix with squares of the eigenvalues on the diagonal.

Repeating from Section 7.4.2, the normalized modal matrix x_n is made up of eigenvectors as defined below:

$$x_{ni} = \frac{x_{mi}}{\left[x_{mi}^T m x_{mi} \right]^{\frac{1}{2}}} = \frac{x_{mi}}{q_i}$$

(10.26)

Where q_i is defined as:

$$q_i = \left[\sum_{j=i}^{n} x_{mji} \left(\sum_{k=1}^{n} m_{jk} x_{mki} \right) \right]^{\frac{1}{2}}$$

(10.27)

For a diagonal mass matrix, simplifying q because all the mjk terms are zero:

$$q_i = \left[\sum_{k=1}^{n} m_k x_{mki}^2 \right]^{\frac{1}{2}}$$

(10.28)

Continuing with code from tdofss_eig.m:

```
%   define the mass and stiffness matrices for normalization of eigenvectors
%   and for checking values in principal coordinates

    m = [m1    0    0
```

```
              0    m2    0
              0     0   m3];

      k = [ kl      -kl       0
           -kl     kl+k2     -k2
            0      -k2       k2];
```

%   define the position eigenvectors by taking the first, third and fifth
%   rows of the original six rows in xmo

    xmop1 = [xmo(1,:); xmo(3,:); xmo(5,:)]

%   define the three eigenvectors for the three degrees of freedom by taking
%   the second, fourth and sixth columns

    xmop = [xmop1(:,2) xmop1(:,4) xmop1(:,6)]

%   normalize with respect to mass

    for mode = 1:3

        xn(:,mode) = xmop(:,mode)/sqrt(xmop(:,mode)'*m*xmop(:,mode));

    end

    xn

%   calculate the normalized mass and stiffness matrices for checking

    mm = xn'*m*xn

    km = xn'*k*xn

%   check that the sqrt of diagonal elements of km are eigenvalues

    p = (diag(km)).^0.5;

    [p abs(imag(omegao(1:2:5,:)))]

%   rename the three eigenvalues for convenience in later calculations

    w1 = abs(imag(omegao(1)));

    w2 = abs(imag(omegao(3)));

    w3 = abs(imag(omegao(5)));

Back to MATLAB output, with comments added in bold type:

Repeating xmo, the full, rearranged eigenvector matrix:

xmo =

Columns 1 through 4

```
    0.5774          0.5774          0.5000          0.5000
    0               -0.0000         0 + 0.5000i     0 - 0.5000i
    0.5774          0.5774          0.0000          0.0000
    0               -0.0000         0 + 0.0000i     0 - 0.0000i
    0.5774          0.5774          -0.5000         -0.5000
    0               -0.0000         0 - 0.5000i     0 + 0.5000i
```

Columns 5 through 6

```
    0.2041          0.2041
    0 + 0.3536i     0 - 0.3536i
    -0.4082         -0.4082
    0 - 0.7071i     0 + 0.7071i
    0.2041          0.2041
    0 + 0.3536i     0 - 0.3536i
```

Taking only the position rows:
xmop1 =

```
    0.5774   0.5774   0.5000    0.5000    0.2041    0.2041
    0.5774   0.5774   0.0000    0.0000   -0.4082   -0.4082
    0.5774   0.5774  -0.5000   -0.5000    0.2041    0.2041
```

Taking every other column to form the 3x3 position eigenvector matrix:
xmop =

```
    0.5774    0.5000    0.2041
    0.5774    0.0000   -0.4082
    0.5774   -0.5000    0.2041
```

Normalizing with respect to mass:
xn =

```
    0.5774    0.7071    0.4082
    0.5774    0.0000   -0.8165
    0.5774   -0.7071    0.4082
```

Checking the mass matrix in principal coordinates, should be the identity matrix:
mm =

```
    1.0000   -0.0000    0.0000
   -0.0000    1.0000   -0.0000
    0.0000   -0.0000    1.0000
```

Checking the stiffness matrix in principal coordinates, should be squares of eigenvalues:
km =

```
    0.0000   -0.0000    0.0000
   -0.0000    1.0000   -0.0000
    0.0000   -0.0000    3.0000
```

Comparing the square root of the diagonal elements of the stiffness matrix in principal coordinates with the eigenvalues:
ans =

```
0.0000      0
1.0000   1.0000
1.7321   1.7321
```

10.6.7 Writing Homogeneous Equations of Motion

Now that we know the eigenvalues, we can write the homogeneous equations of motion in the principal coordinate system by inspection. We can also use the normalized eigenvectors to transform the forcing function and initial conditions to principal coordinates, yielding the complete solution for either transient or frequency domain problems in principal coordinates. We can then back-transform to the physical coordinate system to get the desired results in physical coordinates. Through the modal formulation we can define the contributions of various modes to the total response.

For a problem of this size, there is no need to use the modal formulation. When solving real problems, however, whether they be large MATLAB based problems or ANSYS finite element models, using the modal formulation has advantages. As mentioned earlier, ANSYS gives the eigenvalues and eigenvectors normalized with respect to mass as normal output of an eigenvalue run. Therefore, all one has to do to solve in MATLAB is to take that ANSYS output information and build the equations of motion in state space form and solve, taking advantage of the flexibility, plotting capability and speed of MATLAB to perform other studies. The modal approach is what gives us the capability to create complete state space models of the system mechanical dynamics in a form that can be used by the servo engineers in their state space servo/mechanical models.

10.6.7.1 Equations of Motion – Physical Coordinates

We will start with the equations of motion in physical coordinates with forces as shown in (10.29) and assume zero initial conditions. The reason we are assuming zero initial conditions is that converting initial conditions requires the inverse of the complete modal matrix, which is not convenient when using ANSYS modal results to build a reduced (smaller size) model. Fortunately, a large majority of real life problems can be solved with zero initial conditions.

$$m\ddot{z}_1 + kz_1 - kz_2 = F_1$$
$$m\ddot{z}_2 - kz_1 + 2kz_2 - kz_3 = F_2$$
$$m\ddot{z}_3 - kz_2 + kz_3 = F_3$$
$$\text{IC's}:\ z_1, z_2, z_3, \dot{z}_1, \dot{z}_2, \dot{z}_3 = 0$$

$$(10.29)$$

Knowing the eigenvalues and eigenvectors normalized with respect to mass, we can write the damped homogeneous equations of motion in principal coordinates by inspection.   The forces in principal coordinates, F_{p1}, F_{p2} and F_{p3} are obtained by premultiplying the force vector in physical coordinates by the transpose of the normalized eigenvector:

$$\mathbf{F}_p = \mathbf{x}_n^T \mathbf{F} \tag{10.30}$$

$\mathbf{x}_n$ was defined in (10.20) as a 3x3 matrix of normalized displacement eigenvectors. The multiplication then results in a 3x1 vector of forces in principal coordinates. The resulting elements are entered in the appropriate positions in the equations in principal coordinates below.

10.6.7.2 Equations of Motion – Principal Coordinates

The three equations of motion in principal coordinates become:

$$\ddot{x}_{p1} = F_{p1}$$
$$\ddot{x}_{p2} + 2\zeta_2\omega_2\dot{x}_{p2} + \omega_2^2 x_{p2} = F_{p2} \tag{10.31a,b,c}$$
$$\ddot{x}_{p3} + 2\zeta_3\omega_3\dot{x}_{p3} + \omega_3^2 x_{p3} = F_{p3}$$

where ω_1, ω_2, and ω_3 are the three eigenvalues, with units of radians/sec. The "zeta" terms, ζ_1, ζ_2 and ζ_3, represent the percentages of critical damping for each of the three modes, all of which can be different and are typically obtained from experimental results. For example, 2% of critical damping would give a ζ value of 0.02.

Now we can convert the second order differential equations above to state space form by solving for the highest derivative:

$$\ddot{x}_{p1} = F_{p1}$$
$$\ddot{x}_{p2} = F_{p2} - \omega_2^2 x_{p2} - 2\zeta_2\omega_2\dot{x}_{p2} \tag{10.32a,b,c}$$
$$\ddot{x}_{p3} = F_{p3} - \omega_3^2 x_{p3} - 2\zeta_3\omega_3\dot{x}_{p3}$$

Defining states:

$$x_1 = x_{p1} \quad \text{displacement of mode 1 (not of mass 1)}$$
$$x_2 = \dot{x}_{p1} \quad \text{derivative of displacement of mode 1}$$
$$x_3 = x_{p2} \quad \text{displacement of mode 2}$$
$$x_4 = \dot{x}_{p2} \quad \text{derivative of displacement of mode 2}$$
$$x_5 = x_{p3} \quad \text{displacement of mode 3}$$
$$x_6 = \dot{x}_{p3} \quad \text{derivative of displacement of mode 3}$$

Rewriting the equations of motion using the states:

$$\dot{x}_1 = x_2$$
$$\dot{x}_2 = F_{p1}$$
$$\dot{x}_3 = x_4$$
$$\dot{x}_4 = F_{p2} - \omega_2^2 x_3 - 2\zeta_2 \omega_2 x_4 \qquad \text{(10.33a-f)}$$
$$\dot{x}_5 = x_6$$
$$\dot{x}_6 = F_{p3} - \omega_3^2 x_3 - 2\zeta_3 \omega_3 x_3$$

Rewriting in matrix form:

$$\dot{\mathbf{x}} = \mathbf{A}\mathbf{x} + \mathbf{B}\mathbf{u} \qquad \text{(10.34)}$$

$$\begin{bmatrix} \dot{x}_1 \\ \dot{x}_2 \\ \dot{x}_3 \\ \dot{x}_4 \\ \dot{x}_5 \\ \dot{x}_6 \end{bmatrix} = \begin{bmatrix} 0 & 1 & 0 & 0 & 0 & 0 \\ 0 & 0 & 0 & 0 & 0 & 0 \\ 0 & 0 & 0 & 1 & 0 & 0 \\ 0 & 0 & -\omega_2^2 & -2\zeta_2\omega_2 & 0 & 0 \\ 0 & 0 & 0 & 0 & 0 & 1 \\ 0 & 0 & 0 & 0 & -\omega_3^2 & -2\zeta_3\omega_3 \end{bmatrix} \begin{bmatrix} x_1 \\ x_2 \\ x_3 \\ x_4 \\ x_5 \\ x_6 \end{bmatrix} + \begin{bmatrix} 0 \\ F_{p1} \\ 0 \\ F_{p2} \\ 0 \\ F_{p3} \end{bmatrix} u \quad \text{(10.35)}$$

Now that the complete state space equations of motion are known, the six states in principal coordinates can be solved for their frequency and/or time domain responses.

Let us assume that we are interested in the three displacements and the three velocities. The output matrix equation then becomes, where $\mathbf{y}_p$ is the displacements in principal coordinates:

$$
\mathbf{y}_p = \mathbf{Cx} =
\begin{bmatrix}
1 & 0 & 0 & 0 & 0 & 0 \\
0 & 1 & 0 & 0 & 0 & 0 \\
0 & 0 & 1 & 0 & 0 & 0 \\
0 & 0 & 0 & 1 & 0 & 0 \\
0 & 0 & 0 & 0 & 1 & 0 \\
0 & 0 & 0 & 0 & 0 & 1
\end{bmatrix}
\begin{bmatrix}
x_1 \\ x_2 \\ x_3 \\ x_4 \\ x_5 \\ x_6
\end{bmatrix}
=
\begin{bmatrix}
x_1 \\ x_2 \\ x_3 \\ x_4 \\ x_5 \\ x_6
\end{bmatrix}
\tag{10.36}
$$

With the six desired outputs in principal coordinates, we can back-transform them into physical coordinates by the following transform:

$$
\begin{bmatrix}
z_1 \\ \dot{z}_1 \\ z_2 \\ \dot{z}_2 \\ z_3 \\ \dot{z}_3
\end{bmatrix}
= \mathbf{x}_n \mathbf{y}_p =
\begin{bmatrix}
x_{n11} & 0 & x_{n12} & 0 & x_{n13} & 0 \\
0 & x_{n11} & 0 & x_{n12} & 0 & x_{n13} \\
x_{n21} & 0 & x_{n22} & 0 & x_{n23} & 0 \\
0 & x_{n21} & 0 & x_{n22} & 0 & x_{n23} \\
x_{n31} & 0 & x_{n32} & 0 & x_{n33} & 0 \\
0 & x_{n31} & 0 & x_{n32} & 0 & x_{n33}
\end{bmatrix}
\begin{bmatrix}
y_{p1} \\ y_{p2} \\ y_{p3} \\ y_{p4} \\ y_{p5} \\ y_{p6}
\end{bmatrix}
$$

$$
=
\begin{bmatrix}
x_{n11} y_{p1} + x_{n12} y_{p3} + x_{n13} y_{p5} \\
x_{n11} y_{p2} + x_{n12} y_{p4} + x_{n13} y_{p6} \\
x_{n21} y_{p1} + x_{n22} y_{p3} + x_{n23} y_{p5} \\
x_{n21} y_{p2} + x_{n22} y_{p4} + x_{n23} y_{p6} \\
x_{n31} y_{p1} + x_{n32} y_{p3} + x_{n33} y_{p5} \\
x_{n31} y_{p2} + x_{n32} y_{p4} + x_{n33} y_{p6}
\end{bmatrix}
\tag{10.37}
$$

Instead of doing the two multiplications shown in (10.36) and (10.37), $\mathbf{C}$ times $\mathbf{x}$ to get $\mathbf{y}_p$ and then premultiplying $\mathbf{y}_p$ by $\mathbf{x}_n$ to get the displacements and velocities in physical coordinates, we could have done a single multiplication if $\mathbf{C}$ were defined as shown in (10.38), using eigenvector entries directly in the definition:

$$
C = \begin{bmatrix}
X_{n11} & 0 & X_{n12} & 0 & X_{n13} & 0 \\
0 & X_{n11} & 0 & X_{n12} & 0 & X_{n13} \\
X_{n21} & 0 & X_{n22} & 0 & X_{n23} & 0 \\
0 & X_{n21} & 0 & X_{n22} & 0 & X_{n23} \\
X_{n31} & 0 & X_{n32} & 0 & X_{n33} & 0 \\
0 & X_{n31} & 0 & X_{n32} & 0 & X_{n33}
\end{bmatrix}
\tag{10.38}
$$

Rewriting the output equation using C defined in (10.38) and expanding to see individual terms:

$$
\begin{bmatrix} z_1 \\ \dot{z}_1 \\ z_2 \\ \dot{z}_2 \\ z_3 \\ \dot{z}_3 \end{bmatrix}
= Cx =
\begin{bmatrix}
X_{n11} & 0 & X_{n12} & 0 & X_{n13} & 0 \\
0 & X_{n11} & 0 & X_{n12} & 0 & X_{n13} \\
X_{n21} & 0 & X_{n22} & 0 & X_{n23} & 0 \\
0 & X_{n21} & 0 & X_{n22} & 0 & X_{n23} \\
X_{n31} & 0 & X_{n32} & 0 & X_{n33} & 0 \\
0 & X_{n31} & 0 & X_{n32} & 0 & X_{n33}
\end{bmatrix}
\begin{bmatrix} x_1 \\ x_2 \\ x_3 \\ x_4 \\ x_5 \\ x_6 \end{bmatrix}
$$

$$
= \begin{bmatrix}
X_{n11}x_1 + X_{n12}x_3 + X_{n13}x_5 \\
X_{n11}x_2 + X_{n12}x_4 + X_{n13}x_6 \\
X_{n21}x_1 + X_{n22}x_3 + X_{n23}x_5 \\
X_{n21}x_2 + X_{n22}x_4 + X_{n23}x_6 \\
X_{n31}x_1 + X_{n32}x_3 + X_{n33}x_5 \\
X_{n31}x_2 + X_{n32}x_4 + X_{n33}x_6
\end{bmatrix}
= \begin{bmatrix}
X_{n11}y_{p1} + X_{n12}y_{p3} + X_{n13}y_{p5} \\
X_{n11}y_{p2} + X_{n12}y_{p4} + X_{n13}y_{p6} \\
X_{n21}y_{p1} + X_{n22}y_{p3} + X_{n23}y_{p5} \\
X_{n21}y_{p2} + X_{n22}y_{p4} + X_{n23}y_{p6} \\
X_{n31}y_{p1} + X_{n32}y_{p3} + X_{n33}y_{p5} \\
X_{n31}y_{p2} + X_{n32}y_{p4} + X_{n33}y_{p6}
\end{bmatrix}
\tag{10.39}
$$

10.6.8 Individual Mode Contributions, Modal State Space Form

In Section 8.7 we discussed in detail how individual modes contribute to the overall frequency response. Here we will show how to calculate individual modal contributions in modal state space form.

We start with repeating (10.35), the modal state space equations of motion.

$$
\begin{bmatrix} \dot{x}_1 \\ \dot{x}_2 \\ \dot{x}_3 \\ \dot{x}_4 \\ \dot{x}_5 \\ \dot{x}_6 \end{bmatrix}
=
\begin{bmatrix}
0 & 1 & 0 & 0 & 0 & 0 \\
0 & 0 & 0 & 0 & 0 & 0 \\
0 & 0 & 0 & 1 & 0 & 0 \\
0 & 0 & -\omega_2^2 & -2\zeta_2\omega_2 & 0 & 0 \\
0 & 0 & 0 & 0 & 0 & 1 \\
0 & 0 & 0 & 0 & -\omega_3^2 & -2\zeta_3\omega_3
\end{bmatrix}
\begin{bmatrix} x_1 \\ x_2 \\ x_3 \\ x_4 \\ x_5 \\ x_6 \end{bmatrix}
+
\begin{bmatrix} 0 \\ F_{p1} \\ 0 \\ F_{p2} \\ 0 \\ F_{p3} \end{bmatrix} u
\tag{10.40}
$$

Notice how the three sets of uncoupled first order equations in (10.40) appear as blocks of 2x2 coefficients along the diagonal. Note also that if the eigenvalues, ω_i, and damping ratios, ζ_i, are known, the entire system matrix $\mathbf{A}$ can be filled out by inspection, as we will do in future chapters where ANSYS results are used to automatically build a model.

The first 2x2 block along the diagonal

$$\begin{bmatrix} 0 & 1 \\ 0 & 0 \end{bmatrix} \tag{10.41}$$

represents the response of the first mode, the second 2x2 block

$$\begin{bmatrix} 0 & 1 \\ -\omega_2^2 & -2\zeta_2\omega_2 \end{bmatrix} \tag{10.42}$$

represents the response of the second mode and the third 2x2 block

$$\begin{bmatrix} 0 & 1 \\ -\omega_3^2 & -2\zeta_3\omega_3 \end{bmatrix} \tag{10.43}$$

represents the response of the third mode.

Note that the three modes are not coupled and the equations of motion in state space modal form may be rewritten separately as:

$$\begin{bmatrix} \dot{x}_1 \\ \dot{x}_2 \end{bmatrix} = \begin{bmatrix} 0 & 1 \\ 0 & 0 \end{bmatrix} \begin{bmatrix} x_1 \\ x_2 \end{bmatrix} + \begin{bmatrix} 0 \\ F_{p1} \end{bmatrix} u \qquad \text{mode 1} \tag{10.44}$$

$$\begin{bmatrix} \dot{x}_3 \\ \dot{x}_4 \end{bmatrix} = \begin{bmatrix} 0 & 1 \\ -\omega_2^2 & -2\zeta_2\omega_2 \end{bmatrix} \begin{bmatrix} x_3 \\ x_4 \end{bmatrix} + \begin{bmatrix} 0 \\ F_{p2} \end{bmatrix} u \qquad \text{mode 2} \tag{10.45}$$

$$\begin{bmatrix} \dot{x}_5 \\ \dot{x}_6 \end{bmatrix} = \begin{bmatrix} 0 & 1 \\ -\omega_3^2 & -2\zeta_3\omega_3 \end{bmatrix} \begin{bmatrix} x_5 \\ x_6 \end{bmatrix} + \begin{bmatrix} 0 \\ F_{p3} \end{bmatrix} u \qquad \text{mode 3} \tag{10.46}$$

For the output equation, defining a version of (10.42) which will output only displacements, not velocities:

$$\mathbf{z} = \mathbf{y} = \mathbf{Cx} \tag{10.47}$$

Expanding:

$$
\begin{bmatrix} z_1 \\ z_2 \\ z_3 \end{bmatrix} = \begin{bmatrix} x_{n11} & 0 & x_{n12} & 0 & x_{n13} & 0 \\ x_{n21} & 0 & x_{n22} & 0 & x_{n23} & 0 \\ x_{n31} & 0 & x_{n32} & 0 & x_{n33} & 0 \end{bmatrix} \begin{bmatrix} x_1 \\ x_2 \\ x_3 \\ x_4 \\ x_5 \\ x_6 \end{bmatrix}
\tag{10.48}
$$

Similarly, the output equations can be written separately as (10.49) to (10.51), where the $z_{31,m3}$ subscript notation stands for the displacement of mass 3 due to force at mass 1 contributed by mode 3. Here we are dealing with only the z11 transfer function. The modal contributions to any of the four unique transfer functions can be solved in a similar fashion.

$$
\begin{bmatrix} z_{11,m1} \\ z_{21,m1} \\ z_{31,m1} \end{bmatrix} = \begin{bmatrix} x_{n11} & 0 \\ x_{n21} & 0 \\ x_{n31} & 0 \end{bmatrix} \begin{bmatrix} x_1 \\ x_2 \end{bmatrix} \qquad \text{mode 1} \tag{10.49}
$$

$$
\begin{bmatrix} z_{11,m2} \\ z_{21,m2} \\ z_{31,m2} \end{bmatrix} = \begin{bmatrix} x_{n12} & 0 \\ x_{n22} & 0 \\ x_{n32} & 0 \end{bmatrix} \begin{bmatrix} x_3 \\ x_4 \end{bmatrix} \qquad \text{mode 2} \tag{10.50}
$$

$$
\begin{bmatrix} z_{11,m3} \\ z_{21,m3} \\ z_{31,m3} \end{bmatrix} = \begin{bmatrix} x_{n13} & 0 \\ x_{n23} & 0 \\ x_{n33} & 0 \end{bmatrix} \begin{bmatrix} x_5 \\ x_6 \end{bmatrix} \qquad \text{mode 3} \tag{10.51}
$$

We are familiar with using (10.35) and (10.39) to solve for frequency responses for systems. With the use of (10.44) to (10.51) we can plot and see how each individual mode contributes to the overall response. We will examine this further in the code seen in the next chapter.

10.7   Real Modes – Argand Diagrams, Initial Condition Responses of Individual Modes

In Chapter 5, we introduced the concept of using Argand diagrams to visualize complex modes and to show how the complex eigenvector components combine to create "real" displacements and velocities. We will use the

MATLAB code **tdof_prop_damped.m** to define the eigenvectors for Argand plotting and solve for the transient responses.

The methodology followed is:

1) **Solve** the original undamped system equation for eigenvalues and eigenvectors.

2) **Plot** the eigenvectors normalized to unity using a deformed mode shape plot.

3) **Normalize** the displacement eigenvector entries with respect to mass to convert to principal coordinates for the proportionally damped case.

4) **Form** the system matrix in principal coordinates using proportional damping.

5) **Solve** for the eigenvalues and eigenvectors of the system matrix in principal coordinates.

6) **Plot** the real and imaginary displacements of each of the normal modes separately, since the three modes are uncoupled with proportional damping.

7) **Back transform** to physical coordinates using the normalized displacement eigenvectors.

8) **Plot** the real and imaginary displacements of each of the degrees of freedom separately.

For the undamped case we will use $c1 = c2 = 0$ and the result will be "normal" modes with "real" eigenvectors.

For proportional damping, we will start with the undamped eigenvectors and add a percentage of critical damping to each mode. This will result in "real" eigenvectors since proportional damping satisfies the Rayleigh damping criterion $c = am + bk$ as discussed in Chapter 7.

10.7.1 Undamped Model, Eigenvectors, Real Modes

The code starts with executing **tdofss_eig.m**, which calculates the eigenvalues and eigenvectors for the undamped problem, $c1 = c2 = 0$. The eigenvectors are then normalized with respect to unity for plotting in Argand form.

```
%   tdof_prop_damped.m       proportionally damped tdof model

%   solve for the eigenvalues of the undamped system model
```

```
tdofss_eig;

subplot(1,1,1);
```

%   now normalize the undamped eigenvectors with respect to the position of
%   mass 1, which will be set to 1.0 - for plotting of undamped Argand diagram

```
     for  cnt = 1:length(omegad)

          xmon1(:,cnt) = xmo(:,cnt)/xmo(1,cnt);

     end

     xmon1
```

The eigenvalues and eigenvectors are:

```
omegaro =
     0              (Note the two poles
     0                at the origin)
     0 + 1.0000i
     0 - 1.0000i
     0 + 1.7321i
     0 - 1.7321i
```

```
xmron1 =
   1.0000      1.0000      1.0000      1.0000      1.0000      1.0000
   0           0           0 + 1.0000i  0 - 1.0000i  0 + 1.7321i  0 - 1.7321i
   1.0000      1.0000      0.0000      0.0000     -2.0000     -2.0000
   0           0           0 + 0.0000i  0 - 0.0000i  0 - 3.4641i  0 + 3.4641i
   1.0000      1.0000     -1.0000     -1.0000      1.0000      1.0000
   0           0           0 - 1.0000i  0 + 1.0000i  0 + 1.7321i  0 - 1.7321i
```

Note that the pairs of eigenvalues for each mode are complex conjugates of each other and that the pairs of eigenvectors for each mode are also complex conjugates of each other.

Once again, some eigenvector elements have complex parts. **Why do we call them "real" when they contain imaginary parts?**

"Real" eigenvectors refers to the fact that all of the **position** entries in the eigenvector are not complex numbers [i.e., not of the form (a+jb)], but are real numbers. The fact that the velocity entries are complex numbers by virtue of multiplying the "real" position eigenvector entries by the eigenvalue does not make the eigenvectors "complex" but refers to the fact that in the undamped case velocity is $90°$ out of phase with position.

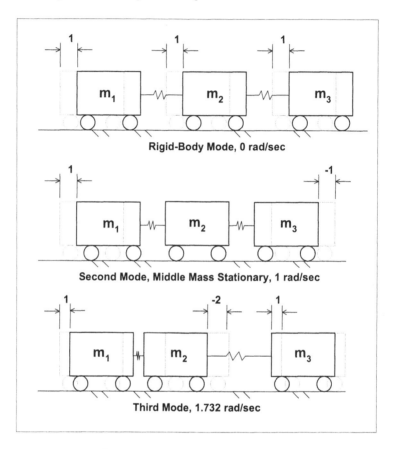

Figure 10.1: Mode shape plots, "real" modes.

For "real" eigenvectors, there are two ways of visualizing the mode shapes and resulting motions. One method we have used several times before, the mode shape plot, shows the deformed shapes of the system for each eigenvector.

Since for real eigenvectors all the degrees of freedom reach their maxima and minima at the same times, any snapshot in time will show the <u>relative</u> displacements, which is why we can plot a deformed mode plot as shown in Figure 10.1.

10.7.2 Principal Coordinate Eigenvalue Problem

The section of code below prompts for the amount of proportional damping, zeta, and then sets up the equations of motion in principal coordinates. After solving the eigenvalue problem, the eigenvalues and eigenvectors are sorted and the magnitude and phase angle of the each eigenvector is defined.

```
%   input proportional damping for equations in principal coordinate system

    zeta = input('input value for zeta, default = 0.02, 2% of critical ... ');

    if (isempty(zeta))
        zeta = 0.02;
    else
    end

%   setup proportionally damped state-space system matrix in principal coordinates

    a_ss = [    0       1       0         0          0        0
                0       0       0         0          0        0
                0       0       0         1          0        0
                0       0     -w2^2   -2*zeta*w2     0        0
                0       0       0         0          0        1
                0       0       0         0        -w3^2   -2*zeta*w3];

%   solve for the eigenvalues of the system matrix with proportional damping

    [xmp,omegap] = eig(a_ss);

%   take the diagonal elements of the generalized eigenvalue matrix omegap

    omegapd = diag(omegap);

%   now reorder the eigenvalues and eigenvectors from low to high frequency,
%   keeping track of how the eigenvalues are ordered in reorder the
%   eigenvectors to match, using indexhz

    [omegaporder,indexhz] = sort(abs(imag(omegapd)));

    for cnt = 1:length(omegapd)

    omegapo(cnt,1) = omegapd(indexhz(cnt));   %  reorder eigenvalues

    xmpo(:,cnt) = xmp(:,indexhz(cnt)); %  reorder eigenvector columns

    end

%   now calculate the magnitude and phase angle of each of the eigenvector
%   entries

    for row = 1:length(omegapd)
```

```
        for  col = 1:length(omegapd)

            xmpomag(row,col) = abs(xmpo(row,col));

            xmpoang(row,col) = (180/pi)*angle(xmpo(row,col));

        end

    end

    omegapo

    xmpo

    xmpomag

    xmpoang
```

10.7.3 Damping Calculation, Eigenvalue Complex Plane Plot

The section below calculates the percentage of critical damping due to the defined amount of input damping, zeta. For example, if 2% of critical damping is defined as input, then we should see that the eigenvalues of the equations of motion in principal coordinates plot as shown in Figure 5.2.

```
%   calculate the percentage of critical damping for each mode

    zeta1 = 0

    theta2 = atan(real(omegapo(3))/imag(omegapo(3)));
    zeta2 = abs(sin(theta2))

    theta3 = atan(real(omegapo(5))/imag(omegapo(5)));
    zeta3 = abs(sin(theta3))

    plot(omegap,'kx')
    grid on
    axis([-3 1 -2 2])
    axis('square')
    title('Proportionally Damped Eigenvalues')
    xlabel('real')
    ylabel('imaginary')
    text(real(omegapo(3))-1,imag(omegapo(3))+0.1,['zeta = ',num2str(zeta2)])
    text(real(omegapo(5))-1,imag(omegapo(5))+0.1,['zeta = ',num2str(zeta3)])

    disp('execution paused to display figure, "enter" to continue'); pause
```

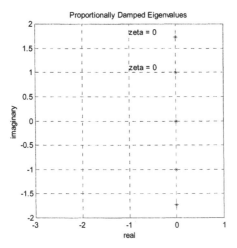

Figure 10.2: Undamped model eigenvalue plot in complex plane.

For the undamped model, we should see that the eigenvalues, poles, should lie on the imaginary axis – and they do.

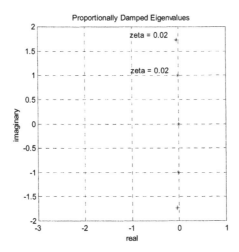

Figure 10.3: Proportionally damped eigenvalue plot, zeta = 2% was input.

The eigenvalues for zeta = 0.02 plot slightly to the left of the imaginary axis.

10.7.4 Principal Displacement Calculations

We showed in (5.54), repeated below, how to calculate the displacements when the system is started with a set of initial conditions which match the eigenvector:

$$\begin{aligned}
\mathbf{x}(t) &= e^{\sigma_{n1}t}\left(e^{j\omega_{n1}t}\mathbf{x}_{n1} + e^{-j\omega_{n2}t}\mathbf{x}_{n2}\right)\\
&= e^{\sigma_{n1}t}\left(e^{j\omega_{n1}t}\mathbf{x}_{n1}\right) + e^{\sigma_{n1}t}\left(e^{-j\omega_{n2}t}\mathbf{x}_{n2}\right)
\end{aligned} \tag{10.52}$$

Since our eigenvalues lie along the imaginary axis, their σ values are zero and $e^{0t} = 1$, the equations can be simplified to:

$$\mathbf{x}(t) = e^{j\omega_{n1}t}\mathbf{x}_{n1} + e^{-j\omega_{n2}t}\mathbf{x}_{n2} \tag{10.53}$$

A time vector from 0 to 15 seconds is defined, and real and imaginary parts are picked from the eigenvalues. Equation (10.52) is used to calculate the motions.

```
%   calculate the motions of the three masses for all three modes - damped case

    t = 0:.12:15;

    sigma11 = real(omegapo(1));   %  sigma for first eigenvalue for mode 1
    omegap11 = imag(omegapo(1));  %  omegap for first eigenvalue for mode 1

    sigma12 = real(omegapo(2));   %  sigma for second eigenvalue for mode 1
    omegap12 = imag(omegapo(2));  %  omegap for second eigenvalue for mode 1

    sigma21 = real(omegapo(3));   %  sigma for first eigenvalue for mode 2
    omegap21 = imag(omegapo(3));  %  omegap for first eigenvalue for mode 2

    sigma22 = real(omegapo(4));   %  sigma for second eigenvalue for mode 2
    omegap22 = imag(omegapo(4));  %  omegap for second eigenvalue for mode 2
    sigma31 = real(omegapo(5));   %  sigma for first eigenvalue for mode 3
    omegap31 = imag(omegapo(5));  %  omegap for first eigenvalue for mode 3

    sigma32 = real(omegapo(6));   %  sigma for second eigenvalue for mode 3
    omegap32 = imag(omegapo(6));  %  omegap for second eigenvalue for mode 3
%   displacements of mode 1 in principal coordinates

    zp111 = exp(sigma11*t).*(exp(i*omegap11*t)*xmpo(1,1));   %  mass 1
    zp112 = exp(sigma12*t).*(exp(i*omegap12*t)*xmpo(1,2));   %  mass 1

%   displacements of mode 2 in principal coordinates

    zp221 = exp(sigma21*t).*(exp(i*omegap21*t)*xmpo(3,3));   %  mass 2
```

```
        zp222 = exp(sigma22*t).*(exp(i*omegap22*t)*xmpo(3,4));    % mass 2

%   displacements of mode 3 in principal coordinates

        zp331 = exp(sigma31*t).*(exp(i*omegap31*t)*xmpo(5,5));    % mass 3
        zp332 = exp(sigma32*t).*(exp(i*omegap32*t)*xmpo(5,6));    % mass 3
```

10.7.5 Transformation to Physical Coordinates

The section of code below sets up the appropriate size matrices to enable back-transforming from principal to physical coordinates.

```
%   calculate the motions of each mass for mode 2
%   define matrix of displacements vs time for each eigenvector

        z221 = [zeros(1,length(t))
                zp221
                zeros(1,length(t))];

        z222 = [zeros(1,length(t))
                zp222
                zeros(1,length(t))];

%   back-transform from principal to physical coordinates

        zmode21 = xn*z221;

        zmode22 = xn*z222;

        z1mode21 = zmode21(1,:);

        z2mode21 = zmode21(2,:);

        z3mode21 = zmode21(3,:);

        z1mode22 = zmode22(1,:);

        z2mode22 = zmode22(2,:);

        z3mode22 = zmode22(3,:);

%   calculate the motions of each mass for mode 3
%   define matrix of displacements vs time for each eigenvector

        z331 = [zeros(1,length(t))
                zeros(1,length(t))
                zp331];

        z332 = [zeros(1,length(t))
                zeros(1,length(t))
                zp332];
```

```
zmode31 = xn*z331;

zmode32 = xn*z332;

z1mode31 = zmode31(1,:);

z2mode31 = zmode31(2,:);

z3mode31 = zmode31(3,:);

z1mode32 = zmode32(1,:);

z2mode32 = zmode32(2,:);

z3mode32 = zmode32(3,:);
```

10.7.6 Plotting Results

The plotting commands for mode 2 are listed below; those for mode 3 have been eliminated for brevity.

```
%   plot principal displacements of mode 2

    plot(t,real(zp221),'k-',t,real(zp222),'k+-',t,imag(zp221),'k.-',t,imag(zp222),'ko-')
    title('principal real and imag disp for mode 2')
    legend('real','real','imag','imag')
    axis([0 max(t) -1 1])
    grid on

    disp('execution paused to display figure, "enter" to continue'); pause

%   plot physical disp of masses for mode 2

    plot(t,real(z1mode21),'k-',t,real(z1mode22),'k+-',t,imag(z1mode21), ...
                        'k.-',t,imag(z1mode22),'ko-')
    title('physical real and imag disp for mass 1, mode 2')
    legend('real','real','imag','imag')
    axis([0 max(t) -0.5 0.5])
    grid on

    disp('execution paused to display figure, "enter" to continue'); pause

    plot(t,real(z2mode21),'k-',t,real(z2mode22),'k+-',t,imag(z2mode21),...
                        'k.-',t,imag(z2mode22),'ko-')
    title('physical real and imag disp for mass 2, mode 2')
    legend('real','real','imag','imag')
    axis([0 max(t) -0.5 0.5])
    grid on

    disp('execution paused to display figure, "enter" to continue'); pause

    plot(t,real(z3mode21),'k-',t,real(z3mode22),'k+-',t,imag(z3mode21), ...
```

```
                                 'k.-',t,imag(z3mode22),'ko-')
      title('physical real and imag disp for mass 3, mode 2')
      legend('real','real','imag','imag')
      axis([0 max(t) -0.5 0.5])
      grid on

      disp('execution paused to display figure, "enter" to continue'); pause

      plot(t,real(z1mode21+z1mode22),'k-',t,real(z2mode21+z2mode22), ...
                                'k+-',t,real(z3mode21+z3mode22),'k.-')
      title('physical disp z1, z2, z3 mode 2')
      legend('mass 1','mass 2','mass 3')
      axis([0 max(t) -1 1])
      grid on

      disp('execution paused to display figure, "enter" to continue'); pause

%   plot subplots for notes

%   plot principal disp of mode 2

      subplot(3,2,1)
      plot(t,real(zp221),'k-',t,real(zp222),'k+-',t,imag(zp221),'k.-',t,imag(zp222),'ko-')
      title('principal disp for mode 2')
      legend('real','real','imag','imag')
      axis([0 max(t) -1 1])
      grid on

%   plot physical disp of masses for mode 2

      subplot(3,2,3)
      plot(t,real(z1mode21),'k-',t,real(z1mode22),'k+-',t,imag(z1mode21), ...
                          'k.-',t,imag(z1mode22),'ko-')
      title('physical real and imag disp for mass 1, mode 2')
      legend('real','real','imag','imag')
      axis([0 max(t) -0.5 0.5])
      grid on

      subplot(3,2,4)
      plot(t,real(z2mode21),'k-',t,real(z2mode22),'k+-',t,imag(z2mode21), ...
                           'k.-',t,imag(z2mode22),'ko-')
      title('physical real and imag disp for mass 2, mode 2')
      legend('real','real','imag','imag')
      axis([0 max(t) -0.5 0.5])
      grid on

      subplot(3,2,5)
      plot(t,real(z3mode21),'k-',t,real(z3mode22),'k+-',t,imag(z3mode21), ...
                            'k.-',t,imag(z3mode22),'ko-')
      title('physical real and imag disp for mass 3, mode 2')
      legend('real','real','imag','imag')
      axis([0 max(t) -0.5 0.5])
      grid on

      subplot(3,2,6)
```

```
    plot(t,real(z1mode21+z1mode22),'k+-',t,real(z2mode21+z2mode22), ...
                    'k.-',t,real(z3mode21+z3mode22),'ko-')
    title('physical disp for z1, z2, z3 mode 2')
    legend('mass 1','mass 2','mass 3')
    axis([0 max(t) -1 1])
    grid on

    disp('execution paused to display figure, "enter" to continue'); pause

    subplot(1,1,1)
```

10.7.7 Undamped / Proportionally Damped Argand Diagram, Mode 2

As in the Argand diagrams explained in Chapter 5, the two complex conjugate eigenvectors for each mode are plotted side by side. The direction of the rotation of the eigenvector is indicated by the arrow associated with the $e^{j\omega t}$ or $e^{-j\omega t}$ terms. The addition of the two counter-rotating complex eigenvectors for an arbitrary time "t" is shown in the middle and below the two individual eigenvector plots for each dof. The addition plot shows how the two imaginary components cancel each other out, leaving only the real portion of the motion.

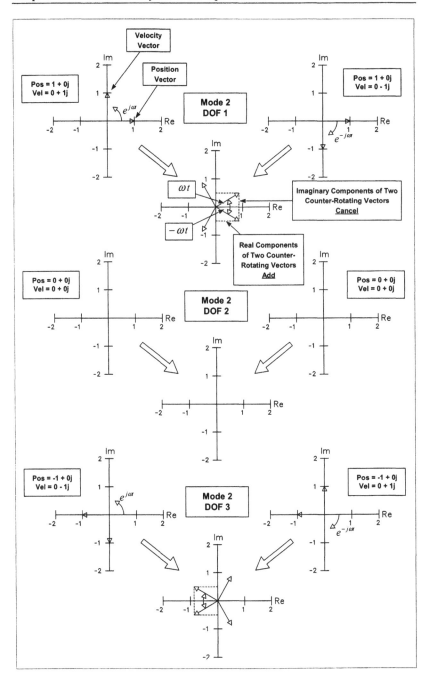

Figure 10.4: Argand diagram for undamped or proportionally damped system, mode 2.

10.7.8 Undamped / Proportionally Damped Argand Diagram, Mode 3

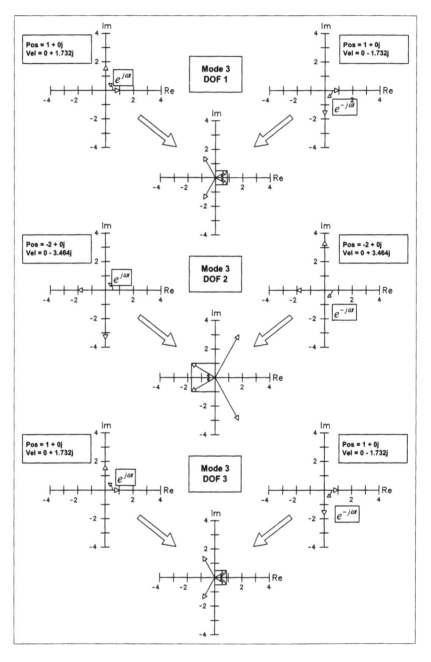

Figure 10.5: Argand diagram for undamped or proportionally damped system, mode 3.

10.7.9 Proportionally Damped Initial Condition Response, Mode 2

Figures 10.6 to 10.10 show the initial condition responses for mode 2 for proportional damping of 2%. Mode 2 is the mode where mass 2 is stationary and masses 1 and 3 are moving out of phase with each other with equal amplitude.

Figure 10.6 shows the real and imaginary components of the two complex eigenvector responses that make up mode 2. Note that the two imaginary components are out of phase and cancel each other while the two real components are overlaid and will add. Figures 10.7 to 10.9 show the real and imaginary components for each of the three masses. The motions of mass 2 are zero, while the motions of masses 1 and 3 are out of phase with each other, consistent with the shape of mode 2 in Figure 10.1. Figure 10.10 shows the physical displacements of the three masses versus time. The Argand diagram vectors for mode 2, Figure 10.4, can be matched with each figure for each degree of freedom.

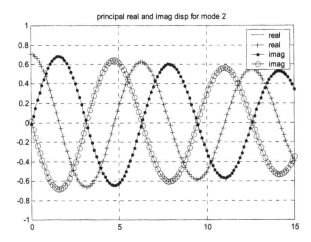

Figure 10.6: Principal real and imaginary displacements, mode 2.

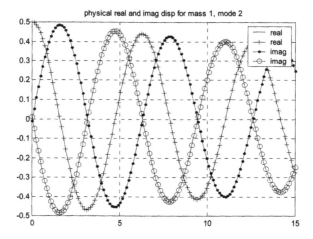

Figure 10.7: Physical real and imaginary displacements for mass 1, mode 2.

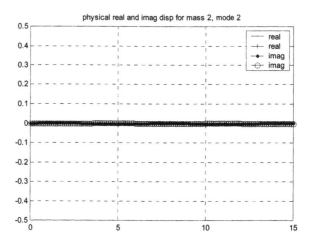

Figure 10.8: Physical real and imaginary displacements for mass 2, mode 2.

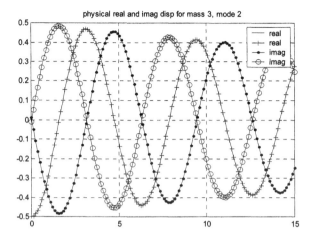

Figure 10.9: Physical real and imaginary displacements for mass 3, mode 2.

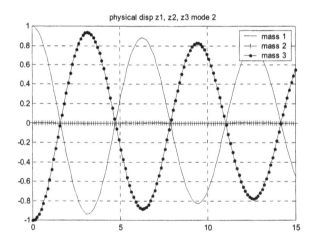

Figure 10.10: Physical displacements for masses 1, 2 and 3, mode 2.

10.7.10 Proportionally Damped Initial Condition Response, Mode 3

Figures 10.11 to 10.15 show the initial condition responses for mode 3 for 2% proportional damping, where mass 2 moves twice as far and out of phase with masses 1 and 3.

Figure 10.11 shows the real and imaginary components of the two complex eigenvector responses that make up mode 2. As in the previous section, note that the two imaginary components are out of phase and cancel each other while the two real components are overlaid and will add. Figures 10.12 to 10.14 display the real and imaginary components for each of the three masses. Figure 10.15 shows the physical displacements of the three masses versus time. The Argand diagram vectors for mode 2, Figure 10.5, can be matched with each figure for each degree of freedom.

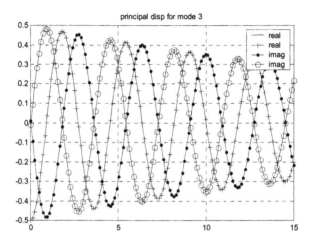

Figure 10.11: Principal real and imaginary displacements, mode 3.

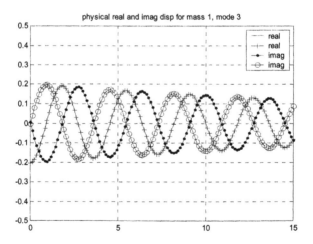

Figure 10.12: Physical real and imaginary displacements for mass 1, mode 3.

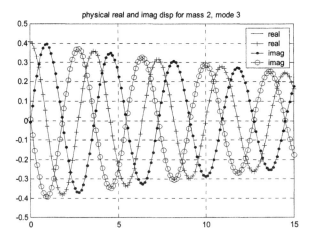

Figure 10.13: Physical real and imaginary displacements for mass 2, mode 3.

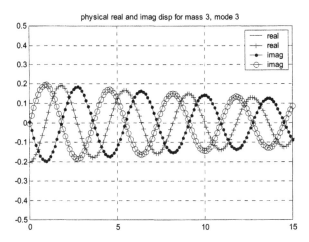

Figure 10.14: Physical real and imaginary displacements for mass 3, mode 3.

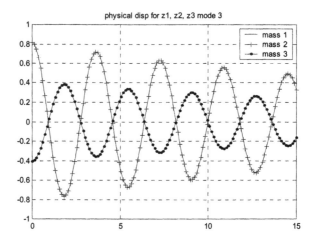

Figure 10.15: Physical real and imaginary displacements for masses 1,2 and 3, mode 3.

Problems

Note: All the problems refer to the two dof system shown in Figure P2.2.

P10.1 Write the homogeneous equations of motion in state space form for the undamped two dof system with $m_1 = m_2 = m = 1$, $k_1 = k_2 = k = 1$. Set up the eigenvalue problem and expand the determinant to reveal the characteristic equation. Compare with the denominator terms from P2.2.

P10.2   Solve for the eigenvalues and eigenvectors in state space form. Compare with the results from P7.1. What is the relationship between the displacement and velocity eigenvector terms?

P10.3 (MATLAB) Modify the **tdofss_eig.m** code for the undamped two dof system with $m_1 = m_2 = m = 1$, $k_1 = k_2 = k = 1$. Print out the eigenvalue and eigenvector results and compare with the results from P10.2. What changes are required to the MATLAB eigenvectors to make them match the P10.2 results? After normalizing with respect to mass, confirm that the equations of motion consist of an identity mass matrix and a stiffness matrix with squares of the eigenvalues along the diagonal.

P10.4 Write the equations of motion in principal coordinates in state space form, knowing only the eigenvalues and eigenvectors, similar to (10.35). Use the displacements of mass 1 and mass 2 as outputs. Show how the output matrix C can be formulated to only require a single multiplication to give outputs (Section 10.6.7.2). Identify the 2x2 submatrices which define the state equations of each mode. Are the individual modes uncoupled?

P10.5 (MATLAB) Modify the **tdof_prop_damped.m** code for the two dof system with $m_1 = m_2 = m = 1$, $k_1 = k_2 = k = 1$. Plot the eigenvalue locations in the s-plane for zero damping and for proportional damping of 2% (0.02). List the eigenvalues and eigenvectors for the undamped and proportional damping cases and note the differences. Plot the initial condition responses when started in initial conditions which match each of the two eigenvectors.

P10.6 Plot Argand diagrams for the undamped system.

CHAPTER 11

FREQUENCY RESPONSE: MODAL STATE SPACE FORM

11.1 Introduction

In Chapter 10 we constructed the modal form of the state equations for the overall frequency response as well as for the individual mode contributions. This short chapter of MATLAB code will carry out both overall and individual mode frequency response calculations. The code will also allow us to plot the different forms of frequency responses covered in Chapter 3.

11.2 Modal State Space Setup, tdofss_modal_xfer_modes.m Listing

After executing the "**tdofss_eig.m**" code to provide eigenvalues and eigenvectors, we enter a section of code that yields similar results to those resulting from an ANSYS simulation. In the ANSYS case, we would have access to the eigenvalues and mass normalized eigenvectors, similar to the "xn" and "w1, w2 and w3" from **tdofss_eig.m**.

Since we can add proportional damping to our modal model, the code prompts for a value for zeta.

Knowing zeta and the eigenvalues, the system matrix can be setup as shown in (10.35), as 2x2 blocks along the diagonal. The three 2x2 submatrices of the system matrix are defined for individual mode contribution calculations.

The next step is to define a 6x3 input matrix, 6 states and three possible inputs representing forces applied to only mass 1, only mass 2 or only mass 3. We start out by defining three separate 3x1 force vectors, one for each mass, F1, F2 and F3. Each of these vectors is transformed from physical to principal coordinates by premultiplying by xn transpose. The three 3x1 vectors are padded with zeros resulting in three 6x1 vectors, which are then inserted as columns in the 6x3 input matrix "b."

The output matrix, "c," is defined in one step as shown in (10.38) by incorporating the appropriate elements of "xn." However, only displacement states are output, giving a 3x6 matrix.

The direct transmission matrix is set to zero.

```
%        tdofss_modal_xfer_modes.m      state-space modal form transfer function analysis
%        of tdof model, proportional damping, modal contribution plotting

         clf;

         clear all;

%        run tdofss_eig.m to provide eigenvalues and eigenvectors

         tdofss_eig;

%        note, this is the point where we would start if we had eigenvalue results from ANSYS,
%        using the eigenvalues and eigenvectors to define state space equations in
%        principal coordinates

%        define damping ratio to be used for proportional damping in the state space equation
%        in principal coordinates

         zeta = input('input zeta, 0.02 = 2% of critical damping (default) ... ');

         if (isempty(zeta))
         zeta = 0.02;
         else
         end

%        setup 6x6 state-space system matrix for all three modes in principal
%        coordinates, a_ss

         a_ss = [0        1        0        0        0        0
                 0        0        0        0        0        0
                 0        0        0        1        0        0
                 0        0      -w2^2   -2*zeta*w2   0        0
                 0        0        0        0        0        1
                 0        0        0        0      -w3^2   -2*zeta*w3];

%        setup three 2x2 state-space matrices, one for each individual mode

         a1_ss = a_ss(1:2,1:2);

         a2_ss = a_ss(3:4,3:4);

         a3_ss = a_ss(5:6,5:6);

%        transform the 3x1 force vectors in physical coordinates to principal coordinates and
%        then insert the principal forces in the appropriate rows in the state-space
%        6x1 input matrix, padding with zeros as appropriate

%        define three force vectors in physical coordinates, where each is for
%        a force applied to a single mass

         F1 = [1 0 0]';

         F2 = [0 1 0]';

         F3 = [0 0 1]';
```

```
%          calculate the three force vectors in principal coordinates by pre-multiplying
%          by the transpose of the normalized modal matrix

           Fp1 = xn'*F1;

           Fp2 = xn'*F2;

           Fp3 = xn'*F3;

%          expand the force vectors in principal coordinates from 3x1 to 6x1, padding with zeros

           b1 = [0 Fp1(1) 0 Fp1(2) 0 Fp1(3)]';        % principal force applied at mass 1

           b2 = [0 Fp2(1) 0 Fp2(2) 0 Fp2(3)]';        % principal force applied at mass 2

           b3 = [0 Fp3(1) 0 Fp3(2) 0 Fp3(3)]';        % principal force applied at mass 3

           b = [b1 b2 b3];

%          the output matrix c is setup in one step, to allow the "bode" command to
%          output the desired physical coordinates directly without having to go
%          through any intermediate steps.

%          setup the output matrix for displacement transfer functions, each row
%          represents the position outputs of mass 1, mass 2 and mass 3
%          velocities not included, so c is only 3x6 instead of 6x6

           c = [xn(1,1)   0   xn(1,2)   0   xn(1,3)   0
                xn(2,1)   0   xn(2,2)   0   xn(2,3)   0
                xn(3,1)   0   xn(3,2)   0   xn(3,3)   0];

%          define direct transmission matrix d

           d = zeros(3,3);
```

11.3 Frequency Response Calculation

We will begin this section by defining the vector of frequencies to be used for the frequency response plot. Then we will define a state space model, using the matrices defined in the section above.

Because we are using a 6x3 input matrix and a 3x6 output matrix, we have access to nine frequency response plots, the displacement for all three degrees of freedom for three different force application points. To plot the four distinct frequency responses, the appropriate indices are used to define magnitude and phase.

```
%          Define a vector of frequencies to use, radians/sec. The logspace command uses
%          the log10 value as limits, i.e. -1 is 10^-1 = 0.1 rad/sec, and 1 is
%          10^1 = 10 rad/sec. The 200 defines 200 frequency points.
```

```
                w = logspace(-1,1,200);

%       define four state-space systems using the "ss" command
%                       sys is for all modes for all 3 forcing functions
%                       sys1 is for mode 1 for all 3 forcing functions
%                       sys2 is for mode 2 for all 3 forcing functions
%                       sys3 is for mode 3 for all 3 forcing functions

                sys = ss(a_ss,b,c,d);

                sys1 = ss(a1_ss,b(1:2,:),c(:,1:2),d);

                sys2 = ss(a2_ss,b(3:4,:),c(:,3:4),d);

                sys3 = ss(a3_ss,b(5:6,:),c(:,5:6),d);

%       use the bode command with left hand magnitude and phase vector arguments
%       to provide values for further analysis/plotting

                [mag,phs] = bode(sys,w);

                [mag1,phs1] = bode(sys1,w);

                [mag2,phs2] = bode(sys2,w);

                [mag3,phs3] = bode(sys3,w);

%       pick out the specific magnitudes and phases for four distinct responses

                z11mag = mag(1,1,:);

                z21mag = mag(2,1,:);

                z31mag = mag(3,1,:);

                z22mag = mag(2,2,:);

                z11magdb = 20*log10(z11mag);

                z21magdb = 20*log10(z21mag);

                z31magdb = 20*log10(z31mag);

                z22magdb = 20*log10(z22mag);

                z11phs = phs(1,1,:);

                z21phs = phs(2,1,:);

                z31phs = phs(3,1,:);

                z22phs = phs(2,2,:);

%       pick out the three individual mode contributions to z11
```

```
        z111mag = mag1(1,1,:);

        z112mag = mag2(1,1,:);

        z113mag = mag3(1,1,:);

        z111magdb = 20*log10(z111mag);

        z112magdb = 20*log10(z112mag);

        z113magdb = 20*log10(z113mag);

        z111phs = phs1(1,1,:);

        z112phs = phs2(1,1,:);

        z113phs = phs3(1,1,:);
```

11.4 Frequency Response Plotting

```
%       truncate peaks for plotting of expanded linear scale

        z11plotmag = z11mag;

        z111plotmag = z111mag;

        z112plotmag = z112mag;

        z113plotmag = z113mag;

        for  cnt = 1:length(z11mag)

                if  z11plotmag(cnt) >= 3.0

                        z11plotmag(cnt) = 3.0;

                end

                if  z111plotmag(cnt) >= 3.0

                        z111plotmag(cnt) = 3.0;

                end

                if  z112plotmag(cnt) >= 3.0

                        z112plotmag(cnt) = 3.0;

                end

                if  z113plotmag(cnt) >= 3.0
```

```
                               z113plotmag(cnt) = 3.0;

                  end

      end

%     plot the four transfer functions separately, in a 2x2 subplot form

      subplot(2,2,1)
      semilogx(w,z11magdb(1,:),'k-')
      title('state space, z11, z33 db magnitude')
      ylabel('magnitude, db')
      axis([.1 10 -150 50])
      grid

      subplot(2,2,2)
      semilogx(w,z21magdb(1,:),'k-')
      title('state space, z21, z12, z23, z32 db magnitude')
      ylabel('magnitude, db')
      axis([.1 10 -150 50])
      grid

      subplot(2,2,3)
      semilogx(w,z31magdb(1,:),'k-')
      title('state space, z31, z13 db magnitude')
      xlabel('frequency, rad/sec')
      ylabel('magnitude, db')
      axis([.1 10 -150 50])
      grid

      subplot(2,2,4)
      semilogx(w,z22magdb(1,:),'k-')
      title('state space, z22 db magnitude')
      xlabel('frequency, rad/sec')
      ylabel('magnitude, db')
      axis([.1 10 -150 50])
      grid

      disp('execution paused to display figure, "enter" to continue'); pause

      subplot(2,2,1)
      semilogx(w,z11phs(1,:),'k-')
      title('state space, z11, z33 phase')
      ylabel('phase, deg')
      grid

      subplot(2,2,2)
      semilogx(w,z21phs(1,:),'k-')
      title('state space, z21, z12, z23, z32 phase')
      ylabel('phase, deg')
      grid

      subplot(2,2,3)
      semilogx(w,z31phs(1,:),'k-')
      title('state space, z31, z13 phase')
```

```
         xlabel('frequency, rad/sec')
         ylabel('phase, deg')
         grid

         subplot(2,2,4)
         semilogx(w,z22phs(1,:),'k-')
         title('state space, z22 phase')
         xlabel('frequency, rad/sec')
         ylabel('phase, deg')
         grid

         disp('execution paused to display figure, "enter" to continue'); pause

%        plot the overall plus individual mode contributions separately

         subplot(2,2,1)
         semilogx(w,z11magdb(1,:),'k-')
         title('State-Space Modal, z11 db magnitude')
         ylabel('magnitude, db')
         axis([.1 10 -60 40])
         grid

         subplot(2,2,2)
         semilogx(w,z111magdb(1,:),'k-')
         title('State-Space Modal, z11 db magnitude of mode 1')
         ylabel('magnitude, db')
         axis([.1 10 -60 40])
         grid

         subplot(2,2,3)
         semilogx(w,z112magdb(1,:),'k-')
         title('State-Space Modal, z11 db magnitude of mode 2')
         xlabel('frequency, rad/sec')
         ylabel('magnitude, db')
         axis([.1 10 -60 40])
         grid

         subplot(2,2,4)
         semilogx(w,z113magdb(1,:),'k-')
         title('State-Space Modal, z11 db magnitude of mode 3')
         xlabel('frequency, rad/sec')
         ylabel('magnitude, db')
         axis([.1 10 -60 40])
         grid

         disp('execution paused to display figure, "enter" to continue'); pause

         subplot(2,2,1)
         semilogx(w,z11phs(1,:),'k-')
         title('State-Space Modal, z11 phase')
         ylabel('phase, deg')
         grid

         subplot(2,2,2)
         semilogx(w,z111phs(1,:),'k-')
```

```
title('State-Space Modal, z11 phase of mode 1')
ylabel('phase, deg')
grid

subplot(2,2,3)
semilogx(w,z112phs(1,:),'k-')
title('State-Space Modal, z11 phase of mode 2')
xlabel('frequency, rad/sec')
ylabel('phase, deg')
grid

subplot(2,2,4)
semilogx(w,z113phs(1,:),'k-')
title('State-Space Modal, z11 phase of mode 3')
xlabel('frequency, rad/sec')
ylabel('phase, deg')
grid

disp('execution paused to display figure, "enter" to continue'); pause

subplot(1,1,1);

%       plot the overlaid transfer function and individual mode contributions

loglog(w,z11mag(1,:),'k+:',w,z111mag(1,:),'k-',w,z112mag(1,:),'k-',w, ...
             z113mag(1,:),'k-')
title('State-Space Modal Mode Contributions, z11 db magnitude')
xlabel('frequency, rad/sec')
ylabel('magnitude, db')
axis([.1 10 .001 100])
grid

disp('execution paused to display figure, "enter" to continue'); pause

semilogx(w,z11mag(1,:),'k+:',w,z111mag(1,:),'k-',w,z112mag(1,:), ...
             'k-',w,z113mag(1,:),'k-')
title('State-Space Modal Mode Contributions, z11 linear magnitude')
xlabel('frequency, rad/sec')
ylabel('magnitude')
grid

disp('execution paused to display figure, "enter" to continue'); pause

semilogx(w,z11plotmag(1,:),'k+:',w,z111plotmag(1,:),'k-', ...
             w,z112plotmag(1,:),'k-',w,z113plotmag(1,:),'k-')
title('State-Space Modal Mode Contributions, z11 linear magnitude')
xlabel('frequency, rad/sec')
ylabel('magnitude')
axis([.1 10 0 3]);
grid

disp('execution paused to display figure, "enter" to continue'); pause

semilogx(w,z11phs(1,:),'k+:',w,z111phs(1,:),'k-',w,z112phs(1,:),'k-', ...
             w,z113phs(1,:),'k-')
```

```
title('State-Space Modal Mode Contributions, z11 phase')
xlabel('frequency, rad/sec')
ylabel('phase, deg')
grid
```

11.5 Code Results – Frequency Response Plots, 2% of Critical Damping

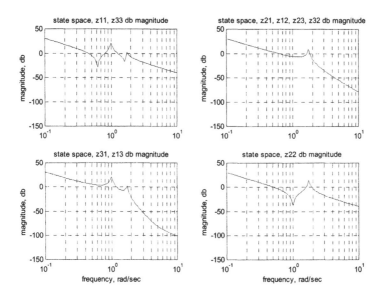

Figure 11.1: Magnitude output for four distinct frequency responses, proportional
damping zeta = 2%.

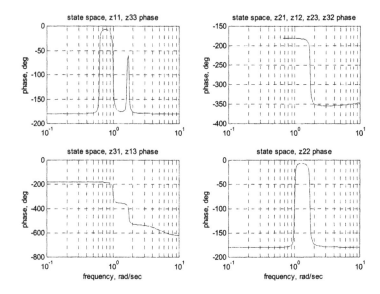

Figure 11.2: **Phase output for four distinct frequency responses, proportional damping zeta = 2%.**

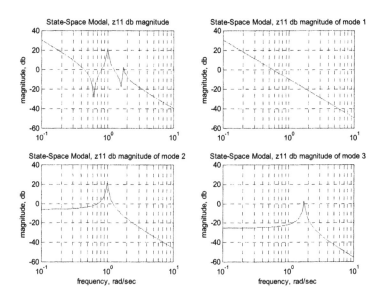

Figure 11.3: **Magnitude output for z11 frequency response and individual mode contributions.**

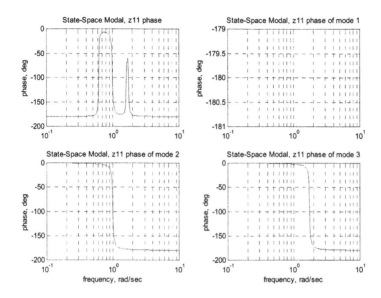

Figure 11.4: Phase output for z11 frequency response and individual mode contributions.

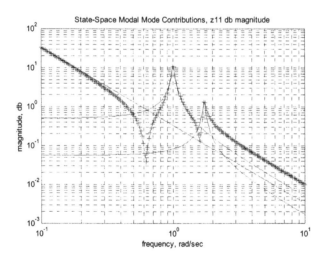

Figure 11.5: Overlaid magnitude output for z11 frequency response and individual mode contributions.

11.6 Forms of Frequency Response Plotting

This section of code is used to plot various forms of frequency responses for the z11 transfer function, as shown in Chapter 3, Section 3.6. All the plots except the Nyquist plot use user-defined damping and 200 frequency points. The Nyquist section recalculates the system matrix to use a damping zeta of 0.02 and 800 frequency points in order to plot in the designated format.

```
%        plot only z11 transfer function in different formats

         orient tall

%        log mag, log freq

         subplot(2,1,1)
         loglog(w,z11mag(1,:),'k-')
         title('z11, z33 log mag versus log freq')
         ylabel('magnitude')
         grid

         subplot(2,1,2)
         semilogx(w,z11phs(1,:),'k-')
         title('z11, z33 phase versus log freq')
         xlabel('frequency, rad/sec')
         ylabel('phase, deg')
         grid

         disp('execution paused to display figure, "enter" to continue'); pause

%        db mag, log freq

         subplot(2,1,1)
         semilogx(w,z11magdb(1,:),'k-')
         title('z11, z33 db mag versus log freq')
         ylabel('magnitude, db')
         grid

         subplot(2,1,2)
         semilogx(w,z11phs(1,:),'k-')
         title('z11, z33 phase versus log freq')
         xlabel('frequency, rad/sec')
         ylabel('phase, deg')
         grid

         disp('execution paused to display figure, "enter" to continue'); pause

%        db mag, lin freq

         subplot(2,1,1)
         plot(w,z11magdb(1,:),'k-')
         title('z11, z33 db mag versus linear freq')
         ylabel('magnitude, db')
         grid

         subplot(2,1,2)
         plot(w,z11phs(1,:),'k-')
         title('z11, z33 phase versus linear freq')
         xlabel('frequency, rad/sec')
```

```
            ylabel('phase, deg')
            grid

            disp('execution paused to display figure, "enter" to continue'); pause

%           lin mag, lin freq

            subplot(2,1,1)
            plot(w,z11mag(1,:),'k-')
            title('z11, z33 linear mag versus linear freq')
            ylabel('magnitude')
            grid

            subplot(2,1,2)
            plot(w,z11phs(1,:),'k-')
            title('z11, z33 phase versus linear freq')
            xlabel('frequency, rad/sec')
            ylabel('phase, deg')
            grid

            disp('execution paused to display figure, "enter" to continue'); pause

%           linear real versus log freq, linear imag versus log freq

            z11real = z11mag.*cos(z11phs*pi/180);       % convert from mag/angle to real

            z11realdb = 20*log10(z11real);

            z11imag = z11mag.*sin(z11phs*pi/180);  %  convert from mag/angle to imag

            z11imagdb = 20*log10(z11imag);

            subplot(2,1,1)
            semilogx(w,z11real(1,:),'k-')
            title('z11, z33 linear real mag versus log freq')
            ylabel('real magnitude')
            grid

            subplot(2,1,2)
            semilogx(w,z11imag(1,:),'k-')
            title('z11, z33 linear imaginary versus log freq')
            xlabel('frequency, rad/sec')
            ylabel('imaginary magnitude');
            grid

            disp('execution paused to display figure, "enter" to continue'); pause

%           linear real versus linear freq, linear imag versus linear freq

            subplot(2,1,1)
            plot(w,z11real(1,:),'k-')
            title('z11, z33 linear real mag versus linear freq')
            ylabel('real magnitude')
            grid
```

```
        subplot(2,1,2)
        plot(w,z11imag(1,:),'k-')
        title('z11, z33 linear imaginary versus linear freq')
        xlabel('frequency, rad/sec')
        ylabel('imaginary magnitude');
        grid

        disp('execution paused to display figure, "enter" to continue'); pause

%       real versus imaginary (Nyquist), redo frequency response with 800 points for
%       finer frequency resolution for Nyquist plot and use zeta = 0.02 to fit on plot

        zeta = 0.02;

a_ss = [0       1       0          0          0          0
        0       0       0          0          0          0
        0       0       0          1          0          0
        0       0     -w2^2    -2*zeta*w2     0          0
        0       0       0          0          0          1
        0       0       0          0        -w3^2    -2*zeta*w3];

        w = logspace(-1,1,800);

        sys = ss(a_ss,b,c,d);

        [mag,phs] = bode(sys,w);

        z11mag = mag(1,1,:);

        z11magdb = 20*log10(z11mag);

        z11phs = phs(1,1,:);

        z11real = z11mag.*cos(z11phs*pi/180);     % convert from mag/angle to real

        z11imag = z11mag.*sin(z11phs*pi/180);   % convert from mag/angle to imag

        subplot(1,1,1)

        plot(z11real(1,:),z11imag(1,:),'k+:')
        title('z11, z33 real versus imaginary, "Nyquist"')
        ylabel('imag')
        axis('square')
        axis([-15 15 -15 15])
        grid
```

Problem

Note: This problem refers to the two dof system shown in Figure P2.2.

P11.1 (MATLAB) Modify the **tdofss_modal_xfer_modes.m** code for the two dof system with $m_1 = m_2 = m = 1$, $k_1 = k_2 = k = 1$ and plot the frequency responses with and without the individual mode contributions overlaid.

CHAPTER 12

TIME DOMAIN: MODAL STATE SPACE FORM

12.1 Introduction

In Chapter 7 we derived the equations of motion in modal form for the system in Figure 12.1. In this chapter we will convert the modal form to state space modal form and obtain the closed form transient solution for the forcing function and initial conditions described in Figure 12.1. MATLAB will then be used to solve the same equations using the ode45 function.

12.2 Equations of Motion – Modal Form

The applied step forces are as shown in Figure 12.1. The initial conditions of position and velocity for each of the three masses are displayed in Table 12.1, the same as Figure 9.1 and Table 9.1.

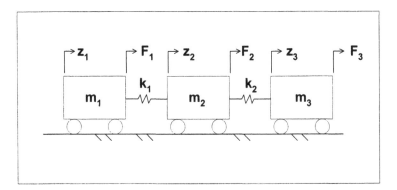

Figure 12.1: Step forces applied to tdof system.

Mass 1	Mass 2	Mass 3
$z_{01} = 0$	$z_{02} = -1$	$z_{03} = 1$
$\dot{z}_{01} = -1$	$\dot{z}_{02} = 2$	$\dot{z}_{03} = -2$

Table 12.1: Initial conditions applied to tdof system.

Repeating results from Chapter 9, where we developed the modal form of the equations of motion:

The force vector in principal coordinates from (9.8) is:

$$\mathbf{F}_p = \mathbf{z}_n^T \mathbf{F} = \begin{bmatrix} F_{p1} \\ F_{p2} \\ F_{p3} \end{bmatrix} = \frac{1}{\sqrt{m}} \begin{bmatrix} \dfrac{1}{\sqrt{3}} & \dfrac{1}{\sqrt{3}} & \dfrac{1}{\sqrt{3}} \\ \dfrac{1}{\sqrt{2}} & 0 & \dfrac{-1}{\sqrt{2}} \\ \dfrac{1}{\sqrt{6}} & \dfrac{-2}{\sqrt{6}} & \dfrac{1}{\sqrt{6}} \end{bmatrix} \begin{bmatrix} 1 \\ 0 \\ -2 \end{bmatrix} = \frac{1}{\sqrt{m}} \begin{bmatrix} \dfrac{-\sqrt{3}}{3} \\ \dfrac{3\sqrt{2}}{2} \\ \dfrac{-\sqrt{6}}{6} \end{bmatrix} \qquad (12.1)$$

With initial conditions from (9.6), (9.7):

$$\mathbf{z}_{po} = \sqrt{m} \begin{bmatrix} 0 \\ \dfrac{-\sqrt{2}}{2} \\ \dfrac{\sqrt{6}}{2} \end{bmatrix}, \quad \dot{\mathbf{z}}_{po} = \sqrt{m} \begin{bmatrix} \dfrac{-\sqrt{3}}{3} \\ \dfrac{\sqrt{2}}{2} \\ \dfrac{-7\sqrt{6}}{6} \end{bmatrix} \qquad (12.2)$$

Using the results of the eigenvalue solution, we can write the homogeneous equations of motion by inspection. The forcing function can be added to the right-hand side, knowing $\mathbf{F}_p$:

$$\dot{\mathbf{x}} = \mathbf{Ax} + \mathbf{Bu} \qquad (12.3)$$

$$\begin{bmatrix} \dot{x}_1 \\ \dot{x}_2 \\ \dot{x}_3 \\ \dot{x}_4 \\ \dot{x}_5 \\ \dot{x}_6 \end{bmatrix} = \begin{bmatrix} 0 & 1 & 0 & 0 & 0 & 0 \\ 0 & 0 & 0 & 0 & 0 & 0 \\ 0 & 0 & 0 & 1 & 0 & 0 \\ 0 & 0 & -\omega_2^2 & -2\zeta_2\omega_2 & 0 & 0 \\ 0 & 0 & 0 & 0 & 0 & 1 \\ 0 & 0 & 0 & 0 & -\omega_3^2 & -2\zeta_3\omega_3 \end{bmatrix} \begin{bmatrix} x_1 \\ x_2 \\ x_3 \\ x_4 \\ x_5 \\ x_6 \end{bmatrix} + \begin{bmatrix} 0 \\ F_{p1} \\ 0 \\ F_{p2} \\ 0 \\ F_{p3} \end{bmatrix} u \qquad (12.4)$$

with initial conditions of:

$$\mathbf{x}_{po} = \begin{bmatrix} z_{po1} \\ \dot{z}_{po1} \\ z_{po2} \\ \dot{z}_{po2} \\ z_{p03} \\ \dot{z}_{p03} \end{bmatrix} = \sqrt{m} \begin{bmatrix} 0 \\ \dfrac{-\sqrt{3}}{3} \\ \dfrac{-\sqrt{2}}{2} \\ \dfrac{\sqrt{2}}{2} \\ \dfrac{\sqrt{6}}{2} \\ \dfrac{-7\sqrt{6}}{6} \end{bmatrix} \qquad (12.5)$$

12.3 Solving Equations of Motion Using Laplace Transforms

Now that we know the complete state space equations of motion in principal coordinates and the initial conditions on the six states in principal coordinates, the equations can be solved in the time domain. The first order equations of motion above are similar in nature to the second order equations of motion in Table 7.2. The three **sets** of first order equations in modal state space form are uncoupled as were the three second order equations of motion in modal form (7.89).

Expanding the three sets of equations:

$$\begin{aligned}
\dot{x}_1 &= x_2 \\
\dot{x}_2 &= F_{p1}u \\
\dot{x}_3 &= x_4 \\
\dot{x}_4 &= -\omega_2^2 x_3 - 2\zeta_2\omega_2 x_4 + F_{p2}u \\
\dot{x}_5 &= x_6 \\
\dot{x}_6 &= -\omega_3^2 x_5 - 2\zeta_3\omega_3 x_6 + F_{p3}u
\end{aligned} \qquad (12.6a\text{-}f)$$

Taking the Laplace transform of the first two equations above:

$$\begin{aligned}
sx_1(s) - x_1(0) &= x_2(s) \\
sx_2(s) - x_2(0) &= F_{p1}u(s) = \frac{F_{p1}}{s}
\end{aligned} \qquad (12.7a,b)$$

Solving for $x_1(s)$:

$$sx_1(s) - x_1(0) = x_2(s)$$

$$s[sx_1(s) - x_1(0)] - x_2(0) = F_{pl}u(s) = \frac{F_{pl}}{s}$$

$$s^2 x_1(s) = \frac{F_{pl}}{s} + sx_1(0) + x_2(0) \qquad (12.8a\text{-}f)$$

$$x_1(s) = \frac{F_{pl}}{s^3} + \frac{sx_1(0)}{s^2} + \frac{x_2(0)}{s^2}$$

$$x_1(s) = \frac{-\sqrt{3}}{3s^3\sqrt{m}} + \frac{0}{s} - \frac{\sqrt{3m}}{3s^2}$$

The three terms on the right-hand side of (12.8f) represent the displacement of the first mode of vibration due to the force, initial displacement and initial velocity, respectively. This equation for $x_1(s)$ is the same as for $z_{pl}(s)$ in (9.17). Using the same back-transformation yields the identical result for the principal displacement as for $z_{pl}(t)$ in (9.20).

$$x_1(t) = \frac{-t^2}{2\sqrt{3m}} + 0 - \frac{\sqrt{3m}\ t}{3}$$

$$= \frac{-t^2}{2\sqrt{3}} + 0 - \frac{\sqrt{3}\ t}{3} \qquad (12.9)$$

The two sets of equations for modes 2 and 3 can be solved for $x_3(t)$ and $x_5(t)$ in a similar fashion, again giving results which are the same as for $z_{p2}(t)$ and $z_{p3}(t)$ in (9.27) and (9.34). The three velocity states in principal coordinates can be defined by differentiating the displacement states.

Summarizing the solution in principal state space coordinates:

$$\mathbf{x}(t) = \begin{bmatrix} x_1 \\ x_2 \\ x_3 \\ x_4 \\ x_5 \\ x_6 \end{bmatrix} = \begin{bmatrix} -\dfrac{t^2}{2\sqrt{3}} - \dfrac{\sqrt{3}\,t}{3} \\[2ex] -\dfrac{t}{\sqrt{3}} - \dfrac{\sqrt{3}}{3} \\[2ex] \dfrac{3\sqrt{2}}{2} - \dfrac{3\sqrt{2}}{2}\cos t - \dfrac{\sqrt{2}}{2}\cos t + \dfrac{\sqrt{2}}{2}\sin t \\[2ex] \dfrac{3\sqrt{2}}{2}\sin t + \dfrac{\sqrt{2}}{2}\sin t + \dfrac{\sqrt{2}}{2}\cos t \\[2ex] -\dfrac{\sqrt{6}}{18} + \dfrac{\sqrt{6}}{18}\cos\sqrt{3}\,t + \dfrac{\sqrt{6}}{2}\cos\sqrt{3}\,t - \dfrac{7}{\sqrt{3}}\sin\sqrt{3}\,t \\[2ex] -\dfrac{\sqrt{6}\sqrt{3}}{18}\sin\sqrt{3}\,t + \dfrac{\sqrt{6}\sqrt{3}}{2}\sin\sqrt{3}\,t - \dfrac{7\sqrt{3}}{\sqrt{3}}\cos\sqrt{3}\,t \end{bmatrix} \qquad (12.10a\text{-}f)$$

Let us assume that we are interested in three displacements and three velocities; the output matrix is shown below in (12.11), repeated from (10.38):

$$\mathbf{C} = \begin{bmatrix} x_{n11} & 0 & x_{n12} & 0 & x_{n13} & 0 \\ 0 & x_{n11} & 0 & x_{n12} & 0 & x_{n13} \\ x_{n21} & 0 & x_{n22} & 0 & x_{n23} & 0 \\ 0 & x_{n21} & 0 & x_{n22} & 0 & x_{n23} \\ x_{n31} & 0 & x_{n32} & 0 & x_{n33} & 0 \\ 0 & x_{n31} & 0 & x_{n32} & 0 & x_{n33} \end{bmatrix} \qquad (12.11)$$

$$
\begin{bmatrix} z_1 \\ \dot{z}_1 \\ z_2 \\ \dot{z}_2 \\ z_3 \\ \dot{z}_3 \end{bmatrix} = \mathbf{Cx} = \begin{bmatrix} x_{n11} & 0 & x_{n12} & 0 & x_{n13} & 0 \\ 0 & x_{n11} & 0 & x_{n12} & 0 & x_{n13} \\ x_{n21} & 0 & x_{n22} & 0 & x_{n23} & 0 \\ 0 & x_{n21} & 0 & x_{n22} & 0 & x_{n23} \\ x_{n31} & 0 & x_{n32} & 0 & x_{n33} & 0 \\ 0 & x_{n31} & 0 & x_{n32} & 0 & x_{n33} \end{bmatrix} \begin{bmatrix} x_1 \\ x_2 \\ x_3 \\ x_4 \\ x_5 \\ x_6 \end{bmatrix}
$$

$$
= \begin{bmatrix} \frac{1}{\sqrt{3}} & 0 & \frac{1}{\sqrt{2}} & 0 & \frac{1}{\sqrt{6}} & 0 \\ 0 & \frac{1}{\sqrt{3}} & 0 & \frac{1}{\sqrt{2}} & 0 & \frac{1}{\sqrt{6}} \\ \frac{1}{\sqrt{3}} & 0 & 0 & 0 & \frac{-2}{\sqrt{6}} & 0 \\ 0 & \frac{1}{\sqrt{3}} & 0 & 0 & 0 & \frac{-2}{\sqrt{6}} \\ \frac{1}{\sqrt{3}} & 0 & \frac{-1}{\sqrt{2}} & 0 & \frac{1}{\sqrt{6}} & 0 \\ 0 & \frac{1}{\sqrt{3}} & 0 & \frac{-1}{\sqrt{2}} & 0 & \frac{1}{\sqrt{6}} \end{bmatrix} \begin{bmatrix} x_1 \\ x_2 \\ x_3 \\ x_4 \\ x_5 \\ x_6 \end{bmatrix}
$$

(12.12)

With (12.12) we have the complete time domain results in physical coordinates.

12.4 MATLAB Code tdofss_modal_time_ode45.m – Time Domain Modal Contributions

12.4.1 Modal State Space Model Setup, Code Listing

This first section executes **tdofss_eig.m** to calculate the eigenvalues and eigenvectors. It then sets up the 6x6 system matrix and defines three individual mode 2x2 submatrices.

The force vector in physical coordinates is defined, applying step forces as defined in Figure 12.1. It is transformed to a forcing function in principal coordinates and expanded to 6x1 size by padding with zeros. To specify the input matrices for each of the three modes, three 2x1 submatrices are defined.

The output matrix is setup as a 3x6 matrix, to calculate displacements. Once again, three submatrices of 3x2 size are defined for the individual modes.

```
%          tdofss_modal_time_ode45.m      state space modal form transfer function analysis
%          of tdof model, proportional damping, modal contribution plotting

           clf;

%          run tdofss_eig.m to provide eigenvalues and eigenvectors

           tdofss_eig;

           global a_ss a1_ss a2_ss a3_ss b b1 b2 b3 u

%          note, this is the point where we would start if we had eigenvalue results from ANSYS,
%          using the eigenvalues and eigenvectors to define state space equations in
%          principal coordinates

%          define damping ratio to be used for proportional damping in the state space equation
%          in principal coordinates

           zeta = input('input zeta, 0.02 = 2% of critical damping (default) ... ');

           if (isempty(zeta))
           zeta = 0.02;
           else
           end

%          setup 6x6 state-space system matrix for all three modes in principal
%          coordinates, a_ss

           a_ss = [    0          1          0          0          0          0
                       0          0          0          0          0          0
                       0          0          0          1          0          0
                       0          0        -w2^2     -2*zeta*w2     0          0
                       0          0          0          0          0          1
                       0          0          0          0        -w3^2     -2*zeta*w3];

%          setup three 2x2 state-space matrices, one for each individual mode

           a1_ss = a_ss(1:2,1:2);

           a2_ss = a_ss(3:4,3:4);

           a3_ss = a_ss(5:6,5:6);

%          transform the 3x1 force vector in physical coordinates to principal coordinates and
%          then insert the principal forces in the appropriate rows in the state-space
%          6x1 input matrix, padding with zeros as appropriate

           F = [1 0 -2]';

           Fp = xn'*F;

%          expand the force vectors in principal coordinates from 3x1 to 6x1, padding with zeros

           b = [0 Fp(1) 0 Fp(2) 0 Fp(3)]';    %  principal forces applied to all masses
```

```
            b1 = b(1:2);

            b2 = b(3:4);

            b3 = b(5:6);
%           the output matrix c is setup in one step, to allow the "bode" command to
%           output the desired physical coordinates directly without having to go
%           through any intermediate steps.

%           setup the output matrix for displacement transfer functions, each row
%           represents the position outputs of mass 1, mass 2 and mass 3
%           velocities not included, so c is only 3x6 instead of 6x6

            c = [xn(1,1)  0  xn(1,2)  0  xn(1,3)  0
                 xn(2,1)  0  xn(2,2)  0  xn(2,3)  0
                 xn(3,1)  0  xn(3,2)  0  xn(3,3)  0];

            c1 = c(:,1:2);

            c2 = c(:,3:4);

            c3 = c(:,5:6);
%           define direct transmission matrix d

            d = 0;
```

12.4.2 Problem Setup, Initial Conditions, Code Listing

Now that the model is in place, we can solve for transient response. The input scalar, "u" is set to "1," for a unity step function. The total time is set and a vector of time span from 0 to 10 seconds (default) is setup for input to the ode routine.

The two 3x1 initial condition displacement and velocity vectors with initial displacements and velocities from Figure 12.1 are set up, then transformed to principal coordinates. Next the 6x1 initial condition vector is constructed from appropriate elements of the two 3x1 vectors. We are now ready to solve the problem.

```
%           transient response using the ode45 command

            u = 1;

            ttotal = input('Input total time for Simulation, default = 10 sec, ... ');

            if (isempty(ttotal))
            ttotal = 10;
            else
```

```
          end

          tspan = [0 ttotal];

%         calculate the initial conditions in principal coordinates using the inverse of the
%         normalized modal matrix

          x0phys = [0 -1 1]';            %         initial condition position, physical coord

          x0dphys = [-1 2 -2]';          %         initial condition velocity, physical coord

          x0 = inv(xn)*x0phys;

          x0d = inv(xn)*x0dphys;

%         create the initial condition state vector

          x0ss = [x0(1) x0d(1) x0(2) x0d(2) x0(3) x0d(3)];

          x0ss1 = x0ss(1:2);

          x0ss2 = x0ss(3:4);

          x0ss3 = x0ss(5:6);
```

12.4.3 Solving Equations Using ode45, Code Listing

The ode45 "options" parameter, which can be used to control many options for use in the solution, is set to a null vector.

Next, the total response in principal coordinates and the three individual mode responses in principal coordinates are calculated using MATLAB's ode45 differential equation solver. Four functions, listed separately in the following sections, are used by ode45 to define the equations to solve.

The responses in principal coordinates are then transformed to physical coordinates.

```
%         use the ode45 non-stiff differential equation solver

          options = [ ];                           % no options specified

%         total response, principal coord, states are modes of vibration

          [t,x] = ode45('tdofssmodalfun',tspan,x0ss,options);

% mode 1 response, principal coord

          [t1,x1] = ode45('tdofssmodal1fun',tspan,x0ss1,options);

% mode 2 response, principal coord
```

```
            [t2,x2] = ode45('tdofssmodal2fun',tspan,x0ss2,options);

% mode 3 response, principal coord

            [t3,x3] = ode45('tdofssmodal3fun',tspan,x0ss3,options);

% total response, physical coord

        z_ode = c*x';

% mode 1 response, physical coord

        z_ode1 = c1*x1';

% mode 2 response, physical coord

        z_ode2 = c2*x2';

% mode 3 response, physical coord

        z_ode3 = c3*x3';
```

12.4.4 Plotting, Code Listing

```
%       plot displacements in principal coordinates

        subplot(1,1,1);

        plot(t1,x1(:,1),'k+-',t2,x2(:,1),'kx-',t3,x3(:,1),'k-')
        title('Displacements in Principal Coordinate System, ode45')
        xlabel('Time, sec')
        ylabel('Displacements')
        legend('zp1','zp2','zp3',2)
        grid

        disp('execution paused to display figure, "enter" to continue'); pause

        axis([0 1 -2 2]);

        disp('execution paused to display figure, "enter" to continue'); pause

%       plot displacements in physical coordinates

        plot(t,z_ode(1,:),'k+-',t,z_ode(2,:),'kx-',t,z_ode(3,:),'k-')
        title('Displacements in Physical Coordinate System, ode45')
        xlabel('Time, sec')
        ylabel('Displacements')
        legend('z1','z2','z3',3)
        grid

        disp('execution paused to display figure, "enter" to continue'); pause

%       load previous closed-form solutions for tplot, z1, z2, z3 if zeta = 0
```

```
        if zeta == 0

        load tdof_modal_time_z1z2z3;

        plot(t,z_ode(1,:),'k-',t,z_ode(2,:),'k-',t,z_ode(3,:),'k-',tplot,z1,'k.-',tplot,z2, ...
                    'k.-',tplot,z3,'k.-')
        title('Displacements in Physical Coordinate System from ode45 (ode) ...
                    and Closed Form (cf)')
        xlabel('Time, sec')
        ylabel('Vibration Displacements')
        legend('ode dof 1','ode dof 2','ode dof 3','cf dof 1','cf dof 2','cf dof 3')
        grid

        disp('execution paused to display figure, "enter" to continue'); pause

        else
        end

%       plot the modal contributions to the motion of masses 1, 2 and 3

        plot(t1,z_ode1(1,:),'k+-',t2,z_ode2(1,:),'kx-',t3,z_ode3(1,:),'k-')
        title('Displacement of dof 1 for Modes 1, 2 and 3, ode45')
        xlabel('Time, sec')
        ylabel('Displacements')
        legend('Mode 1','Mode 2','Mode 3')
        grid

        disp('execution paused to display figure, "enter" to continue'); pause

        plot(t1,z_ode1(2,:),'k+-',t2,z_ode2(2,:),'kx-',t3,z_ode3(2,:),'k-')
        title('Displacement of dof 2 for Modes 1, 2 and 3, ode45')
        xlabel('Time, sec')
        ylabel('Displacements')
        legend('Mode 1','Mode 2','Mode 3')
        grid

        disp('execution paused to display figure, "enter" to continue'); pause

        plot(t1,z_ode1(3,:),'k+-',t2,z_ode2(3,:),'kx-',t3,z_ode3(3,:),'k-')
        title('Displacement of dof 3 for Modes 1, 2 and 3, ode45')
        xlabel('Time, sec')
        ylabel('Displacements')
        legend('Mode 1','Mode 2','Mode 3')
        grid
```

12.4.5 Functions Called: tdofssmodalfun.m, tdofssmodal1fun.m, tdofssmodal2fun.m, tdofssmodal3fun.m

The ode45 differential equation solver calls function files depending on which solution is being performed. The four functions for calculating the system response as well as individual responses of modes 1, 2 and 3 are listed below. Each simply defines the state equation where the derivative of the state vector is equal to the system matrix times the states plus the input matrix times the input: $\dot{\mathbf{x}} = \mathbf{Ax} + \mathbf{Bu}$. The "global" assignments make all the variables defined available both to the calling program and to the function.

System response:

```
function xprime = tdofssmodalfun(t,x)

%       function for calculating the transient response of tdof_ss_modal_time_ode45.m

        global a_ss a1_ss a2_ss a3_ss b b1 b2 b3 u

        xprime = a_ss*x + b*u;
```

Mode 1 response:

```
function xprime = tdofssmodal1fun(t1,x1)

%       function for calculating the transient response of tdof_ss_modal_time_ode45.m

        global a_ss a1_ss a2_ss a3_ss b b1 b2 b3 u

        xprime = a1_ss*x1 + b1*u;
```

Mode 2 response:

```
function xprime = tdofssmodal2fun(t2,x2)

%       function for calculating the transient response of tdof_ss_modal_time_ode45.m

        global a_ss a1_ss a2_ss a3_ss b b1 b2 b3 u

        xprime = a2_ss*x2 + b2*u;
```

Mode 3 response:

```
function xprime = tdofssmodal3fun(t3,x3)

%       function for calculating the transient response of tdof_ss_modal_time_ode45.m

        global a_ss a1_ss a2_ss a3_ss b b1 b2 b3 u

        xprime = a3_ss*x3 + b1*u;
```

12.5 Plotted Results

The following figures should be compared with Figures 9.2 through 9.7, which were plotted using the closed form modal solutions.

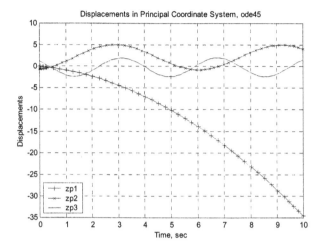

Figure 12.2: Displacements in principal coordinate system using ode45.

The motions of the rigid body and two oscillatory modes are clearly seen.

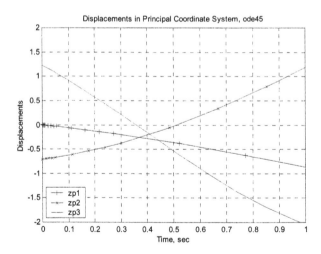

Figure 12.3: Displacements in principal coordinate system, expanded scales to see initial conditions.

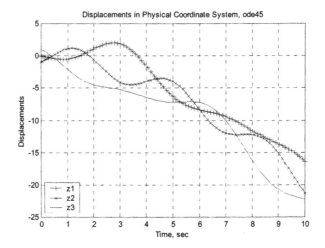

Figure 12.4: Displacements in physical coordinate system.

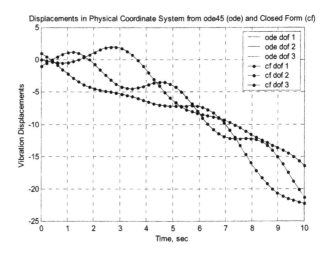

Figure 12.5: Displacements in physical coordinate system – comparing closed form solution from Chapter 7.

The three plots below show how one can study the motions of degrees of freedom due to individual modes. Use zeta = 0 when running **tdofss_modal_time_ode45.m** in order to plot the closed form solution.

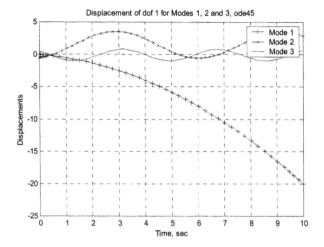

Figure 12.6: Displacement of mass 1 for modes 1, 2 and 3.

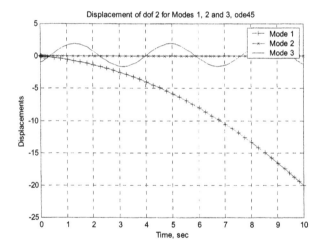

Figure 12.7: Displacement of mass 2 for modes 1, 2 and 3.

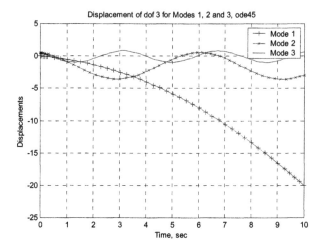

Figure 12.8: Displacement of mass 3 for modes 1, 2 and 3.

Problem

P12.1 Using the initial conditions and forcing functions from P7.4, solve for the time domain response of the states in principal coordinates in closed form using Laplace transforms. Define the output matrix if the outputs required are the displacements of both masses.

CHAPTER 13

FINITE ELEMENTS: STIFFNESS MATRICES

13.1 Introduction

The purpose of this chapter is to use two simple examples to explain the basics of how finite element stiffness matrices are formulated and how static finite element analysis is performed.

Chapter 2 discussed building global stiffness matrices column by column, giving a unit displacement to the dof associated with each column and entering constraint forces for each dof along the column. This chapter will show another method of building global stiffness matrices, based on using **element** stiffness matrices, combining them in an orderly way to generate the global stiffness matrix. The first example uses the lumped parameter 6dof example seen in Section 2.2.4. The second example uses a two-element cantilever. Static condensation is used to prepare for a development of Guyan reduction in the next chapter.

The next chapter will use element mass matrices to assemble global mass matrices and will introduce dynamics using finite elements.

13.2 Six dof Model – Element and Global Stiffness Matrices

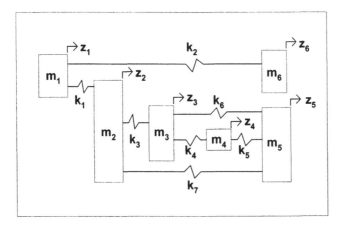

Figure 13.1: Six dof stiffness matrix model.

13.2.1 Overview

The global stiffness matrix for the model in Figure 13.1 was defined previously by inspection (Table 2.2). Each column of the matrix was defined by giving a unit displacement to the dof associated with that column and then defining the constraints required to hold the system in that configuration. This method works very well for hand calculations, but creating stiffness and mass matrices with computers requires a different, more systematic approach, where individual element stiffness matrices are developed and combined to give the global stiffness matrix.

We can define an element stiffness matrix for each of the springs in the figure, where the size of the element stiffness matrix is (nxn), and n is the total number of degrees of freedom associated with the element. For a uni-axial spring, there are two degrees of freedom, the displacements in the "z" direction at both ends, hence a 2x2 stiffness matrix.

Each element stiffness matrix can be set up using the "inspection" method, by displacing first the left-hand dof for the first column, and then the right-hand dof for the second column as shown in Figure 13.2.

13.2.2 Element Stiffness Matrix

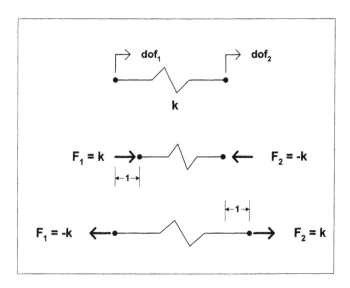

Figure 13.2: Spring element stiffness matrix development.

The resulting element stiffness matrix, $\mathbf{k}_{el}$, for a general uni-axial spring element is then:

$$\mathbf{k}_{el,i} = \begin{bmatrix} k_i & -k_i \\ -k_i & k_i \end{bmatrix} \tag{13.1}$$

For spring element 3, for example, the element stiffness matrix would be:

$$\mathbf{k}_{el,3} = \begin{bmatrix} k_3 & -k_3 \\ -k_3 & k_3 \end{bmatrix} \tag{13.2}$$

13.2.3 Building Global Stiffness Matrix Using Element Stiffness Matrices

The total number of degrees of freedom for the problem is 6, so the complete system stiffness matrix, the **global** stiffness matrix, is a 6x6 matrix. Each row and column of every element stiffness matrix can be associated with a **global degree of freedom**.

For element 1, which is connected to degrees of freedom 1 and 2:

1st and 2nd columns of global stiffness matrix
$\tag{13.3}$

$$\mathbf{k}_{el,1} = \begin{array}{cc} z_1 & z_2 \\ \begin{bmatrix} k_1 & -k_1 \\ -k_1 & k_1 \end{bmatrix} & \begin{array}{l} z_1 \\ z_2 \end{array} \end{array} \quad \begin{array}{l} 1^{st} \text{ row of global stiffness matrix} \\ 2^{nd} \text{ row of global stiffness matrix} \end{array}$$

For element 2, which is connected to degrees of freedom 1 and 6:

1st and 6th columns of global stiffness matrix
$\tag{13.4}$

$$\mathbf{k}_{el,2} = \begin{array}{cc} z_1 & z_6 \\ \begin{bmatrix} k_2 & -k_2 \\ -k_2 & k_2 \end{bmatrix} & \begin{array}{l} z_1 \\ z_6 \end{array} \end{array} \quad \begin{array}{l} 1^{st} \text{ row of global stiffness matrix} \\ 6^{th} \text{ row of global stiffness matrix} \end{array}$$

For element 3, which is connected to degrees of freedom 2 and 3:

2nd and 3rd columns of global stiffness matrix
$\tag{13.5}$

$$\mathbf{k}_{el,3} = \begin{array}{cc} z_2 & z_3 \\ \begin{bmatrix} k_3 & -k_3 \\ -k_3 & k_3 \end{bmatrix} & \begin{array}{l} z_2 \\ z_3 \end{array} \end{array} \quad \begin{array}{l} 2^{nd} \text{ row of global stiffness matrix} \\ 3^{rd} \text{ row of global stiffness matrix} \end{array}$$

For element 4, which is connected to degrees of freedom 3 and 4:

$$3^{rd} \text{ and } 4^{th} \text{ columns of global stiffness matrix}$$ (13.6)

$$\mathbf{k}_{el,4} = \begin{matrix} z_3 & z_4 \\ \begin{bmatrix} k_4 & -k_4 \\ -k_4 & k_4 \end{bmatrix} & \begin{matrix} z_3 & 3^{rd} \text{ row of global stiffness matrix} \\ z_4 & 4^{th} \text{ row of global stiffness matrix} \end{matrix} \end{matrix}$$

For element 5, which is connected to degrees of freedom 4 and 5:

$$4^{th} \text{ and } 5^{th} \text{ columns of global stiffness matrix}$$ (13.7)

$$\mathbf{k}_{el,5} = \begin{matrix} z_4 & z_5 \\ \begin{bmatrix} k_5 & -k_5 \\ -k_5 & k_5 \end{bmatrix} & \begin{matrix} z_4 & 4^{th} \text{ row of global stiffness matrix} \\ z_5 & 5^{th} \text{ row of global stiffness matrix} \end{matrix} \end{matrix}$$

For element 6, which is connected to degrees of freedom 3 and 5:

$$3^{rd} \text{ and } 5^{th} \text{ columns of global stiffness matrix}$$ (13.8)

$$\mathbf{k}_{el,6} = \begin{matrix} z_3 & z_5 \\ \begin{bmatrix} k_6 & -k_6 \\ -k_6 & k_6 \end{bmatrix} & \begin{matrix} z_3 & 3^{rd} \text{ row of global stiffness matrix} \\ z_5 & 5^{th} \text{ row of global stiffness matrix} \end{matrix} \end{matrix}$$

For element 7, which is connected to degrees of freedom 2 and 5:

$$2^{nd} \text{ and } 5^{th} \text{ columns of global stiffness matrix}$$ (13.9)

$$\mathbf{k}_{el,7} = \begin{matrix} z_2 & z_5 \\ \begin{bmatrix} k_7 & -k_7 \\ -k_7 & k_7 \end{bmatrix} & \begin{matrix} z_2 & 2^{nd} \text{ row of global stiffness matrix} \\ z_5 & 5^{th} \text{ row of global stiffness matrix} \end{matrix} \end{matrix}$$

The global stiffness matrix starts out as a 6x6 null matrix, then each element is cycled through and its elements added to the previous matrix. The initial null matrix is:

$$\mathbf{k}_g = \begin{bmatrix} 0 & 0 & 0 & 0 & 0 & 0 \\ 0 & 0 & 0 & 0 & 0 & 0 \\ 0 & 0 & 0 & 0 & 0 & 0 \\ 0 & 0 & 0 & 0 & 0 & 0 \\ 0 & 0 & 0 & 0 & 0 & 0 \\ 0 & 0 & 0 & 0 & 0 & 0 \end{bmatrix} \tag{13.10}$$

After adding the element stiffness matrix for element 1:

$$\mathbf{k}_g = \begin{bmatrix} k_1 & -k_1 & 0 & 0 & 0 & 0 \\ -k_1 & k_1 & 0 & 0 & 0 & 0 \\ 0 & 0 & 0 & 0 & 0 & 0 \\ 0 & 0 & 0 & 0 & 0 & 0 \\ 0 & 0 & 0 & 0 & 0 & 0 \\ 0 & 0 & 0 & 0 & 0 & 0 \end{bmatrix} \tag{13.11}$$

After adding the element stiffness matrices for elements 1 to 2:

$$\mathbf{k}_g = \begin{bmatrix} k_1+k_2 & -k_1 & 0 & 0 & 0 & -k_2 \\ -k_1 & k_1 & 0 & 0 & 0 & 0 \\ 0 & 0 & 0 & 0 & 0 & 0 \\ 0 & 0 & 0 & 0 & 0 & 0 \\ 0 & 0 & 0 & 0 & 0 & 0 \\ -k_2 & 0 & 0 & 0 & 0 & k_2 \end{bmatrix} \tag{13.12}$$

After adding the element stiffness matrices for elements 1 to 3:

$$\mathbf{k}_g = \begin{bmatrix} k_1+k_2 & -k_1 & 0 & 0 & 0 & -k_2 \\ -k_1 & k_1+k_3 & -k_3 & 0 & 0 & 0 \\ 0 & -k_3 & k_3 & 0 & 0 & 0 \\ 0 & 0 & 0 & 0 & 0 & 0 \\ 0 & 0 & 0 & 0 & 0 & 0 \\ -k_2 & 0 & 0 & 0 & 0 & k_2 \end{bmatrix} \tag{13.13}$$

After adding the element stiffness matrices for elements 1 to 4:

$$\mathbf{k}_g = \begin{bmatrix} k_1+k_2 & -k_1 & 0 & 0 & 0 & -k_2 \\ -k_1 & k_1+k_3 & -k_3 & 0 & 0 & 0 \\ 0 & -k_3 & k_3+k_4 & -k_4 & 0 & 0 \\ 0 & 0 & -k_4 & k_4 & 0 & 0 \\ 0 & 0 & 0 & 0 & 0 & 0 \\ -k_2 & 0 & 0 & 0 & 0 & k_2 \end{bmatrix} \tag{13.14}$$

After adding the element stiffness matrices for elements 1 to 5:

$$\mathbf{k}_g = \begin{bmatrix} k_1+k_2 & -k_1 & 0 & 0 & 0 & -k_2 \\ -k_1 & k_1+k_3 & -k_3 & 0 & 0 & 0 \\ 0 & -k_3 & k_3+k_4 & -k_4 & 0 & 0 \\ 0 & 0 & -k_4 & k_4+k_5 & -k_5 & 0 \\ 0 & 0 & 0 & -k_5 & k_5 & 0 \\ -k_2 & 0 & 0 & 0 & 0 & k_2 \end{bmatrix} \tag{13.15}$$

After adding the element stiffness matrices for elements 1 to 6:

$$\mathbf{k}_g = \begin{bmatrix} k_1+k_2 & -k_1 & 0 & 0 & 0 & -k_2 \\ -k_1 & k_1+k_3 & -k_3 & 0 & 0 & 0 \\ 0 & -k_3 & k_3+k_4+k_6 & -k_4 & -k_6 & 0 \\ 0 & 0 & -k_4 & k_4+k_5 & -k_5 & 0 \\ 0 & 0 & -k_6 & -k_5 & k_5+k_6 & 0 \\ -k_2 & 0 & 0 & 0 & 0 & k_2 \end{bmatrix} \tag{13.16}$$

After adding the element stiffness matrices for elements 1 to 7 we have the final global stiffness matrix.

$$\mathbf{k}_g = \begin{bmatrix} k_1+k_2 & -k_1 & 0 & 0 & 0 & -k_2 \\ -k_1 & k_1+k_3+k_7 & -k_3 & 0 & -k_7 & 0 \\ 0 & -k_3 & k_3+k_4+k_6 & -k_4 & -k_6 & 0 \\ 0 & 0 & -k_4 & k_4+k_5 & -k_5 & 0 \\ 0 & -k_7 & -k_6 & -k_5 & k_5+k_6+k_7 & 0 \\ -k_2 & 0 & 0 & 0 & 0 & k_2 \end{bmatrix}$$

$$\tag{13.17}$$

This checks against the original global stiffness matrix defined by inspection in Table 2.2 and fulfills the symmetry requirement.

$$
\begin{array}{c}
\begin{array}{cccccc} 1 & \quad 2 & \quad 3 & \quad 4 & \quad 5 & \quad 6 \end{array}\\
\begin{array}{c} 1\\2\\3\\4\\5\\6 \end{array}
\begin{bmatrix}
(k_1+k_2) & -k_1 & 0 & 0 & 0 & -k_2\\
-k_1 & (k_1+k_3+k_7) & -k_3 & 0 & -k_7 & 0\\
0 & -k_3 & (k_3+k_4+k_6) & -k_4 & -k_6 & 0\\
0 & 0 & -k_4 & (k_4+k_5) & -k_5 & 0\\
0 & -k_7 & -k_6 & -k_5 & (k_5+k_6+k_7) & 0\\
-k_2 & 0 & 0 & 0 & 0 & k_2
\end{bmatrix}
\end{array}
$$

$$(13.18)$$

13.3 Two-Element Cantilever Beam

We will now do a static finite element displacement analysis of a two-element cantilever beam. We start by showing the original model and defining the degrees of freedom for the idealized beam, Figure 13.3.

Note that even though the left-hand side node is grounded in the actual beam, there are degrees of freedom associated with the node to allow generating global stiffness and mass matrices for all nodes. The constrained degrees of freedom will be accounted for once the complete global stiffness matrix is available. For this model, each of the three nodes has two degrees of freedom, a translation and a rotation.

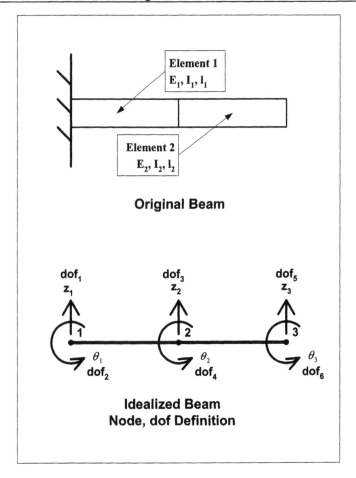

Figure 13.3: Two-element cantilever beam model and node definition.

13.3.1 Element Stiffness Matrix

The element stiffness matrix can be developed by using basic strength of materials techniques to analyze the forces required to displace each degree of freedom a unit value in the positive direction:

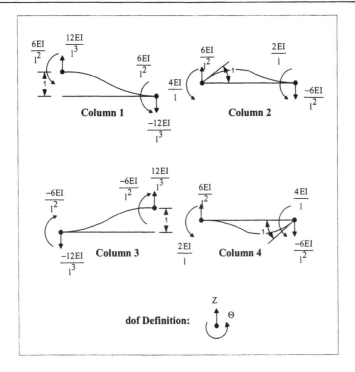

Figure 13.4: Beam element stiffness matrix terms.

13.3.2 Degree of Freedom Definition – Beam Stiffness Matrix

Using the degrees of freedom in Figure 13.5 results in the following element stiffness matrix:

$$
\mathbf{k}_{el,i} = E_i I_i
\begin{bmatrix}
\dfrac{12}{l_i^3} & \dfrac{6}{l_i^2} & \dfrac{-12}{l_i^3} & \dfrac{6}{l_i^2} \\[2mm]
\dfrac{6}{l_i^2} & \dfrac{4}{l_i} & \dfrac{-6}{l_i^2} & \dfrac{2}{l_i} \\[2mm]
\dfrac{-12}{l_i^3} & \dfrac{-6}{l_i^2} & \dfrac{12}{l_i^3} & \dfrac{-6}{l_i^2} \\[2mm]
\dfrac{6}{l_i^2} & \dfrac{2}{l_i} & \dfrac{-6}{l_i^2} & \dfrac{4}{l_i}
\end{bmatrix}
\tag{13.19}
$$

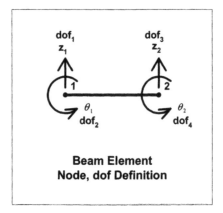

Figure 13.5: Beam element node and degree of freedom definition.

13.3.3 Building Global Stiffness Matrix Using Element Stiffness Matrices

To build the global stiffness matrix, we start with a 6x6 null matrix, with the six degrees of freedom being the translation and rotation of each of the three nodes, again including the constrained node 1 degrees of freedom:

$$
\mathbf{k}_g =
\begin{bmatrix}
0 & 0 & 0 & 0 & 0 & 0 \\
0 & 0 & 0 & 0 & 0 & 0 \\
0 & 0 & 0 & 0 & 0 & 0 \\
0 & 0 & 0 & 0 & 0 & 0 \\
0 & 0 & 0 & 0 & 0 & 0 \\
0 & 0 & 0 & 0 & 0 & 0
\end{bmatrix}
\begin{matrix}
\text{displacement of node 1} \\
\text{rotation of node 1} \\
\text{displacement of node 2} \\
\text{rotation of node 2} \\
\text{displacement of node 3} \\
\text{rotation of node 3}
\end{matrix}
\qquad (13.20)
$$

The two 4x4 element stiffness matrices are:

$$
\mathbf{k}_{el,1} = E_1 I_1
\begin{bmatrix}
\dfrac{12}{l_1^3} & \dfrac{6}{l_1^2} & \dfrac{-12}{l_1^3} & \dfrac{6}{l_1^2} \\[2ex]
\dfrac{6}{l_1^2} & \dfrac{4}{l_1} & \dfrac{-6}{l_1^2} & \dfrac{2}{l_1} \\[2ex]
\dfrac{-12}{l_1^3} & \dfrac{-6}{l_1^2} & \dfrac{12}{l_1^3} & \dfrac{-6}{l_1^2} \\[2ex]
\dfrac{6}{l_1^2} & \dfrac{2}{l_1} & \dfrac{-6}{l_1^2} & \dfrac{4}{l_1}
\end{bmatrix}
\qquad (13.21)
$$

$$\mathbf{k}_{el,2} = E_2 I_2 \begin{bmatrix} \dfrac{12}{l_2^3} & \dfrac{6}{l_2^2} & \dfrac{-12}{l_2^3} & \dfrac{6}{l_2^2} \\[2mm] \dfrac{6}{l_2^2} & \dfrac{4}{l_2} & \dfrac{-6}{l_2^2} & \dfrac{2}{l_2} \\[2mm] \dfrac{-12}{l_2^3} & \dfrac{-6}{l_2^2} & \dfrac{12}{l_2^3} & \dfrac{-6}{l_2^2} \\[2mm] \dfrac{6}{l_2^2} & \dfrac{2}{l_2} & \dfrac{-6}{l_2^2} & \dfrac{4}{l_2} \end{bmatrix} \qquad (13.22)$$

Building up the global stiffness matrix, element by element, inserting element 1 first:

$$\mathbf{k}_g = \begin{bmatrix} \dfrac{12E_1 I_1}{l_1^3} & \dfrac{6E_1 I_1}{l_1^2} & \dfrac{-12E_1 I_1}{l_1^3} & \dfrac{6E_1 I_1}{l_1^2} & 0 & 0 \\[2mm] \dfrac{6E_1 I_1}{l_1^2} & \dfrac{4E_1 I_1}{l_1} & \dfrac{-6E_1 I_1}{l_1^2} & \dfrac{2E_1 I_1}{l_1} & 0 & 0 \\[2mm] \dfrac{-12E_1 I_1}{l_1^3} & \dfrac{-6E_1 I_1}{l_1^2} & \dfrac{12E_1 I_1}{l_1^3} & \dfrac{-6E_1 I_1}{l_1^2} & 0 & 0 \\[2mm] \dfrac{6E_1 I_1}{l_1^2} & \dfrac{2E_1 I_1}{l_1} & \dfrac{-6E_1 I_1}{l_1^2} & \dfrac{4E_1 I_1}{l_1} & 0 & 0 \\[2mm] 0 & 0 & 0 & 0 & 0 & 0 \\[2mm] 0 & 0 & 0 & 0 & 0 & 0 \end{bmatrix} \qquad (13.23)$$

Inserting the element 2 terms leaves $\mathbf{k}_g$:

$$\begin{bmatrix}
\dfrac{12E_1I_1}{l_1^3} & \dfrac{6E_1I_1}{l_1^2} & \dfrac{-12E_1I_1}{l_1^3} & \dfrac{6E_1I_1}{l_1^2} & 0 & 0 \\[4mm]
\dfrac{6E_1I_1}{l_1^2} & \dfrac{4E_1I_1}{l_1} & \dfrac{-6E_1I_1}{l_1^2} & \dfrac{2E_1I_1}{l_1} & 0 & 0 \\[4mm]
\dfrac{-12E_1I_1}{l_1^3} & \dfrac{-6E_1I_1}{l_1^2} & \left(\dfrac{12E_1I_1}{l_1^3}+\dfrac{12E_2I_2}{l_2^3}\right) & \left(\dfrac{-6E_1I_1}{l_1^2}+\dfrac{6E_2I_2}{l_2^2}\right) & \dfrac{-12E_2I_2}{l_2^3} & \dfrac{6E_2I_2}{l_2^2} \\[4mm]
\dfrac{6E_1I_1}{l_1^2} & \dfrac{2E_1I_1}{l_1} & \left(\dfrac{-6E_1I_1}{l_1^2}+\dfrac{6E_2I_2}{l_2^2}\right) & \left(\dfrac{4E_1I_1}{l_1}+\dfrac{4E_2I_2}{l_2}\right) & \dfrac{-6E_2I_2}{l_2^2} & \dfrac{2E_2I_2}{l_2} \\[4mm]
0 & 0 & \dfrac{-12E_2I_2}{l_2^3} & \dfrac{-6E_2I_2}{l_2^2} & \dfrac{12E_2I_2}{l_2^3} & \dfrac{-6E_2I_2}{l_2^2} \\[4mm]
0 & 0 & \dfrac{6E_2I_2}{l_2^2} & \dfrac{2E_2I_2}{l_2^2} & \dfrac{-6E_2I_2}{l_2^2} & \dfrac{4E_2I_2}{l_2}
\end{bmatrix}$$

$$(13.24)$$

Note how the contributions for the stiffness elements for node 2 from the left-hand and right-hand beams add together.

13.3.4 Eliminating Constraint Degrees of Freedom from Stiffness Matrix

We now have the entire global stiffness matrix, including the degrees of freedom which are constrained, the translation and rotation of node 1 (the first two rows and columns of k_g). To eliminate the constrained degrees of freedom, we eliminate the rows and columns which correspond to the constrained global degrees of freedom, reducing the global stiffness matrix to a 4x4 matrix:

$$k_g = \begin{bmatrix}
\left(\dfrac{12E_1I_1}{l_1^3}+\dfrac{12E_2I_2}{l_2^3}\right) & \left(\dfrac{-6E_1I_1}{l_1^2}+\dfrac{6E_2I_2}{l_2^2}\right) & \dfrac{-12E_2I_2}{l_2^3} & \dfrac{6E_2I_2}{l_2^2} \\[4mm]
\left(\dfrac{-6E_1I_1}{l_1^2}+\dfrac{6E_2I_2}{l_2^2}\right) & \left(\dfrac{4E_1I_1}{l_1}+\dfrac{4E_2I_2}{l_2}\right) & \dfrac{-6E_2I_2}{l_2^2} & \dfrac{2E_2I_2}{l_2} \\[4mm]
\dfrac{-12E_2I_2}{l_2^3} & \dfrac{-6E_2I_2}{l_2^2} & \dfrac{12E_2I_2}{l_2^3} & \dfrac{-6E_2I_2}{l_2^2} \\[4mm]
\dfrac{6E_2I_2}{l_2^2} & \dfrac{2E_2I_2}{l_2^2} & \dfrac{-6E_2I_2}{l_2^2} & \dfrac{4E_2I_2}{l_2}
\end{bmatrix} \quad (13.25)$$

To facilitate hand calculations, we will make the two-beam elements identical, with the same E, I and lengths, l. The global stiffness matrix can then be rewritten as:

$$
\mathbf{k_g} = EI \begin{bmatrix}
\dfrac{24}{l^3} & 0 & \dfrac{-12}{l^3} & \dfrac{6}{l^2} \\[2mm]
0 & \dfrac{8}{l} & \dfrac{-6}{l^2} & \dfrac{2}{l} \\[2mm]
\dfrac{-12}{l^3} & \dfrac{-6}{l^2} & \dfrac{12}{l^3} & \dfrac{-6}{l^2} \\[2mm]
\dfrac{6}{l^2} & \dfrac{2}{l} & \dfrac{-6}{l^2} & \dfrac{4}{l}
\end{bmatrix}
\qquad (13.26)
$$

13.3.5 Static Solution: Force Applied at Tip

We have all the information required to solve a static problem. For example, we could solve for the displacements of the system for a z direction force applied at the tip of the beam. The equation for static equilibrium of the system is:

$$
\mathbf{k}_g \mathbf{z} = \mathbf{F} \qquad (13.27)
$$

Expanding:

$$
\begin{bmatrix}
k_{g11} & k_{g12} & k_{g13} & k_{g14} \\
k_{g21} & k_{g22} & k_{g23} & k_{g24} \\
k_{g31} & k_{g32} & k_{g33} & k_{g34} \\
k_{g41} & k_{g42} & k_{g43} & k_{g44}
\end{bmatrix}
\begin{bmatrix} z_1 \\ z_2 \\ z_3 \\ z_4 \end{bmatrix}
=
\begin{bmatrix} F_1 \\ F_2 \\ F_3 \\ F_4 \end{bmatrix}
\qquad (13.28)
$$

Where:

z_1 is translation of node 2

z_2 is rotation of node 2

z_3 is translation of node 3

z_4 is rotation of node 3

F_1 is z force applied to node 2

F_2 is y moment applied to node 2

F_3 is z force applied to node 3

F_4 is y moment applied to node 3

13.4 Static Condensation

13.4.1 Derivation

Solving the static equation is trivial using a computer, but doing a 4x4 inverse by hand is difficult, so we will reduce the problem to a 2x2 problem using static condensation. Static condensation is not typically used for static problems, but is the precursor for Guyan reduction (dynamic condensation), which will be introduced in the eigenvalue analysis in the next chapter.

Static condensation involves separating the degrees of freedom into "master" and "slave" degrees of freedom. If master dof's are chosen such that they include all degrees of freedom where forces/moments are applied and also degrees of freedom where displacements are desired, the resulting solution is exact. If the slave dof set includes dof's where forces/moments are applied and/or where displacements are desired, the technique will create errors.

For an **exact static solution**, master dof's are chosen as dof's where forces/moments are applied and where displacements/rotations are desired.

For dynamic problems master degrees of freedom are typically chosen as displacements of the higher mass nodes and rotations of the higher mass moment of inertia nodes, with slave degrees of freedom being the displacements and rotations of the relatively lower inertia nodes.

For the two-element cantilever, we will solve for the two translations of node 2 and node 3 as master degrees of freedom, and will condense (reduce out) the two rotations. We will develop the theory first, then will substitute our cantilever example.

The first step is to rearrange the degrees of freedom, rows and columns of the stiffness matrix, into dependent (slave) displacements to be reduced, z_a, and independent (master) displacements, z_b. This involves moving the second and fourth rows and columns of the cantilever stiffness matrix up to become the first and second rows and columns, which moves the first and third rows and columns down to the second and fourth positions.

$$kz = F \tag{13.29}$$

$$\begin{bmatrix} k_{aa} & k_{ab} \\ k_{ba} & k_{bb} \end{bmatrix} \begin{bmatrix} z_a \\ z_b \end{bmatrix} = \begin{bmatrix} F_a \\ F_b \end{bmatrix} \tag{13.30}$$

Multiplying out the first matrix equation:

$$k_{aa}z_a + k_{ab}z_b = F_a \tag{13.31}$$

Solving for z_a :

$$z_a = k_{aa}^{-1}\left(F_a - k_{ab}z_b\right) \tag{13.32}$$

If no forces (moments) are applied at the dependent (slave) degrees of freedom, $F_a = [0]$, and the equation above becomes:

$$z_a = k_{aa}^{-1}\left(-k_{ab}z_b\right) = -k_{aa}^{-1}k_{ab}z_b \tag{13.33}$$

We can now rewrite the displacement vector in terms of z_b only:

$$z = \begin{bmatrix} z_a \\ z_b \end{bmatrix} = \begin{bmatrix} -k_{aa}^{-1}k_{ab} \\ I \end{bmatrix} z_b = \begin{bmatrix} -k_{aa}^{-1}k_{ab}z_b \\ z_b \end{bmatrix} \tag{13.34}$$

Defining a transformation matrix for brevity:

$$z = \begin{bmatrix} z_a \\ z_b \end{bmatrix} = \begin{bmatrix} -k_{aa}^{-1}k_{ab} \\ I \end{bmatrix} z_b = \begin{bmatrix} T_{ab} \\ I \end{bmatrix} z_b = Tz_b \tag{13.35}$$

Where:

$$T_{ab} = -k_{aa}^{-1}k_{ab} \tag{13.36}$$

Substituting back into the original static equilibrium equation:

$$kz = k(Tz_b) = F \tag{13.37}$$

Multiplying both sides by T^T to reduce the number of degrees of freedom from $(a + b)$ to b:

$$(\mathbf{T}^T\mathbf{k}\mathbf{T})\mathbf{z}_b = \mathbf{T}^T\mathbf{F} \tag{13.38}$$

Expanding the term in parentheses above, and redefining it to be $\mathbf{k}_{bb}^*$:

$$\mathbf{k}_{bb}^* = \mathbf{T}^T\mathbf{k}\mathbf{T} = \begin{bmatrix} \mathbf{T}_{ab}^T & \mathbf{I} \end{bmatrix} \begin{bmatrix} \mathbf{k}_{aa} & \mathbf{k}_{ab} \\ \mathbf{k}_{ba} & \mathbf{k}_{bb} \end{bmatrix} \begin{bmatrix} \mathbf{T}_{ab} \\ \mathbf{I} \end{bmatrix}$$

$$= \begin{bmatrix} (\mathbf{T}_{ab}^T\mathbf{k}_{aa} + \mathbf{k}_{ba}) & (\mathbf{T}_{ab}^T\mathbf{k}_{ab} + \mathbf{k}_{bb}) \end{bmatrix} \begin{bmatrix} \mathbf{T}_{ab} \\ \mathbf{I} \end{bmatrix}$$

$$= (\mathbf{T}_{ab}^T\mathbf{k}_{aa} + \mathbf{k}_{ba})\mathbf{T}_{ab} + (\mathbf{T}_{ab}^T\mathbf{k}_{ab} + \mathbf{k}_{bb})\mathbf{I}$$

$$= \mathbf{T}_{ab}^T\mathbf{k}_{aa}\mathbf{T}_{ab} + \mathbf{k}_{ba}\mathbf{T}_{ab} + \mathbf{T}_{ab}^T\mathbf{k}_{ab} + \mathbf{k}_{bb} \tag{13.39}$$

$$= (-\mathbf{k}_{ba}\mathbf{k}_{aa}^{-1})\mathbf{k}_{aa}(-\mathbf{k}_{aa}^{-1}\mathbf{k}_{ab}) + \mathbf{k}_{ba}(-\mathbf{k}_{aa}^{-1}\mathbf{k}_{ab}) + (-\mathbf{k}_{ba}\mathbf{k}_{aa}^{-1})\mathbf{k}_{ab} + \mathbf{k}_{bb}$$

$$= \mathbf{k}_{ba}\mathbf{k}_{aa}^{-1}\mathbf{k}_{ab} - \mathbf{k}_{ba}\mathbf{k}_{aa}^{-1}\mathbf{k}_{ab} - \mathbf{k}_{ba}\mathbf{k}_{aa}^{-1}\mathbf{k}_{ab} + \mathbf{k}_{bb}$$

$$= \mathbf{k}_{bb} - \mathbf{k}_{ba}\mathbf{k}_{aa}^{-1}\mathbf{k}_{ab}$$

where: $\mathbf{T}_{ab} = -\mathbf{k}_{aa}^{-1}\mathbf{k}_{ab}$ and $\mathbf{T}_{ab}^T = -\mathbf{k}_{ba}\mathbf{k}_{aa}^{-1}$.

So, the original (a + b) degree of freedom problem now can be transformed to a "b" degree of freedom problem by partitioning into dependent and independent degrees of freedom, and solving for the reduced stiffness matrix $\mathbf{k}_{bb}^*$ and reduced force vector $\mathbf{F}_b^*$:

$$\mathbf{F}_b^* = \mathbf{T}^T\mathbf{F}$$

$$= \begin{bmatrix} \mathbf{T}_{ab}^T & \mathbf{I} \end{bmatrix} \begin{bmatrix} \mathbf{F}_a \\ \mathbf{F}_b \end{bmatrix} = \mathbf{T}_{ba}\mathbf{F}_a + \mathbf{F}_b \tag{13.40}$$

$$= \mathbf{F}_b - \mathbf{k}_{ba}\mathbf{k}_{aa}^{-1}\mathbf{F}_a$$

Then the reduced problem becomes:

$$\mathbf{k}^*_{bb}\mathbf{z}_b = \mathbf{F}^*_b \tag{13.41}$$

After the $\mathbf{z}_b$ degrees of freedom are known, the $\mathbf{z}_a$ degrees of freedom can be expanded from the $\mathbf{z}_b$ masters using, if $\mathbf{F}_a = [0]$:

$$\mathbf{z}_a = -\mathbf{k}^{-1}_{aa}\mathbf{k}_{ab}\mathbf{z}_b \tag{13.42}$$

13.4.2 Solving Two-Element Cantilever Beam Static Problem

We will now solve the example cantilever for a force applied at the tip. Earlier we showed that the stiffness matrix is:

$$\mathbf{k}_g = EI \begin{bmatrix} \dfrac{24}{l^3} & 0 & \dfrac{-12}{l^3} & \dfrac{6}{l^2} \\[2mm] 0 & \dfrac{8}{l} & \dfrac{-6}{l^2} & \dfrac{2}{l} \\[2mm] \dfrac{-12}{l^3} & \dfrac{-6}{l^2} & \dfrac{12}{l^3} & \dfrac{-6}{l^2} \\[2mm] \dfrac{6}{l^2} & \dfrac{2}{l} & \dfrac{-6}{l^2} & \dfrac{4}{l} \end{bmatrix} \tag{13.43}$$

Rearranging rows, 1 to 3, 2 to 1, 3 to 4 and 4 to 2:

$$\mathbf{k}_g = EI \begin{bmatrix} 0 & \dfrac{8}{l} & \dfrac{-6}{l^2} & \dfrac{2}{l} \\[2mm] \dfrac{6}{l^2} & \dfrac{2}{l} & \dfrac{-6}{l^2} & \dfrac{4}{l} \\[2mm] \dfrac{24}{l^3} & 0 & \dfrac{-12}{l^3} & \dfrac{6}{l^2} \\[2mm] \dfrac{-12}{l^3} & \dfrac{-6}{l^2} & \dfrac{12}{l^3} & \dfrac{-6}{l^2} \end{bmatrix} \tag{13.44}$$

Rearranging columns, 1 to 3, 2 to 1, 3 to 4 and 4 to 2:

$$\mathbf{k}_g = EI \begin{bmatrix} \dfrac{8}{1} & \dfrac{2}{1} & 0 & \dfrac{-6}{1^2} \\ \dfrac{2}{1} & \dfrac{4}{1} & \dfrac{6}{1^2} & \dfrac{-6}{1^2} \\ 0 & \dfrac{6}{1^2} & \dfrac{24}{1^3} & \dfrac{-12}{1^3} \\ \dfrac{-6}{1^2} & \dfrac{-6}{1^2} & \dfrac{-12}{1^3} & \dfrac{12}{1^3} \end{bmatrix} \tag{13.45}$$

Breaking out and identifying the four submatrices of dependent (a) and independent (b) degrees of freedom:

$$\mathbf{k}_{aa} = \dfrac{EI}{1} \begin{bmatrix} 8 & 2 \\ 2 & 4 \end{bmatrix} \qquad \mathbf{k}_{ab} = \dfrac{EI}{1^2} \begin{bmatrix} 0 & -6 \\ 6 & -6 \end{bmatrix}$$

$$\mathbf{k}_{ba} = \dfrac{EI}{1^2} \begin{bmatrix} 0 & 6 \\ -6 & -6 \end{bmatrix} \qquad \mathbf{k}_{bb} = \dfrac{EI}{1^3} \begin{bmatrix} 24 & -12 \\ -12 & 12 \end{bmatrix}$$

$$\text{(13.46a-d)}$$

Finding the inverse of $\mathbf{k}_{aa}$:

$$\mathbf{k}_{aa}^{-1} = \dfrac{1}{14EI} \begin{bmatrix} 2 & -1 \\ -1 & 4 \end{bmatrix} \tag{13.47}$$

$$-\mathbf{k}_{aa}^{-1}\mathbf{k}_{ab} = \dfrac{-1}{141} \begin{bmatrix} -6 & -6 \\ 24 & -18 \end{bmatrix} \tag{13.48}$$

$$\mathbf{k}_{ba}\mathbf{k}_{aa}^{-1}\mathbf{k}_{ab} = \dfrac{EI}{141^3} \begin{bmatrix} 144 & -108 \\ -108 & 144 \end{bmatrix} \tag{13.49}$$

$$\mathbf{k}^*_{bb} = \mathbf{k}_{bb} - \mathbf{k}_{ba}\mathbf{k}^{-1}_{aa}\mathbf{k}_{ab}$$

$$= \frac{EI}{14l^3}\left\{\begin{bmatrix} 336 & -168 \\ -168 & 168 \end{bmatrix} - \begin{bmatrix} 144 & -108 \\ -108 & 144 \end{bmatrix}\right\} \qquad (13.50)$$

$$= \frac{EI}{14l^3}\begin{bmatrix} 192 & -60 \\ -60 & 24 \end{bmatrix}$$

$$\mathbf{k}^{*-1}_{bb} = \frac{14l^3}{1008 EI}\begin{bmatrix} 24 & 60 \\ 60 & 192 \end{bmatrix} = \frac{l^3}{72 EI}\begin{bmatrix} 24 & 60 \\ 60 & 192 \end{bmatrix} \qquad (13.51)$$

Solving for the two displacements, $\mathbf{z}_b$ for a tip force of magnitude P:

$$\mathbf{z}_b = \begin{bmatrix} z_2 \\ z_3 \end{bmatrix} = \mathbf{k}^{*-1}_{bb}\,\mathbf{F}^*_{bb}$$

$$= \frac{l^3}{72 EI}\begin{bmatrix} 24 & 60 \\ 60 & 192 \end{bmatrix}\begin{bmatrix} 0 \\ P \end{bmatrix} \qquad (13.52)$$

$$= \frac{Pl^3}{72 EI}\begin{bmatrix} 60 \\ 192 \end{bmatrix} = \frac{Pl^3}{EI}\begin{bmatrix} \dfrac{60}{72} \\ \dfrac{192}{72} \end{bmatrix} = \frac{Pl^3}{EI}\begin{bmatrix} \dfrac{5}{6} \\ \dfrac{8}{3} \end{bmatrix}$$

The tip displacement is:

$$z_3 = \frac{8Pl^3}{3EI} \qquad (13.53)$$

The well-known solution for the displacement of the tip of a cantilever is:

$$z_{tip} = \frac{PL^3}{3EI} \qquad (13.54)$$

Knowing that the total length of the cantilever, L, is 2l:

$$z_{tip} = \frac{PL^3}{3EI} = \frac{P(2l)^3}{3EI} = \frac{8Pl^3}{3EI} \qquad (13.55)$$

The reduced problem has provided the correct solution. Once again, normally we would not solve a reduced static problem except during a hand calculation, but the derivation of static condensation will be useful in the next chapter when dynamic condensation, Guyan reduction, is introduced.

Problems

P13.1   Assemble the global mass and stiffness matrices for Figure P2.1 element by element. Compare results with P2.1 results.

P13.2   In Section 13.4.2 we solved for the displacements of a two-element cantilever beam with a tip load by reducing out the rotations of the beam. Solve the problem by reducing out the rotations of the middle and tip nodes and the displacement of the middle node. Use a symbolic algebra program to invert the 3x3 $\mathbf{k}_{aa}$ matrix.

CHAPTER 14

FINITE ELEMENTS: DYNAMICS

14.1 Introduction

The chapter starts out with discussions of various mass matrix formulations. The 6dof lumped mass example from Chapter 2 is used for the lumped mass matrix example. A two-element cantilever is used to develop the consistent mass example. Using the same technique as in the previous chapter, the global mass matrix is built up as an assemblage of element mass matrices. A method analogous to static condensation, Guyan reduction, is developed and used to reduce the size of the two-element cantilever problem. The cantilever is then solved for its eigenvalues by hand using Guyan reduction. The same cantilever is solved for eigenvalues and eigenvectors using MATLAB and results are compared to the hand calculations.

Following the two-element cantilever example, a second MATLAB code allows solving for eigenvalues and eigenvectors for a uniform cantilever beam with user-defined number of elements. The results of the MATLAB code are compared with the results from an ANSYS model for the same 10-element cantilever.

This 10-element cantilever will be the last eigenvalue analysis in the book using MATLAB. Further chapters will start with eigenvalue results from ANSYS models, which will be used to build state space MATLAB models. These MATLAB models are then used for frequency and time domain analyses. This chapter serves as a bridge between carrying out analyses completely in MATLAB and using ANSYS results as the starting point for state space MATLAB models. Hence, we will reintroduce ANSYS eigenvalue/eigenvector results and start becoming familiar with their form and interpretation.

14.2 Six dof Global Mass Matrix

The lumped mass matrix is simple to construct because there is only a single degree of freedom associated with each mass element. This leads to the 6x6 diagonal mass matrix below, which can be constructed in the same manner as the 6dof stiffness matrix in the previous chapter.

$$
\mathbf{m_g} = \begin{bmatrix} m_1 & 0 & 0 & 0 & 0 & 0 \\ 0 & m_2 & 0 & 0 & 0 & 0 \\ 0 & 0 & m_3 & 0 & 0 & 0 \\ 0 & 0 & 0 & m_4 & 0 & 0 \\ 0 & 0 & 0 & 0 & m_5 & 0 \\ 0 & 0 & 0 & 0 & 0 & m_6 \end{bmatrix} \quad (14.1)
$$

14.3 Cantilever Dynamics

14.3.1 Overview – Mass Matrix Forms

In order to solve for the dynamics of the cantilever beam, we need to develop a mass matrix to complete the equations of motion. For a beam finite element, there are a number of different mass matrix formulations, each of which will be covered below:

  1) Lumped mass, displacements only

  2) Lumped mass, displacements and rotations both included

  3) Consistent mass – distributed mass effect

14.3.2 Lumped Mass

Beam-element lumped parameter mass and inertia terms in the mass matrix relate point inertial loads to point accelerations and give only diagonal terms. Equation (14.2) below shows the lumped mass matrix including both displacements and rotations:

$$
\mathbf{m_1} = \begin{bmatrix} \left(\dfrac{ml}{2}\right) & 0 & 0 & 0 \\ 0 & \left(\dfrac{ml^3}{24}+\dfrac{mlI_y}{2A}\right) & 0 & 0 \\ 0 & 0 & \left(\dfrac{ml}{2}\right) & 0 \\ 0 & 0 & 0 & \left(\dfrac{ml^3}{24}+\dfrac{mlI_y}{2A}\right) \end{bmatrix} \quad (14.2)
$$

For the lumped mass for displacement terms only, the (2,2) and (4,4) terms in (14.2) would be set to zero. Notation is as follows: m is mass per unit length, l is the element length, I_y is the cross-sectional moment of inertia about the y axis and A is the cross section area. This lumped mass formulation assumes a prismatic beam (same area and moment of inertia along the length) and effectively lumps half of the mass and inertia at each end (Archer 1963).

14.3.3 Consistent Mass

Lumped mass formulations were state of the art in structural dynamics until Archer's classic paper introduced the consistent mass matrix in 1963.

We will see in the development below that the consistent mass matrix for a beam element is a filled matrix. The filled matrix can be combined with other consistent mass matrices of other elements of the structure, in the same manner as the element stiffness matrices are combined, to yield the final global mass matrix.

The element consistent mass matrix for a prismatic beam is, with mass per unit length m and length l (Weaver 1990):

$$\mathbf{m}_e = \frac{ml}{420} \begin{bmatrix} 156 & 221 & 54 & -131 \\ 221 & 41^2 & 131 & -31^2 \\ 54 & 131 & 156 & -221 \\ -131 & -31^2 & -221 & 41^2 \end{bmatrix} \tag{14.3}$$

Figure 14.1 shows the unit accelerations of each of the four degrees of freedom which correspond to the four columns of the consistent mass matrix, analogous to the beam element stiffness description in Chapter 13.

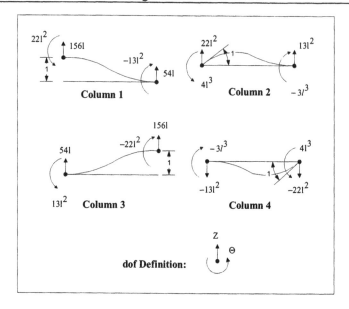

Figure 14.1: Beam element consistent mass matrix terms.

14.4 Dynamics of Two-Element Cantilever – Consistent Mass Matrix

We already have the global stiffness matrix for the two-element cantilever beam from (13.26):

$$\mathbf{k}_g = EI \begin{bmatrix} \dfrac{24}{l^3} & 0 & \dfrac{-12}{l^3} & \dfrac{6}{l^2} \\[2mm] 0 & \dfrac{8}{l} & \dfrac{-6}{l^2} & \dfrac{2}{l} \\[2mm] \dfrac{-12}{l^3} & \dfrac{-6}{l^2} & \dfrac{12}{l^3} & \dfrac{-6}{l^2} \\[2mm] \dfrac{6}{l^2} & \dfrac{2}{l} & \dfrac{-6}{l^2} & \dfrac{4}{l} \end{bmatrix} \qquad (14.4)$$

The global mass matrix (using consistent mass) can also be built by combining the terms from each of the beam elements as follows:

$$\mathbf{m}_g = \frac{1}{420} \begin{bmatrix} 156m_1l_1 & 22m_1l_1^2 & 54m_1l_1 \\ 22m_1l_1^2 & 4m_1l_1^3 & 13m_1l_1^2 \\ 54m_1l_1 & 13m_1l_1^2 & (156m_1l_1 + 156m_2l_2) \\ -13m_1l_1^2 & -3m_1l_1^3 & (-22m_1l_1^2 + 22m_2l_2^2) \\ 0 & 0 & 54m_2l_2 \\ 0 & 0 & -13m_2l_2^2 \end{bmatrix}$$

$$\begin{bmatrix} -13m_1l_1^2 & 0 & 0 \\ -3m_1l_1^3 & 0 & 0 \\ (-22m_1l_1^2 + 22m_2l_2^2) & 54m_2l_2 & -13m_2l_2^2 \\ (4m_1l_1^3 + 4m_2l_2^3) & 13m_2l_2^2 & -3m_2l_2^3 \\ 13m_2l_2^2 & 156m_2l_2 & -22m_2l_2^2 \\ -3m_2l_2^3 & -22m_2l_2^2 & 4m_2l_2^3 \end{bmatrix}$$

$$(14.5)$$

Once again, assuming the two elements have the same properties and lengths, the global mass matrix becomes:

$$\mathbf{m}_g = \frac{1}{420} \begin{bmatrix} 156ml & 22ml^2 & 54ml & -13ml^2 & 0 & 0 \\ 22ml^2 & 4ml^3 & 13ml^2 & -3ml^3 & 0 & 0 \\ 54ml & 13ml^2 & 312ml & 0 & 54ml & -13ml^2 \\ -13ml^2 & -3ml^3 & 0 & 8ml^3 & 13ml^2 & -3ml^3 \\ 0 & 0 & 54ml & 13ml^2 & 156ml & -22ml^2 \\ 0 & 0 & -13ml^2 & -3ml^3 & -22ml^2 & 4ml^3 \end{bmatrix} \quad (14.6)$$

Taking into account the two constrained degrees of freedom at the built in end, we can eliminate the first two rows and columns:

$$\mathbf{m}_g = \frac{1}{420} \begin{bmatrix} 312ml & 0 & 54ml & -13ml^2 \\ 0 & 8ml^3 & 13ml^2 & -3ml^3 \\ 54ml & 13ml^2 & 156ml & -22ml^2 \\ -13ml^2 & -3ml^3 & -22ml^2 & 4ml^3 \end{bmatrix} \quad (14.7)$$

Having the mass and stiffness matrices allows us to solve the eigenvalue problem for the homogeneous equations of motion:

$$\mathbf{m}_g \ddot{\mathbf{z}} + \mathbf{k}_g \mathbf{z} = [0] \quad (14.8)$$

In order to solve the problem by hand, we will need to find several inverses, so we will again see if we can cut the 4x4 problem down to 2x2 size. We will now use Guyan reduction to reduce the size of the problem.

14.5 Guyan Reduction

Guyan reduction is a method of decreasing the number of degrees of freedom in a dynamics problem, similar to the process of static condensation in a statics problem. Unlike static condensation, however, Guyan reduction introduces errors due to the approximations made. The magnitude of the errors introduced depends upon the choice of degrees of freedom to be reduced, the dependent or slave degrees of freedom. The most popular choice of degrees of freedom to be reduced are translations of nodes with relatively lower masses and rotations of nodes with relatively lower mass moment of inertia. This leaves translations of relatively larger mass nodes and rotations of relatively larger mass moment of inertia nodes as the independent degrees of freedom. In a typical finite element problem, the analyst will define masters as degrees of freedom where forces/moment are applied, where displacements or rotations are required for output, or where known large masses/mass moments of inertia occur. The finite element program will then be allowed to choose an additional set of degrees of freedom and add them to the master set. Typically the program sorts along the diagonal of the mass matrix, adding degrees of freedom associated with the larger terms.

14.5.1 Guyan Reduction Derivation

Starting with the undamped equations of motion:

$$m\ddot{z} + kz = [0] \tag{14.9}$$

Rearranging and partitioning into displacements to be reduced, z_a, and independent displacements, z_b:

$$\begin{bmatrix} m_{aa} & m_{ab} \\ m_{ba} & m_{bb} \end{bmatrix} \begin{bmatrix} \ddot{z}_a \\ \ddot{z}_b \end{bmatrix} + \begin{bmatrix} k_{aa} & k_{ab} \\ k_{ba} & k_{bb} \end{bmatrix} \begin{bmatrix} z_a \\ z_b \end{bmatrix} = \begin{bmatrix} F_a \\ F_a \end{bmatrix} \tag{14.10}$$

Multiplying out the first matrix equation:

$$m_{aa}\ddot{z}_a + m_{ab}\ddot{z}_b + k_{aa}z_a + k_{ab}z_b = F_a \tag{14.11}$$

Solving the above for z_a:

$$z_a = k_{aa}^{-1}\left(F_a - k_{ab}z_b - m_{aa}\ddot{z}_a - m_{ab}\ddot{z}_b\right)$$
$$= -k_{aa}^{-1}k_{ab}z_b + k_{aa}^{-1}\left(F_a - m_{aa}\ddot{z}_a - m_{ab}\ddot{z}_b\right) \qquad (14.12)$$

Instead of letting z_a depend upon the entire right-hand side of (14.13), the approximation of static equilibrium is introduced:

$$z_a = -k_{aa}^{-1}k_{ab}z_b \qquad (14.13)$$

Typically the choice of degrees of freedom to be reduced does not include any degrees of freedom to which forces are applied, thus $F_a = 0$. The static equilibrium approximation basically sets the term in brackets in (14.12) to zero. Setting $F_a = 0$ and using the second derivative of (14.13), we can see the form of m_{ab}:

$$0 = F_a - m_{aa}\ddot{z}_a - m_{ab}\ddot{z}_b$$
$$= -m_{aa}\ddot{z}_a - m_{ab}\ddot{z}_b$$
$$= -m_{aa}(-k_{aa}^{-1}k_{ab}\ddot{z}_b) - m_{ab}\ddot{z}_b \qquad (14.14a,b)$$
$$= m_{aa}k_{aa}^{-1}k_{ab} - m_{ab}$$

$$m_{ab} = m_{aa}k_{aa}^{-1}k_{ab}$$

We assume that the $m_{aa}\ddot{z}_a$ terms are zero and that m_{aa} and m_{ab} are related as in (14.14b). The force transmission between the $\ddot{z}_a$ and $\ddot{z}_b$ degrees of freedom is related only to the stiffnesses as denoted in (14.14), hence the "static equilibrium" approximation.

Assuming (14.13) holds, the displacement vector z can be written in terms of z_b only:

$$z = \begin{bmatrix} z_a \\ z_b \end{bmatrix} = \begin{bmatrix} -k_{aa}^{-1}k_{ab} \\ I \end{bmatrix} z_b = \begin{bmatrix} T_{ab} \\ I \end{bmatrix} z_b = T z_b \qquad (14.15)$$

where:

$$T_{ab} = -k_{aa}^{-1}k_{ab} \qquad (14.16)$$

$$\mathbf{T} = \begin{bmatrix} \mathbf{T}_{ab} \\ \mathbf{I} \end{bmatrix} \qquad (14.17)$$

Substitution of (14.14), with derivatives, into (14.9) yields:

$$\mathbf{mT\ddot{z}}_b + \mathbf{kTk}_b = \mathbf{F} \qquad (14.18)$$

Equation (14.18) still contains $(a + b)$ degrees of freedom, so premultiplication by $\mathbf{T}^T$ is required to reduce to (b) degrees of freedom and to return symmetry to the reduced mass and stiffness matrices:

$$\left(\mathbf{T}^T\mathbf{mT}\right)\ddot{z}_b + \left(\mathbf{T}^T\mathbf{kT}\right)z_b = \mathbf{T}^T\mathbf{F} \qquad (14.19)$$

Rewriting in a more compact form:

$$\mathbf{m}^*_{bb}\ddot{z}_b + \mathbf{k}^*_{bb}z_b = \mathbf{F}^*_b \qquad (14.20)$$

Equation (14.20) is the final reduced equation of motion which can be solved for the displacements of type b. Displacements of type a (assuming static equilibrium) can then be solved for using (14.13).

$\mathbf{k}^*_{bb}$ can be shown to be the same as that derived in the static condensation Section 13.4.1, (13.39):

$$\mathbf{k}^*_{bb} = \begin{bmatrix} \mathbf{T}^T_{ab} & \mathbf{I} \end{bmatrix} \begin{bmatrix} \mathbf{k}_{aa} & \mathbf{k}_{ab} \\ \mathbf{k}_{ba} & \mathbf{k}_{bb} \end{bmatrix} \begin{bmatrix} \mathbf{T}_{ab} \\ \mathbf{I} \end{bmatrix}$$

$$= \begin{bmatrix} \left(\mathbf{T}^T_{ab}\mathbf{k}_{aa} + \mathbf{k}_{ba}\right) & \left(\mathbf{T}^T_{ab}\mathbf{k}_{ab} + \mathbf{k}_{bb}\right) \end{bmatrix} \begin{bmatrix} \mathbf{T}_{ab} \\ \mathbf{I} \end{bmatrix}$$

$$= \mathbf{T}^T_{ab}\mathbf{k}_{aa}\mathbf{T}_{ab} + \mathbf{k}_{ba}\mathbf{T}_{ab} + \mathbf{T}^T_{ab}\mathbf{k}_{ab} + \mathbf{k}_{bb}$$

$$= \mathbf{k}_{ba}\mathbf{k}^{-1}_{aa}\mathbf{k}_{aa}\mathbf{k}^{-1}_{aa}\mathbf{k}_{ab} - \mathbf{k}_{ba}\mathbf{k}^{-1}_{aa}\mathbf{k}_{ab} - \mathbf{k}_{ba}\mathbf{k}^{-1}_{aa}\mathbf{k}_{ab} + \mathbf{k}_{bb}$$

$$= \mathbf{k}_{bb} - \mathbf{k}_{ba}\mathbf{k}^{-1}_{aa}\mathbf{k}_{ab} \qquad (14.21)$$

14.5.2 Two-Element Cantilever Eigenvalues Closed Form Solution Using Guyan Reduction

Repeating the rearranged global stiffness matrix from the static run, (13.45):

$$k_g = EI \begin{bmatrix} \dfrac{8}{1} & \dfrac{2}{1} & 0 & \dfrac{-6}{1^2} \\[2mm] \dfrac{2}{1} & \dfrac{4}{1} & \dfrac{6}{1^2} & \dfrac{-6}{1^2} \\[2mm] 0 & \dfrac{6}{1^2} & \dfrac{24}{1^3} & \dfrac{-12}{1^3} \\[2mm] \dfrac{-6}{1^2} & \dfrac{-6}{1^2} & \dfrac{-12}{1^3} & \dfrac{12}{1^3} \end{bmatrix} \tag{14.22}$$

Breaking out and identifying the four submatrices of dependent (a) and independent (b) degrees of freedom:

$$k_{aa} = \frac{EI}{1}\begin{bmatrix} 8 & 2 \\ 2 & 4 \end{bmatrix} \qquad k_{ab} = \frac{EI}{1^2}\begin{bmatrix} 0 & -6 \\ 6 & -6 \end{bmatrix}$$

$$\tag{14.23a-d}$$

$$k_{ba} = \frac{EI}{1^2}\begin{bmatrix} 0 & 6 \\ -6 & -6 \end{bmatrix} \qquad k_{bb} = \frac{EI}{1^3}\begin{bmatrix} 24 & -12 \\ -12 & 12 \end{bmatrix}$$

Finding the inverse of k_{aa} :

$$k_{aa}^{-1} = \frac{1}{14EI}\begin{bmatrix} 2 & -1 \\ -1 & 4 \end{bmatrix} \tag{14.24}$$

$$-k_{aa}^{-1}k_{ab} = \frac{-1}{141}\begin{bmatrix} -6 & -6 \\ 24 & -18 \end{bmatrix} \tag{14.25}$$

$$k_{ba}k_{aa}^{-1}k_{ab} = \frac{EI}{141^3}\begin{bmatrix} 144 & -108 \\ -108 & 144 \end{bmatrix} \tag{14.26}$$

$$\mathbf{k}_{bb}^{*} = \mathbf{k}_{bb} - \mathbf{k}_{ba}\mathbf{k}_{aa}^{-1}\mathbf{k}_{ab}$$

$$= \frac{EI}{14l^3}\left\{\begin{bmatrix} 336 & -168 \\ -168 & 168 \end{bmatrix} - \begin{bmatrix} 144 & -108 \\ -108 & 144 \end{bmatrix}\right\}$$

$$= \frac{EI}{14l^3}\begin{bmatrix} 192 & -60 \\ -60 & 24 \end{bmatrix}$$

$$(14.27)$$

The transformation matrix T is given by:

$$\mathbf{T} = \begin{bmatrix} \mathbf{T}_{ab} \\ \mathbf{I} \end{bmatrix} = \begin{bmatrix} -\mathbf{k}_{aa}^{-1}\mathbf{k}_{ab} \\ \mathbf{I} \end{bmatrix} = \begin{bmatrix} \dfrac{6}{141} & \dfrac{6}{141} \\ \dfrac{-24}{141} & \dfrac{18}{141} \\ 1 & 0 \\ 0 & 1 \end{bmatrix} \qquad (14.28)$$

The mass matrix now needs to be rearranged into "a" and "b" submatrices and then transformed to $\mathbf{m}_{bb}^{*}$:

$$\mathbf{m}_g = \frac{1}{420}\begin{bmatrix} 312ml & 0 & 54ml & -13ml^2 \\ 0 & 8ml^3 & 13ml^2 & -3ml^3 \\ 54ml & 13ml^2 & 156ml & -22ml^2 \\ -13ml^2 & -3ml^3 & -22ml^2 & 4ml^3 \end{bmatrix} \qquad (14.29)$$

Rearranging rows 1 to 3, 2 to 1, 3 to 4 and 4 to 2:

$$\mathbf{m}_g = \frac{1}{420}\begin{bmatrix} 0 & 8ml^3 & 13ml^2 & -3ml^3 \\ -13ml^2 & -3ml^3 & -22ml^2 & 4ml^3 \\ 312ml & 0 & 54ml & -13ml^2 \\ 54ml & 13ml^2 & 156ml & -22ml^2 \end{bmatrix} \qquad (14.30)$$

Rearranging columns 1 to 3, 2 to 1, 3 to 4 and 4 to 2:

$$\mathbf{m}_g = \frac{m}{420} \begin{bmatrix} 8l^3 & -3l^3 & 0 & 13l^2 \\ -3l^3 & 4l^3 & -13l^2 & -22l^2 \\ 0 & -13l^2 & 3121 & 541 \\ 13l^2 & -22l^2 & 541 & 1561 \end{bmatrix} \qquad (14.31)$$

Separating into submatrices:

$$\mathbf{m}_{aa} = \frac{ml^3}{420} \begin{bmatrix} 8 & -3 \\ -3 & 4 \end{bmatrix} \qquad\qquad \mathbf{m}_{ab} = \frac{ml^2}{420} \begin{bmatrix} 0 & 13 \\ -13 & -22 \end{bmatrix}$$

$$(14.32\text{a-d})$$

$$\mathbf{m}_{ba} = \frac{ml^2}{420} \begin{bmatrix} 0 & -13 \\ 13 & -22 \end{bmatrix} \qquad\qquad \mathbf{m}_{bb} = \frac{ml}{420} \begin{bmatrix} 312 & 54 \\ 54 & 156 \end{bmatrix}$$

Calculating $\mathbf{m}_{bb}^*$:

$$\mathbf{m}_{bb}^* = \mathbf{T}^{\mathrm{T}} \mathbf{m} \mathbf{T} \qquad (14.33)$$

Carrying out the multiplications:

$$\mathbf{m}_{bb}^* = ml \begin{bmatrix} 1528 & 241 \\ \hline 1715 & 1372 \\ 241 & 471 \\ \hline 1372 & 1715 \end{bmatrix} \qquad (14.34)$$

14.6 Eigenvalues of Reduced Equations for Two-Element Cantilever, State Space Form

The second order reduced equation of motion is shown in (14.35), (14.36), using the 2x2 stiffness matrix from static condensation, (13.50). We will now generate the state space form of the second order reduced equations. It is useful to see how to convert a second order set of differential equations with a filled (not diagonal) mass matrix to state space form. Once we have the equations of motion in state space form, we will use a symbolic algebra program to solve for the eigenvalues.

$$\mathbf{m}_{bb}^* \ddot{\mathbf{z}}_b + \mathbf{k}_{bb}^* \mathbf{z}_b = [0] \qquad (14.35)$$

$$
ml \begin{bmatrix} \dfrac{1528}{1715} & \dfrac{241}{1372} \\ \dfrac{241}{1372} & \dfrac{471}{1715} \end{bmatrix} \begin{bmatrix} \ddot{z}_{b1} \\ \ddot{z}_{b2} \end{bmatrix} + \dfrac{EI}{14l^3} \begin{bmatrix} 192 & -60 \\ -60 & 24 \end{bmatrix} \begin{bmatrix} z_{b1} \\ z_{b2} \end{bmatrix} = \begin{bmatrix} 0 \\ 0 \end{bmatrix} \qquad (14.36)
$$

z_{b1} and z_{b2} are the first two reduced degrees of freedom, the displacements of nodes 2 and 3.

Normally, we would solve each of the equations of motion for the highest derivative and then convert to state space form, but we cannot do that here because the mass matrix is filled, meaning that there is more than one second derivative in each equation. To get around this problem, we will first convert the equation to state space form. We will then take the inverse of the mass matrix and premultiply, leaving only the identity matrix to multiply with the derivative vector.

Converting to state space form, where x_1 and x_2 are displacement and velocity of node 2 and x_3 and x_4 are the displacement and velocity of node 3, respectively:

$$
\mathbf{m}_{ss}\dot{\mathbf{x}} + \mathbf{k}_{ss}\mathbf{x} = \begin{bmatrix} 0 \end{bmatrix} \qquad (14.37)
$$

$$
\begin{bmatrix} 1 & 0 & 0 & 0 \\ 0 & \dfrac{1528ml}{1715} & 0 & \dfrac{241ml}{1372} \\ 0 & 0 & 1 & 0 \\ 0 & \dfrac{241ml}{1372} & 0 & \dfrac{471ml}{1715} \end{bmatrix} \begin{bmatrix} \dot{x}_1 \\ \dot{x}_2 \\ \dot{x}_3 \\ \dot{x}_4 \end{bmatrix} + \begin{bmatrix} 0 & -1 & 0 & 0 \\ \dfrac{192EI}{14l^3} & 0 & \dfrac{-60EI}{14l^3} & 0 \\ 0 & 0 & 0 & -1 \\ \dfrac{-60EI}{14l^3} & 0 & \dfrac{24EI}{14l^3} & 0 \end{bmatrix} \begin{bmatrix} x_1 \\ x_2 \\ x_3 \\ x_4 \end{bmatrix} = \begin{bmatrix} 0 \\ 0 \\ 0 \\ 0 \end{bmatrix}
$$

$$(14.38)$$

Note that the "1" terms are on the diagonal in the mass matrix and the "-1" terms are off diagonal in the stiffness matrix. Taking the inverse of $\mathbf{m}_{ss}$:

$$
\mathbf{m}_{ss}^{-1} = \begin{bmatrix} 1 & 0 & 0 & 0 \\ 0 & \dfrac{263760}{205367ml} & 0 & \dfrac{-168700}{205367ml} \\ 0 & 0 & 1 & 0 \\ 0 & \dfrac{-168700}{205367ml} & 0 & \dfrac{855680}{205367ml} \end{bmatrix}
\tag{14.39}
$$

Premultiplying the equation of motion by $\mathbf{m}_{ss}^{-1}$:

$$
\begin{bmatrix} 1 & 0 & 0 & 0 \\ 0 & 1 & 0 & 0 \\ 0 & 0 & 1 & 0 \\ 0 & 0 & 0 & 1 \end{bmatrix}\begin{bmatrix} \dot{x}_1 \\ \dot{x}_2 \\ \dot{x}_3 \\ \dot{x}_4 \end{bmatrix} + \begin{bmatrix} 0 & -1 & 0 & 0 \\ \dfrac{4340280EI}{205367ml^4} & 0 & \dfrac{-1419600EI}{205367ml^4} & 0 \\ 0 & 0 & 0 & -1 \\ \dfrac{-5980800EI}{205367ml^4} & 0 & \dfrac{2189880EI}{205367ml^4} & 0 \end{bmatrix}\begin{bmatrix} x_1 \\ x_2 \\ x_3 \\ x_4 \end{bmatrix} = \begin{bmatrix} 0 \\ 0 \\ 0 \\ 0 \end{bmatrix}
$$

$$
\tag{14.40}
$$

Rewriting without the identity matrix:

$$
\begin{bmatrix} \dot{x}_1 \\ \dot{x}_2 \\ \dot{x}_3 \\ \dot{x}_4 \end{bmatrix} + \begin{bmatrix} 0 & -1 & 0 & 0 \\ \dfrac{4340280EI}{205367ml^4} & 0 & \dfrac{-1419600EI}{205367ml^4} & 0 \\ 0 & 0 & 0 & -1 \\ \dfrac{-5980800EI}{205367ml^4} & 0 & \dfrac{2189880EI}{205367ml^4} & 0 \end{bmatrix}\begin{bmatrix} x_1 \\ x_2 \\ x_3 \\ x_4 \end{bmatrix} = \begin{bmatrix} 0 \\ 0 \\ 0 \\ 0 \end{bmatrix}
\tag{14.41}
$$

Converting to standard state space form, $\dot{\mathbf{x}} = \mathbf{Ax} + \mathbf{Bu}$:

$$
\begin{bmatrix} \dot{x}_1 \\ \dot{x}_2 \\ \dot{x}_3 \\ \dot{x}_4 \end{bmatrix} = \begin{bmatrix} 0 & 1 & 0 & 0 \\ \dfrac{-4340280EI}{205367ml^4} & 0 & \dfrac{1419600EI}{205367ml^4} & 0 \\ 0 & 0 & 0 & 1 \\ \dfrac{5980800EI}{205367ml^4} & 0 & \dfrac{-2189880EI}{205367ml^4} & 0 \end{bmatrix}\begin{bmatrix} x_1 \\ x_2 \\ x_3 \\ x_4 \end{bmatrix}
\tag{14.42}
$$

Using a symbolic algebra program to solve for the eigenvalues:

$$f_{1,2} = \left(\frac{1}{2\pi}\right)\left(\frac{2}{205367}\right)\frac{\sqrt{43127070}\sqrt{EIm(3887 \pm 20\sqrt{34178})}}{ml^2} \qquad (14.43)$$

14.7 MATLAB Code cant_2el_guyan.m –
Two-element Cantilever Eigenvalues/Eigenvectors

14.7.1 Code Description

The MATLAB code **cant_2el_guyan.m** solves for the eigenvalues and eigenvectors of a two-element steel cantilever with dimensions of 0.2 x 2 x 20mm. The code does the following, where each time MATLAB calculates a result it is compared to the hand-calculated result:

1) builds mass and stiffness matrices element by element

2) deletes degrees of freedom associated with constrained left-hand end

3) reorders the matrices and performs Guyan reduction

4) converts to state space form

5) calculates eigenvalues/eigenvectors

The code for **cant_2el_guyan.m** is not listed as similar code is used in **cantbeam_guyan.m**, which is listed below.

14.7.2 Code Results

Substituting for E, I, m and l in (14.43) as shown in the code results in eigenvalues of 398.55 and 2521.1 Hz. The first two eigenvalues for a 10-element model using ANSYS (following section) are calculated to be 397.86 and 2493.2 Hz, giving differences between the two-element and 10-element beams of 0.17% and +1.11%, respectively. The differences between the two-element and theoretical values are +0.1697% and +0.0095%, respectively. Archer's consistent mass paper stated that in order to calculate accurate eigenvalues using consistent mass we only needed one more element than the number of accurate modes desired. In this case we found the frequency of the first mode very accurately using only two elements, and the second mode was only off by 1.11%, even with the errors inherent in the Guyan reduction method.

14.8 MATLAB Code cantbeam_guyan.m –
User-Defined Cantilever Eigenvalues/Eigenvectors

This MATLAB code solves for the eigenvalues and eigenvectors of a cantilever with user-defined dimensions, material properties, number of elements and number of mode shapes to plot. The code is similar to that in **cant_2el_guyan.m** except that Guyan reduction is an option for this code. If Guyan reduction is chosen, all rotations are reduced, leaving only translations as master degrees of freedom. The code is listed below, but is not broken down and commented because the comments integrated with the code should be sufficient.

In order to compare results with the ANSYS run below, a 10-element beam with the following properties is used: width = 2mm, thickness = 0.2, length = 20mm, modulus = $190e^6 \, mN/mm^2$, density = $7.83e^{-6} \, Kg/mm^3$.

14.9 ANSYS Code cantbeam.inp, Code Description

The ANSYS code solves for the eigenvalues and eigenvectors of the same beam as cantbeam_guyan.m.

14.10   MATLAB cantbeam_guyan.m / ANSYS cantbeam.inp Results
Summary

14.10.1 10-Element Beam Frequency Comparison

The Table 14.1 shows the eigenvalues from the 10-element ANSYS and MATLAB runs, both with Guyan reduction, along with theoretical values calculated using the MATLAB code cantbeam_ss_freq_craig.m (Chang 1969). The errors for the first five modes are quite small, with the maximum error (for the ninth mode) being only 6.5%.

Mode No.	MATLAB Cantbeam_ guyan.m	ANSYS Cantbeam.inp	Theoretical	Percent Error, Cantbeam_guyan.m and Theoretical
1	397.88	397.86	397.874572279	-0.0001
2	2493.6	2493.2	2493.437382146	-0.0051
3	6984.5	6982.2	6981.696870181	-0.0408
4	13703	13696	13681.339375292	-0.1646
5	22727	22705	22616.234284744	-0.4887
6	34194	34145	33784.737867762	-1.2113
7	48420	48234	47186.94828572	-2.6126
8	65831	65657	62822.86012645	-4.7893
9	85987	85697	80692.473674351	-6.5619
10	104570	101392	100795.788914948	-3.7445

Table 14.1: 10-element beam frequency comparisons.

14.10.2 20-Element Beam Mode Shape Plots, Modes 1 to 5

Instead of plotting the mode shapes for the 10-element model, we will use a 20-element model to give better resolution and smoother plots. The first five mode shape plots are shown in Figures 14.2 through 14.6 below. Note that for the third and fifth modes the displacements of the middle node are quite small relative to the maximum 1.0. In other words, there is a "node" of the mode near the midpoint of the beam. This meaning for "node" of a mode is not that of a finite element "node," but is a location along the beam where displacement goes to zero for that mode of vibration.

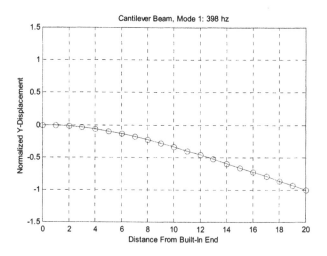

Figure 14.2: Cantilever beam first mode.

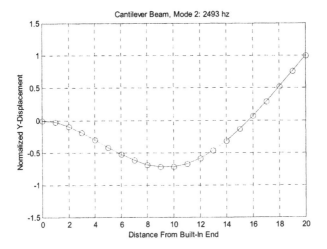

Figure 14.3: Cantilever beam second mode.

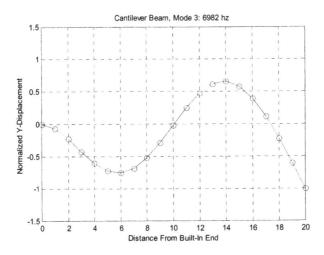

Figure 14.4: Cantilever beam third mode. Note "node" near the beam middle.

We are focusing on "nodes" located near the middle of the beam because in the next chapter we will solve for the frequency responses of a cantilever with a force at the center and output displacement at the tip. We will see that modes with small eigenvector entries for input or output (or both) degrees of freedom are able to be removed from the model, as they contribute little to the input or output of the system.

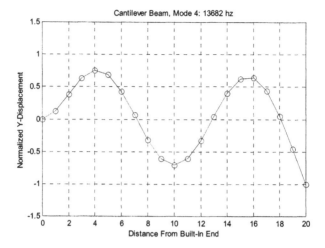

Figure 14.5: Cantilever beam fourth mode.

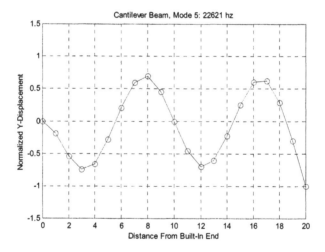

Figure 14.6: Cantilever beam fifth mode. Note the "node" near the midpoint of the beam, and two additional "nodes" to the left and right of the midpoint.

The 10 eigenvectors from the 10-element **cantbeam_guyan.m**, normalized to unity, are shown in Table 14.2. The displacement entry for the built-in left-hand end of the beam is not shown, the 10 rows represent the nodes from left to right, starting with the second node from the end.

Mode: 1	2	3	4	5	6	7	8	9	10
-0.0168	-0.0926	-0.2280	0.3841	-0.5331	-0.6485	0.7129	0.7310	0.7418	-0.6239
-0.0639	-0.3010	-0.6042	0.7519	-0.6535	-0.3274	-0.1055	-0.4942	-0.7714	0.7719
-0.1365	-0.5261	-0.7558	0.4324	0.2109	0.6574	-0.5480	0.0107	0.6458	-0.9023
-0.2299	-0.6834	-0.5256	-0.3153	0.6906	0.1048	0.6100	0.4831	-0.3565	0.9797
-0.3395	-0.7136	**-0.0195**	-0.7053	**-0.0028**	-0.6931	**-0.0029**	-0.6863	**-0.0222**	-1.0000
-0.4611	-0.5894	0.4737	-0.3249	-0.6948	0.1125	-0.6070	0.4771	0.3953	0.9618
-0.5909	-0.3170	0.6571	0.3971	-0.2215	0.6607	0.5534	0.0186	-0.6692	-0.8674
-0.7255	0.0701	0.3945	0.6411	0.5965	-0.3025	0.1160	-0.5089	0.7788	0.7247
-0.8624	0.5238	-0.2288	0.0504	0.2884	-0.4706	-0.5885	0.6466	-0.6636	-0.5252
-1.0000	1.0000	-1.0000	-1.0000	-1.0000	1.0000	1.0000	-1.0000	1.0000	0.7913

Table 14.2: 10-element beam eigenvectors normalized to unity. Note small values for
third, fifth, seventh and ninth mode displacements for midpoint node, in bold type.

The presence of a "node" of a mode can be seen numerically for the 10-element MATLAB model by looking at the fifth row (midpoint of beam) of the eigenvector listing in Table 14.2 and noting the small values for the third, fifth, seventh and ninth modes, highlighted in bold type. Getting a good mental picture of the relationship between the plotted mode shape and the eigenvector listing is quite useful. We will see in the next chapter that the small value of node displacements for certain modes of vibration will mean that for certain transfer functions the modes are less important to include in the reduced (smaller number of states used) state space model, and therefore, can be eliminated.

For eigenvector comparison with the ANSYS results, which are normalized with respect to mass instead of unity, the first two eigenvectors for the 10-element MATLAB beam model, are shown below. Compare with the "UZ" columns in the ANSYS listing below.

4.2387	-23.4098
14.1402	-76.0842
34.4892	-132.9666
58.0918	-172.7285
85.7975	-180.3585
116.5287	-148.9709
149.3145	-80.1210
183.3282	17.7069
217.9284	132.3727
252.7000	252.7326

Table 14.3: MATLAB 10-element beam model, first and second eigenvectors normalized
with respect to mass.

A listing for the first two modes from the ANSYS code **cantbeam.eig** is shown below. The listing displays the title, resonant frequency (eigenvalue) and a listing of eigenvector entries for each degree of freedom. Even though we used Guyan reduction on the ANSYS model, ANSYS back-calculates the eigenvector values of the reduced dof's so there are eigenvector values for both the UZ and ROTY degrees of freedom below. Since we constrained all the degrees of freedom except the displacement in the z-direction and rotation about the y axis, all other degree of freedom entries for the eigenvectors are zero.

```
*DO LOOP ON PARAMETER= I     FROM  1.0000   TO  10.000   BY  1.0000


USE LOAD STEP   1 SUBSTEP   1 FOR LOAD CASE 0

SET COMMAND GOT LOAD STEP=   1 SUBSTEP=   1 CUMULATIVE ITERATION=
1
  TIME/FREQUENCY= 397.86
TITLE= cantbeam.inp, 0.2 thick x 2 wide x 20mm long steel cantilever beam, 10

PRINT DOF  NODAL SOLUTION PER NODE

***** POST1 NODAL DEGREE OF FREEDOM LISTING *****

LOAD STEP=   1  SUBSTEP=   1
FREQ=   397.86     LOAD CASE=  0

THE FOLLOWING DEGREE OF FREEDOM RESULTS ARE IN GLOBAL COORDINATES

NODE  UX        UY        UZ       ROTX     ROTY      ROTZ
   1  0.0000    0.0000    0.0000   0.0000   0.0000    0.0000
   2  0.0000    0.0000    4.2385   0.0000   -4.1366   0.0000
   3  0.0000    0.0000    16.140   0.0000   -7.6631   0.0000
   4  0.0000    0.0000    34.488   0.0000   -10.586   0.0000
   5  0.0000    0.0000    58.090   0.0000   -12.920   0.0000
   6  0.0000    0.0000    85.796   0.0000   -14.695   0.0000
   7  0.0000    0.0000    116.53   0.0000   -15.954   0.0000
   8  0.0000    0.0000    149.31   0.0000   -16.761   0.0000
   9  0.0000    0.0000    183.32   0.0000   -17.198   0.0000
  10  0.0000    0.0000    217.92   0.0000   -17.366   0.0000
  11  0.0000    0.0000    252.70   0.0000   -17.396   0.0000

MAXIMUM ABSOLUTE VALUES
NODE      0        0        11        0       11        0
VALUE  0.0000   0.0000   252.70   0.0000   -17.396   0.0000

*ENDDO INDEX= I

***** POST1 NODAL DEGREE OF FREEDOM LISTING *****

LOAD STEP=   1  SUBSTEP=   2
```

```
FREQ=  2493.2    LOAD CASE=  0

THE FOLLOWING DEGREE OF FREEDOM RESULTS ARE IN GLOBAL COORDINATES

NODE  UX        UY        UZ        ROTX      ROTY      ROTZ
  1   0.0000    0.0000    0.0000    0.0000    0.0000    0.0000
  2   0.0000    0.0000   -23.405    0.0000    21.188    0.0000
  3   0.0000    0.0000   -76.071    0.0000    29.354    0.0000
  4   0.0000    0.0000   -132.95    0.0000    25.705    0.0000
  5   0.0000    0.0000   -172.71    0.0000    12.776    0.0000
  6   0.0000    0.0000   -180.34    0.0000    -5.7217   0.0000
  7   0.0000    0.0000   -148.96    0.0000    -25.506   0.0000
  8   0.0000    0.0000   -80.124    0.0000    -42.575   0.0000
  9   0.0000    0.0000    17.689    0.0000    -54.169   0.0000
 10   0.0000    0.0000    132.34    0.0000    -59.449   0.0000
 11   0.0000    0.0000    252.69    0.0000    -60.537   0.0000

MAXIMUM ABSOLUTE VALUES
NODE       0         0        11         0        11         0
VALUE  0.0000    0.0000    252.69    0.0000    -60.537   0.0000
```

14.11 MATLAB Code cantbeam_guyan.m Listing

```
            echo off
%           cantbeam_guyan.m    cantilever beam finite element program,
%           selectable number of elements.  Solves for eigenvalues and
%           eigenvectors of a cantilever with user-defined dimensions,
%           material properties, number of elements and number of mode shapes
%           to plot.  Guyan reduction is an option.  A 10 element beam is used
%           as an example.  Default beam is 2mm wide by 20mm long by 0.2mm thick.

            clf;

            clear all;

            inp = input('Input "1" to enter beam dimensions, "Enter" to use default ... ');

            if (isempty(inp))
                    inp = 0;
            else
            end

            if  inp == 0

                    wbeam = 2.0
                    tbeam = 0.2
                    lbeam = 20.0
                    E = 190e6
                    density = 7.83e-6

            else

%           input size of beam and material
```

```
                wbeam = input('Input width of beam, default 2mm, ... ');

                if (isempty(wbeam))
                        wbeam = 2.0;
                else
                end

                tbeam = input('Input thickness of beam, default 0.2mm, ... ');

                if (isempty(tbeam))
                        tbeam = 0.2;
                else
                end

                lbeam = input('Input length of beam, default 20mm, ... ');

                if (isempty(lbeam))
                        lbeam = 20.0;
                else
                end

                E = input('Input modulus of material, mN/mm^2, default stainless steel 190e6 ... ');

                if (isempty(E))
                        E = 190e6;
                else
                end

                density = input('Input density of material, Kg/mm^3, default stainless steel 7.83e-6 ...
');

                if (isempty(density))
                        density = 7.83e-6;
                else
                end

                end

%       input number of elements

        num_elements = input('Input number of elements for beam, minimum 2, default 10 ...
');

                if (isempty(num_elements))
                        num_elements = 10;
                else
                end

%       define whether or not to do Guyan Reduction

        guyan = input('enter "1" to do Guyan elimination of rotations, ...
                        "enter" to not do Guyan ... ');

                if (isempty(guyan))
```

```
                           guyan = 0;
        else
        end

        if  guyan == 0

                        num_plot_max = 2*num_elements;

                        num_plot_default = num_elements;

        else

                        num_plot_max = num_elements;

                        num_plot_default = num_elements;

        end

        num_plot = input(['enter the number of modes to plot, max', ...
                        num2str(num_plot_max),', default ',num2str(num_plot_default),' ... ']);

        if (isempty(num_plot))
                        num_plot = 9;
        else
        end
```

```
%       define length of each element, uniform lengths

        l = lbeam/num_elements;
```

```
%       define length vector for plotting, right-to-left numbering

        lvec = 0:l:lbeam;
```

```
%       define the node numbers

        n = 1:num_elements+1;
```

```
%       number the nodes for the elements

        node1 = 1:num_elements;

        node2 = 2:num_elements+1;
```

```
%       size the stiffness and mass matrices to have 2 times the number of nodes
%       to allow for translation and rotation dof's for each node, including built-
%       in end

        max_node1 = max(node1);

        max_node2 = max(node2);

        max_node_used = max([max_node1 max_node2]);

        mnu = max_node_used;
```

```
           k = zeros(2*mnu);

           m = zeros(2*mnu);

%          now build up the global stiffness and consistent mass matrices, element by element


%          calculate I, area and mass per unit length of beam

           I = wbeam*tbeam^3/12;

           area = wbeam*tbeam;

           mpl = density*area;

           for  i = 1:num_elements

                      dof1 = 2*node1(i)-1;
                      dof2 = 2*node1(i);
                      dof3 = 2*node2(i)-1;
                      dof4 = 2*node2(i);

                      k(dof1,dof1) = k(dof1,dof1)+(12*E*I/l^3);
                      k(dof2,dof1) = k(dof2,dof1)+(6*E*I/l^2);
                      k(dof3,dof1) = k(dof3,dof1)+(-12*E*I/l^3);
                      k(dof4,dof1) = k(dof4,dof1)+(6*E*I/l^2);

                      k(dof1,dof2) = k(dof1,dof2)+(6*E*I/l^2);
                      k(dof2,dof2) = k(dof2,dof2)+(4*E*I/l);
                      k(dof3,dof2) = k(dof3,dof2)+(-6*E*I/l^2);
                      k(dof4,dof2) = k(dof4,dof2)+(2*E*I/l);

                      k(dof1,dof3) = k(dof1,dof3)+(-12*E*I/l^3);
                      k(dof2,dof3) = k(dof2,dof3)+(-6*E*I/l^2);
                      k(dof3,dof3) = k(dof3,dof3)+(12*E*I/l^3);
                      k(dof4,dof3) = k(dof4,dof3)+(-6*E*I/l^2);

                      k(dof1,dof4) = k(dof1,dof4)+(6*E*I/l^2);
                      k(dof2,dof4) = k(dof2,dof4)+(2*E*I/l);
                      k(dof3,dof4) = k(dof3,dof4)+(-6*E*I/l^2);
                      k(dof4,dof4) = k(dof4,dof4)+(4*E*I/l);

                      m(dof1,dof1) = m(dof1,dof1)+(mpl/420)*(156*l);
                      m(dof2,dof1) = m(dof2,dof1)+(mpl/420)*(22*l^2);
                      m(dof3,dof1) = m(dof3,dof1)+(mpl/420)*(54*l);
                      m(dof4,dof1) = m(dof4,dof1)+(mpl/420)*(-13*l^2);

                      m(dof1,dof2) = m(dof1,dof2)+(mpl/420)*(22*l^2);
                      m(dof2,dof2) = m(dof2,dof2)+(mpl/420)*(4*l^3);
                      m(dof3,dof2) = m(dof3,dof2)+(mpl/420)*(13*l^2);
                      m(dof4,dof2) = m(dof4,dof2)+(mpl/420)*(-3*l^3);

                      m(dof1,dof3) = m(dof1,dof3)+(mpl/420)*(54*l);
                      m(dof2,dof3) = m(dof2,dof3)+(mpl/420)*(13*l^2);
```

```
                    m(dof3,dof3) = m(dof3,dof3)+(mpl/420)*(156*l);
                    m(dof4,dof3) = m(dof4,dof3)+(mpl/420)*(-22*l^2);

                    m(dof1,dof4) = m(dof1,dof4)+(mpl/420)*(-13*l^2);
                    m(dof2,dof4) = m(dof2,dof4)+(mpl/420)*(-3*l^3);
                    m(dof3,dof4) = m(dof3,dof4)+(mpl/420)*(-22*l^2);
                    m(dof4,dof4) = m(dof4,dof4)+(mpl/420)*(4*l^3);

            end

%           now that stiffness and mass matrices are defined for all dof's, including
%           constrained dof's, need to delete rows and columns of the matrices that
%           correspond to constrained dof's, in the left-to-right case, the first two
%           rows and columns

            k(1:2,:) = [];          % translation/rotation of node 1
            k(:,1:2) = [];

            m(1:2,:) = [];
            m(:,1:2) = [];

            if  guyan == 1

%           Guyan Reduction - reduce out the rotation dof's, leaving displacement dof's
%           re-order the matrices

%           re-order the columns of k

            kr = zeros(2*(mnu-1));

            krr = zeros(2*(mnu-1));

%   rearrange columns, rotation and then displacement dof's

            mkrcolcnt = 0;

            for  mkcolcnt = 2:2:2*(mnu-1)

                    mkrcolcnt = mkrcolcnt + 1;

                    kr(:,mkrcolcnt) = k(:,mkcolcnt);

                    mr(:,mkrcolcnt) = m(:,mkcolcnt);

            end

            mkrcolcnt = num_elements;

            for  mkcolcnt = 1:2:2*(mnu-1)

                    mkrcolcnt = mkrcolcnt + 1;

                    kr(:,mkrcolcnt) = k(:,mkcolcnt);

                    mr(:,mkrcolcnt) = m(:,mkcolcnt);
```

```
            end

%  rearrange rows, rotation and then displacement dofs

        mkrrowcnt = 0;

        for  mkrowcnt = 2:2:2*(mnu-1)

                mkrrowcnt = mkrrowcnt + 1;

                krr(mkrrowcnt,:) = kr(mkrowcnt,:);

                mrr(mkrrowcnt,:) = mr(mkrowcnt,:);

        end

        mkrrowcnt = num_elements;

        for  mkrowcnt = 1:2:2*(mnu-1)

                mkrrowcnt = mkrrowcnt + 1;

                krr(mkrrowcnt,:) = kr(mkrowcnt,:);

                mrr(mkrrowcnt,:) = mr(mkrowcnt,:);

        end

%         define sub-matrices and transformation matrix T

        kaa = krr(1:num_elements,1:num_elements);

        kab = krr(1:num_elements,num_elements+1:2*num_elements);

        T = [-inv(kaa)*kab
                        eye(num_elements,num_elements)]

%         calculate reduced mass and stiffness matrices

        kbb = T'*krr*T

        mbb = T'*mrr*T

        else

        kbb = k;

        mbb = m;

        end

%         define the number of dof for state-space version, 2 times dof left after
%         removing constrained dofs
```

```
        [dof,dof] = size(kbb);
%       define the sizes of mass and stiffness matrices for state-space

        ssdof = 2*dof;

        aud = zeros(ssdof);                      % creates a ssdof x ssdof null matrix

%       divide the negative of the stiffness matrix by the mass matrix

        ksm = inv(mbb)*(-kbb);

%       now expand to state space size
%       fill out unit values in mass and stiffness matrices

        for  row = 1:2:ssdof

                aud(row,row+1) = 1;

        end

%       fill out mass and stiffness terms from m and k

        for  row = 2:2:ssdof

                for  col = 2:2:ssdof

                        aud(row,col-1) = ksm(row/2,col/2);

                end

        end

%       calculate the eigenvalues/eigenvectors of the undamped matrix for plotting
%       and for calculating the damping matrix c

        [evec1,evalu] = eig(aud);

        evalud = diag(evalu);

        evaludhz = evalud/(2*pi);

        num_modes = length(evalud)/2;

%       now reorder the eigenvalues and eigenvectors from low to high freq

        [evalorder,indexhz] = sort(abs((evalud)));

        for  cnt = 1:length(evalud)

                eval(cnt,1) = evalud(indexhz(cnt));

                evalhzr(cnt,1) = round(evaludhz(indexhz(cnt)));

                evec(:,cnt) = evec1(:,indexhz(cnt));
```

```
           end

%          now check for any imaginary eigenvectors and convert to real

           for  cnt = 1:length(evalud)

                     if (imag(evec(1,cnt)) □ imag(evec(3,cnt)) □ imag(evec(5,cnt))) ~= 0

                              evec(:,cnt) = imag(evec(:,cnt));

                 else

                 end

           end

           if  guyan == 0

%          now separate the displacement and rotations in the eigenvectors
%          for plotting mode shapes

           evec_disp = zeros(ceil(dof/2),ssdof);

           rownew = 0;

           for  row = 1:4:ssdof

                     rownew = rownew+1;

                     evec_disp(rownew,:) = evec(row,:);

           end

           evec_rotation = zeros(ceil(dof/2),ssdof);

           rownew = 0;

           for  row = 3:4:ssdof

                     rownew = rownew+1;

                     evec_rotation(rownew,:) = evec(row,:);

           end

           else

           evec_disp = zeros(ceil(dof/4),ssdof);

           rownew = 0;

           for  row = 1:2:ssdof

                     rownew = rownew+1;
```

```
                    evec_disp(rownew,:) = evec(row,:);

        end

        end

%       normalize the displacement eigenvectors wrt one for plotting

        for  col = 1:ssdof

                    evec_disp(:,col) = evec_disp(:,col)/max(abs(real(evec_disp(:,col))));

                    if  evec_disp(floor(dof/2),col) >= 0

                            evec_disp(:,col) = -evec_disp(:,col);

                    else
                    end

        end

%       list eigenvalues, hz

        format long          e

        evaludhz_list = sort(evaludhz(1:2:2*num_modes))

        format short

%       list displacement (not velocity) eigenvectors

        evec_disp(:,1:2:2*num_plot)

        if  guyan == 0

%       plot mode shapes

        for  mode_cnt = 1:num_plot

                    evec_cnt = 2*mode_cnt -1;

                    plot(lvec,[0; evec_disp(:,evec_cnt)],'ko-')
                    title(['Cantilever Beam, Mode ', ...
                            num2str(mode_cnt),': ',num2str(abs(evalhzr(evec_cnt))),' hz']);
                    xlabel('Distance From Built-In End')
                    ylabel('Normalized Y-Displacement')
                    axis([0 lbeam -1.5 1.5])
                    grid on

        disp('execution paused to display figure, "enter" to continue'); pause

        end

        else
```

```
%          plot mode shapes, Guyan Reduced

           for mode_cnt = 1:num_plot

                    evec_cnt = 2*mode_cnt -1;

                    plot(lvec,[0; evec_disp(:,evec_cnt)],'ko-')
                    title(['Cantilever Beam, Mode ', ...
                             num2str(mode_cnt),': ',num2str(abs(evalhzr(evec_cnt))),' hz']);
                    xlabel('Distance From Built-In End')
                    ylabel('Normalized Y-Displacement')
                    axis([0 lbeam -1.5 1.5])
                    grid on

           disp('execution paused to display figure, "enter" to continue'); pause

           end

           end

%          normalization with respect to mass on a filled (not diagonal) mass matrix

%          calculate the displacement (displacement and rotation) eigenvectors
%          to be used for the modal model eigenvectors

           xm = zeros(dof);

           col = 0;

           for mode = 1:2:ssdof

                    col = col + 1;

                    row = 0;

                    for       ndof = 1:2:ssdof

                             row = row + 1;

                             xm(row,col) = evec(ndof,mode);

                    end

           end

%          normalize with respect to mass

           for mode = 1:dof

                    xn(:,mode) = xm(:,mode)/sqrt(xm(:,mode)'*mbb*xm(:,mode));

           end

%          calculate the normalized mass and stiffness matrices for checking
```

```
                mm = xn'*mbb*xn;

                km = xn'*kbb*xn;

%               check that the sqrt of diagonal elements of km are eigenvalues

                p = (diag(km)).^0.5;

                row = 0;

                for cnt = 1:2:ssdof

                        row = row + 1;

                        evalrad(row) = abs((eval(cnt)));

                end

                [p evalrad']/(2*pi)

                evalhz = evalrad/(2*pi);

                semilogy(evalhz)
                title('Resonant Frequencies, Hz')
                xlabel('Mode Number')
                ylabel('Frequency, hz')
                grid
                disp('execution paused to display figure, "enter" to continue'); pause
```

14.12 ANSYS Code cantbeam.inp Listing

```
/title, cantbeam.inp, 0.2 thick x 2 wide x 20mm long steel cantilever beam, 10 elements

/prep7

et,1,4                  ! element type for beam

! steel

ex,1,190e6                      ! mN/mm^2
dens,1,7.83e-6                  ! kg/mm^3
nuxy,1,.293

! real value to define beam characteristics

r,1,0.4,0.1333,0.0013333,0.2,2                  ! area, Izz, Iyy, TKz, TKy

! define plotting characteristics

/view,1,1,-1,1    ! iso view
/angle,1,-60      ! iso view
/pnum,mat,1       ! color by material
```

```
/num,1          ! numbers off
/type,1,0       ! hidden plot
/pbc,all,1      ! show all boundary conditions

csys,0                              ! define global coordinate system

! nodes

n,1,0,0,0                    ! left-hand node
n,11,20,0,0                      ! right-hand node

fill,1,11                    ! interior nodes

nplo

! elements

type,1
mat,1
real,1
e,1,2
egen,10,1,-1

! constrain left-hand end

d,1,all,0                    ! constrain node 1, all dof's

! constrain all but uz and roty for all other nodes to allow only those dof's
! this will give 10 nodes, node 2 through node 11, each with 2 dof, giving a total of 20 dof
! can calculate a maximum of 20 eigenvalues if don't use Guyan reduction to reduce size of
! eigenvalue problem, maximum of 10 eigenvalues if use Guyan reduction

nall
nsel,s,node,,2,11
d,all,ux
d,all,uy
d,all,rotx
d,all,rotz

allsel
nplo
eplo

! ****************** eigenvalue run ******************

fini        ! fini just in case not in begin

/solu       ! enters the solution processor, needs to be here to do editing below

allsel      ! default selects all items of specified entity type, typically nodes, elements

nsel,s,node,,2,11
m,all,uz

antype,modal,new
```

```
modopt,reduc,10        ! method - reduced Householder, number of modes to extract
expass,off             ! key = off, no expansion pass, key = on, do expansion
mxpand,10,,,no         ! nummodes to expand
total,10,1             ! total masters, 10 to be used, exclude rotational dofs

allsel

solve                  ! starts the solution of one load step of a solution sequence, modal here

fini

! plot first mode

/post1

set,1,1

pldi,1


! ****************** output frequencies *********************

/output,cantbeam,frq      ! write out frequency list to ascii file .frq

set,list

/output,term                                        ! returns output to terminal

! **************** output eigenvectors ****************

! define nodes for output:  forces applied or output displacements

nall
!nsel,s,node,,11        ! cantilever tip

/output,cantbeam,eig      ! write out eigenvectors to ascii file .eig

*do,i,1,10
            set,,i
            prdisp
*enddo

/output,term
**************** plot modes *****************

! pldi plots

/show,cantbeam,grp,0
allsel

/view,1,,-1,,          ! side view for plotting
/angle,1,0
/auto

*do,i,1,10
```

```
          set,1,i
   pldi
*enddo

/show,term
```

Problem

P14.1 Modify the **cantbeam_guyan.m** code to allow variable material and geometry properties along the beam by converting the following scalar quantities into user defined vector quantities: wbeam, tbeam, E, density.

Run the modified code for a 20mm long beam with the twice the default values for the left half of the beam and the default parameters for the right-hand side. Plot eigenvalues in hz versus mode number. Plot the first five mode shapes.

CHAPTER 15

SISO STATE SPACE MATLAB MODEL FROM ANSYS MODEL

15.1 Introduction

This chapter will develop a SISO state space MATLAB model from an ANSYS cantilever beam model. The cantilever is admittedly a trivial example, but like the tdof model used in the first part of the book, will serve as a good model to develop a fundamental understanding of the process. As we are going through the simple cantilever example we should be thinking about applying the process to a model of an actual device, for example a complete model of a disk drive, with hundreds of thousands of nodes and up to hundreds of modes in the frequency range of interest. Our objective for the model will be to provide the smallest MATLAB state space model that accurately represents the pertinent dynamics.

The model cantilever is shown in Figure 15.1. It is a 2mm wide by 0.075mm thick by 20mm long steel beam. The coordinate system is indicated on the figure. A z direction force is applied at the midpoint of the beam and z displacement at the tip is the output. Only x-z plane motion is allowed; all other degrees of freedom are constrained.

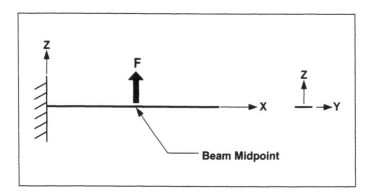

Figure 15.1: Cantilever beam with forcing function at midpoint.

We will begin by analyzing the major issues all finite element analysts face when setting up a model: defining the number of elements to use and calculating the effects of Guyan reduction, if used. We will analyze the cantilever with different numbers of elements. We will also analyze with and without Guyan reduction and compare the resulting resonant frequencies with theoretical results. Knowing the frequency range of interest for the model, typically defined by servo bandwidth considerations, we will define a model (number of elements) that accurately predicts eigenvalues in the range of interest. In this theoretical example we have the luxury of knowing the exact values for the eigenvalues. However, in real life problems, we know that a finite element model is accurate only if we build another model with finer resolution and compare results, and/or have good experimental mode shape data with which to compare.

While Guyan reduction prior to conducting an eigenvalue analysis has been in the main replaced by the Block Lanczos eigenvalue extraction method, Guyan reduction will be presented because it is still used in creating "superelements" for large models (which are then solved using Block Lanczos) and is also used in correlating finite element and experimental model models.

For some problems, the time to perform frequency response calculations using Block Lanczos is of the same order of time as the eigenvalue extraction, which makes using MATLAB for state space frequency response models an efficient adjunct to ANSYS. We will review how to have MATLAB build a state space model given only the eigenvalues and required eigenvector information (eigenvector entries for all modes for only input and output degrees of freedom). This technique will be used for all following models, in conjunction with various mode elimination/truncation techniques.

The problem to be solved in this chapter is: Determine the smallest state space model which accurately constructs the frequency response characteristics through a given frequency range. We will assume for our problem that the servo system requires all significant modes through 20khz be included. The servo system will apply inputs in the z direction at the node located at the mid-length of the cantilever, with z direction displacement of the tip being the output.

The first step in defining the smallest model is to define the eigenvector elements for all modes for only the input and output degrees of freedom. The second step is to analyze the modal contributions of all the modes and sort them to define which ones have the greatest contribution.

One method for reducing the size of a modal model is to simply truncate the higher frequency modes. If this truncation is performed without understanding the contributions of each of the modes to the response, several problems could arise. One problem is that a high frequency mode that could alias to a lower frequency in a sampled servo system may be missed. Another hazard with arbitrarily truncating higher frequency modes is that a mode with a significant dc gain contribution may be eliminated, adversely affecting the model. Typically the contributions of modes decrease as their frequencies increase; however, this is not always the case. In Chapter 16 we will see a cantilever model with an additional tip mass and a tip spring all mounted on a "shaker" base. It is used as an example of how excluding a specific higher frequency mode can result in a model with less than desired accuracy.

15.2 ANSYS Eigenvalue Extraction Methods

ANSYS has a number of different eigenvalue extraction techniques, but for most problems only two methods are commonly used. The first method, Block Lanczos, is the fastest and calculates all the eigenvalues or eigenvalues in a specific frequency range. Most practical models require knowledge of the modes from dc through a specified higher frequency.

The second method, Reduced, performs a Guyan reduction on the model to reduce its size, then calculates all the eigenvalues for the reduced model. All of the "master" degree of freedom eigenvector components are available immediately for use. Obtaining eigenvector components for the reduced degrees of freedom requires an additional calculation step in ANSYS.

For very large models, Block Lanczos has shown to be significantly faster than the Reduced method. If MATLAB state space models are used to calculate frequency responses using Block Lanczos results the total time to get model results can be quite satisfactory. Typically, the Reduced method is used only for small- to medium-size problems.

15.3   Cantilever Model, ANSYS Code cantbeam_ss.inp, MATLAB Code cantbeam_ss_freq.m

The ANSYS code **cantbeam_ss.inp,** listed in Section 15.7, is designed to allow the user to easily change the number of elements "num_elem" as well as the eigenvalue extraction technique "eigext."

The model was run for 2, 4, 6, 8, 10, 12, 16, 32 and 64 elements for both eigenvalue extraction methods. The Lanczos method resulted in twice the number of eigenvalues as the Reduced method because both translations and rotations are degrees of freedom for Lanczos, while the Reduced method has the rotations reduced out.

For those interested, the MATLAB code **cantbeam_ss_freq.m** plots the results of the ANSYS runs along with the theoretical frequencies for up to the first 16 modes (Chang 1969).

Figures 15.2 and 15.3 show the percentage frequency differences between the first 10 modes of the ANSYS Block Lanczos and Reduced runs and the theoretical prediction.

The maximum frequency difference for the Block Lanczos method is 2% and for the Reduced method it is 5%. For the frequency range of interest in our problem, 20 khz, the maximum frequency errors are 1% and 3%, which is deemed satisfactory. We will use the 10-element model with the Reduced method for the rest of the chapter. Real life models will have greater deviations because they have imperfect geometry, joints and connections to ground which are difficult to model accurately, and variations in material and mass properties.

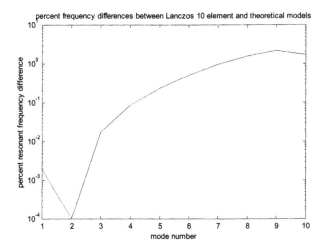

Figure 15.2: Percent resonant frequency differences between 10-element Block Lanczos ANSYS model and theoretical versus mode number.

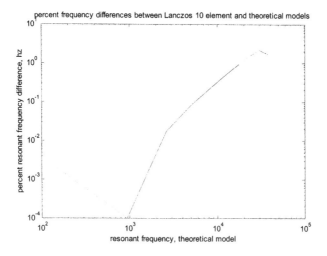

Figure 15.3: Percent resonant frequency differences between 10-element Block Lanczos
ANSYS model and theoretical versus frequency.

15.4 ANSYS 10-element Model Eigenvalue/Eigenvector Summary

```
***** INDEX OF DATA SETS ON RESULTS FILE *****

SET   TIME/FREQ   LOAD STEP   SUBSTEP   CUMULATIVE
 1      149.20        1           1          1
 2      935.05        1           2          2
 3     2619.0         1           3          3
 4     5138.4         1           4          4
 5     8521.2         1           5          5
 6    12820.          1           6          6
 7    18152.          1           7          7
 8    24677.          1           8          8
 9    32229.          1           9          9
10    39191.          1          10         10
```

Table 15.1: Frequency listing from cantbeam10red.frq file – frequencies for all 10 modes,
hz.

In Table 15.2 we can see the eigenvector listing for the first two modes from
the edited cantbeam10red.eig file, which contains information for all nodes for
all 10 modes. As discussed in Section 7.4.2, ANSYS normalizes eigenvectors
with respect to mass by default. Since our problem has input applied at the
middle node (node 7), and output at the tip node (node 11), only those two
nodes are required for the MATLAB model. We can choose to use ANSYS to
output only the eigenvectors for nodes 7 and 11 or we can input the complete

modal matrix below in MATLAB and choose the appropriate rows of data within MATLAB.

```
SET COMMAND GOT LOAD STEP=   1 SUBSTEP=   1 CUMULATIVE ITERATION=
1
 TIME/FREQUENCY= 149.20
TITLE= cantbeam, 10, red

PRINT DOF  NODAL SOLUTION PER NODE

***** POST1 NODAL DEGREE OF FREEDOM LISTING *****

LOAD STEP=   1 SUBSTEP=   1
FREQ=   149.20    LOAD CASE=  0

THE FOLLOWING DEGREE OF FREEDOM RESULTS ARE IN GLOBAL COORDINATES

   NODE    UX        UY        UZ       ROTX      ROTY      ROTZ
    1     0.0000    0.0000    0.0000    0.0000    0.0000    0.0000
    2     0.0000    0.0000    6.9217    0.0000   -6.7553    0.0000
    3     0.0000    0.0000   26.357     0.0000  -12.514     0.0000
    4     0.0000    0.0000   56.320     0.0000  -17.287     0.0000
    5     0.0000    0.0000   94.863     0.0000  -21.099     0.0000
    6     0.0000    0.0000  140.11      0.0000  -23.997     0.0000
    7     0.0000    0.0000  190.29      0.0000  -26.054     0.0000
    8     0.0000    0.0000  243.83      0.0000  -27.371     0.0000
    9     0.0000    0.0000  299.37      0.0000  -28.085     0.0000
   10     0.0000    0.0000  355.87      0.0000  -28.358     0.0000
   11     0.0000    0.0000  412.66      0.0000  -28.407     0.0000

MAXIMUM ABSOLUTE VALUES
NODE      0        0        11        0        11        0
VALUE   0.0000   0.0000   412.66   0.0000   -28.407   0.0000

*ENDDO  INDEX= I

***** POST1 NODAL DEGREE OF FREEDOM LISTING *****

LOAD STEP=   1 SUBSTEP=   2
FREQ=   935.05    LOAD CASE=  0

THE FOLLOWING DEGREE OF FREEDOM RESULTS ARE IN GLOBAL COORDINATES

   NODE    UX        UY        UZ       ROTX      ROTY      ROTZ
    1     0.0000    0.0000    0.0000    0.0000    0.0000    0.0000
    2     0.0000    0.0000  -38.227     0.0000   34.605     0.0000
    3     0.0000    0.0000 -124.24      0.0000   47.942     0.0000
    4     0.0000    0.0000 -217.13      0.0000   41.980     0.0000
    5     0.0000    0.0000 -282.06      0.0000   20.864     0.0000
    6     0.0000    0.0000 -294.52      0.0000   -9.3483    0.0000
    7     0.0000    0.0000 -243.27      0.0000  -41.660     0.0000
    8     0.0000    0.0000 -130.84      0.0000  -69.535     0.0000
    9     0.0000    0.0000   28.911     0.0000  -88.467     0.0000
   10     0.0000    0.0000  216.16      0.0000  -97.088     0.0000
```

11	0.0000	0.0000	**412.70**	0.0000	-98.864	0.0000

MAXIMUM ABSOLUTE VALUES
NODE	0	0	11	0	11	0
VALUE	0.0000	0.0000	412.70	0.0000	-98.864	0.0000

Table 15.2: Eigenvector listing for first two modes from the edited cantbeam10red.eig file.

15.5 Modal Matrix

The ANSYS output file cantbeam10red.eig can be sorted for only the UZ component for all 10 modes and put into a modal matrix form using **ext56uz.m** (see Appendix 1 for usage), as shown in Table 15.3. Each of the 10 columns in Table 5.3 represents the eigenvector for that mode, normalized with respect to mass. Compare the first two columns below with the bold "UZ" entries in the eigenvector listings in Table 15.2.

Columns 1 through 7

0	**0**	0	0	0	0	0
6.9217	-38.2270	94.1860	-159.3800	223.8100	-279.2100	320.1800
26.3570	-124.2400	249.6400	-311.9600	274.3700	-141.0000	-47.3120
56.3200	-217.1300	312.2800	-179.4100	-88.5050	283.0200	-246.1700
94.8630	-282.0600	217.1400	130.8000	-289.9000	45.1810	273.9500
140.1100	-294.5200	8.0768	292.6500	1.1237	-298.4100	-1.2392
190.2900	-243.2700	-195.6800	134.8400	291.6800	48.3890	-272.6900
243.8300	-130.8400	-271.4700	-164.7400	93.0350	284.4600	248.4900
299.3700	28.9110	-162.9800	-266.0000	-250.3900	-130.1600	52.2360
355.8700	216.1600	94.5080	-20.9260	-121.0900	-202.6200	-264.3300
412.6600	412.7000	413.1400	414.9000	419.7700	430.4900	449.0700

Columns 8 through 10

0	0	0
-341.5400	326.0200	223.4200
230.8300	-338.9500	-276.3500
-4.9143	283.7200	323.0000
-225.7900	-156.5300	-350.7100
320.6200	**-9.8888**	357.9400
-222.7800	173.8500	-344.2300
-8.8232	-294.1600	310.4200
237.8700	342.2200	-259.2800
-302.0500	-291.4900	187.8300
467.0400	439.1200	-282.9400

Table 15.3: Eigenvectors for UZ component of cantbeam10red.eig file.

The 11 rows represent the normalized displacements for the 11 nodes, starting with node 1 at the built-in end and node 11 at the tip. Editing the modal matrix to use only the required degrees of freedom (nodes 7 and 11) will take place in MATLAB.

15.6   MATLAB State Space Model from ANSYS Eigenvalue Run – cantbeam_ss_modred.m

In this section we will create a MATLAB state space model using the eigenvalue and eigenvector results from the previous ANSYS run. We discussed in Section 7.9 how to decrease the size of the model by including only degrees of freedom actually used in the particular frequency response or time domain calculations. The new material deals with **how to rank the relative importance of the contributions of each of the individual modes**. In this chapter, we will use a **ranking of dc gains** of individual modes to select the modes to be used.

Once the modes are ranked, the most important can be selected for use, with modes with lower dc gains (typically, but not always, the higher frequency modes) eliminated from the model. When these modes are eliminated from the model their dc gain contributions are not included in the overall dc gain, so there is error in the low frequency gain. In order to eliminate this error, the MATLAB function "modred" is introduced and the theory behind the code is discussed. Using "modred" is analogous to using Guyan reduction to reduce some less important degrees of freedom, in that assumptions are made about some modes being more important than others. This allows reducing the size of the problem to that of the "important" modes, while adjusting the overall dc gain to account for the dc gains of the eliminated modes.

We will find that the simple cantilever beam used for an example in this chapter is not very sensitive to the elimination of higher frequency modes. Including a few modes is sufficient for creating a state space model with good accuracy for both frequency response and step response. Whether "modred" is used is not critical for this example. However, we will see that the example in the next chapter is extremely sensitive to dc gain, and will serve as a good model of the benefits of selecting modes to be eliminated judiciously or by using "modred."

Once the model is created, we will solve for frequency response and step response using various combinations of truncating and sorting modes.

The MATLAB code **cantbeam_ss_modred.m** will be discussed and listed in detail in the following sections.

15.6.1 Input

The code in this section asks the user to define how many elements will be used for the analysis. ANSYS runs have been made for 2, 4, 6, 8, 10, 12, 16, 32 and 64 elements. The ANSYS eigenvector results for each have been stripped out of the ANSYS format and put into frequency vector, "freqvec," and modal matrix, "evr," form and stored as MATLAB .mat files.

```
%          cantbeam_ss_modred.m

           clear all;

           hold off;

           clf;

%          load the .mat file cantbeamXXred, containing evr - the modal matrix, freqvec -
%          the frequency vector and node_numbers - the vector of node numbers for the modal
%          matrix

           model = menu('choose which finite element model to use ... ', ....
                                    '2 beam elements', ...
                                    '4 beam elements', ...
                                    '6 beam elements', ...
                                    '8 beam elements', ...
                                    '10 beam elements', ...
                                    '12 beam elements', ...
                                    '16 beam elements', ...
                                    '32 beam elements', ...
                                    '64 beam elements');

           if  model == 1
                      load cantbeam2red;
           elseif  model == 2
                      load cantbeam4red;
           elseif  model == 3
                      load cantbeam6red;
           elseif  model == 4
                      load cantbeam8red;
           elseif  model == 5
                      load cantbeam10red;
           elseif  model == 6
                      load cantbeam12red;
           elseif  model == 7
                      load cantbeam16red;
           elseif  model == 8
                      load cantbeam32red;
           elseif  model == 9
                      load cantbeam64red;
           end
```

15.6.2 Defining Degrees of Freedom and Number of Modes

The code below checks the size of the modal matrix, where the number of rows indicates how many degrees of freedom are used and the number of columns indicates the number of modes. Since all of the models have an even number of elements, there is always a node at the midpoint of the beam and it is possible to define which row of the modal matrix corresponds to that middle node. The modal matrix row which corresponds to the tip is the last degree of freedom in the matrix. The code also defines a new variable, "xn," the normalized modal matrix.

```
%       define the number of degrees of freedom and number of modes from size
%       of modal matrix

        [numdof,num_modes_total] = size(evr);

%       define rows for middle and  tip nodes

        mid_node_row = 0.5*(numdof-1)+1;

        tip_node_row = numdof;

        xn = evr;
```

15.6.3 Sorting Modes by dc Gain and Peak Gain, Selecting Modes Used

The next step in creating the model is to sort modes of vibration so that only the most important modes are kept. We will discuss in this section two methods of sorting, one which is applicable for models with the same value of damping for all modes, $\zeta_i = \zeta = $ constant ("uniform" damping), and another which is applicable for models with different damping values for each mode ("non-uniform" damping).

Repeating from (8.54a,b) the general equation for the overall transfer function of undamped and damped systems:

$$\frac{z_j}{F_k} = \sum_{i=1}^{m} \frac{z_{nji} z_{nki}}{s^2 + \omega_i^2}$$

$$\frac{z_j}{F_k} = \sum_{i=1}^{m} \frac{z_{nji} z_{nki}}{s^2 + 2\zeta_i \omega_i s + \omega_i^2}$$

(15.1a,b)

This equation shows that in general every transfer function is made up of additive combinations of single degree of freedom systems, with each system having its residue determined by the appropriate input/output eigenvector entries, $z_{nji} z_{nki}$, and with resonant frequency defined by the eigenvalue, ω_i. Substituting $s = j\omega = j0 = 0$ to obtain the ith mode frequency response at dc, the **dc gain**, which is the same for the undamped and damped cases is:

$$\frac{z_{ji}}{F_{ki}} = \frac{z_{nji} z_{nki}}{\omega_i^2}, \qquad (15.2)$$

where $z_{nji} z_{nki}$ is the product of the jth (output) row and kth (force applied) row terms of the ith eigenvector divided by the square of the eigenvalue for the ith mode.

At resonance, the **peak gain** amplitude of each mode is given by substituting $s = j\omega_i$, $s^2 = -\omega_i^2$ into (15.1b):

$$
\begin{aligned}
\frac{z_{ji}}{F_{ki}} &= \frac{z_{nji} z_{nki}}{s^2 + 2\zeta_i \omega_i s + \omega_i^2} \\[6pt]
&= \frac{z_{nji} z_{nki}}{-\omega_i^2 + 2\zeta_i \omega_i^2 \, j + \omega_i^2} \\[6pt]
&= \frac{z_{nji} z_{nki}}{2\zeta_i \omega_i^2 \, j} \\[6pt]
&= \frac{-j \, z_{nji} z_{nki}}{2\zeta_i \omega_i^2} \\[6pt]
&= \frac{-j}{2\zeta_i}\left(\frac{z_{nji} z_{nki}}{\omega_i^2}\right) \\[6pt]
&= \frac{-j}{2\zeta_i}(\text{dc gain})
\end{aligned}
\qquad (15.3)
$$

Comparing (15.2) and (15.3) it is evident that the relationship between the dc gain and peak gain for a mode is that the dc gain term is divided by 2ζ and multiplied by "$-j$," which gives a $-90°$ phase shift at resonance. Since ζ values for mechanical structures are typically small, a few percent of critical damping, 2ζ is a small number, which serves to amplify the response by virtue of the division, thus the resonant "peak" in the response.

If the same value of ζ is used for all modes, then all the dc gain terms are divided by the same 2ζ terms and the relative amplitudes of the dc gains and peak gains are the same, so there is no difference between sorting a uniform damping model using dc gain or peak gain.

However, if the modes have different damping the relationship between the dc gain and peak gain for all the modes is not a constant $1/2\zeta$ value and peak gain must be used to rank modes for importance. In this case, the MATLAB damping parameter "zeta" would not be a scalar but would be a vector with entries corresponding to damping in each mode.

We will use dc gain to rank the relative importance of the modes until Chapter 18, where a technique named "balanced reduction" will be introduced. The code shown below, and throughout the book, is easily modified to sort for peak gain instead of dc gain using (15.3) instead of (15.2) and entering a vector of damping values instead of a scalar.

The code below carries out the calculation of the dc gain and sorts from smallest to largest, keeping track of the new column locations in "index_sort." It then uses the "fliplr" command to list them from largest to smallest, so that the first mode has the highest dc gain. Various plots are then shown to indicate the relative importance of each mode. After plotting the dc gains, the user is asked to define the number of modes to be used in the frequency response, from 1 to all the available modes.

```
%   calculate the dc amplitude of the displacement of each mode by
%   multiplying the forcing function row of the eigenvector by the output row

            omega2 = (2*pi*freqvec)'.^2;    % convert to radians and square

            dc_gain = abs(xn(mid_node_row,:).*xn(tip_node_row,:))./omega2;

            [dc_gain_sort,index_sort] = sort(dc_gain);

            dc_gain_sort = fliplr(dc_gain_sort);

            index_sort = fliplr(index_sort)

            dc_gain_nosort = dc_gain;

            index_orig = 1:num_modes_total;

            semilogy(index_orig,freqvec,'k-');
            title('frequency versus mode number')
            xlabel('mode number')
            ylabel('frequency, hz')
            grid
            pause

            semilogy(index_orig,dc_gain_nosort,'k-')
            title('dc value of each mode contribution versus mode number')
```

```
xlabel('mode number')
ylabel('dc value')
grid off
pause

loglog(freqvec,dc_gain_nosort,'k-')
title('dc value of each mode contribution versus frequency')
xlabel('frequency, hz')
ylabel('dc value')
grid off
pause

semilogy(index_orig,dc_gain_sort,'k-')
title('sorted dc value of each mode versus number of modes included')
xlabel('modes included')
ylabel('sorted dc value')
grid off
pause

num_modes_used = input(['enter how many modes to include ...
            , ',num2str(num_modes_total),' default, max ... ']);

if (isempty(num_modes_used))
        num_modes_used = num_modes_total;
end
```

The first step in any finite element analysis is to understand the resonant frequencies of the model and how they relate to the frequency range of interest for the problem at hand.

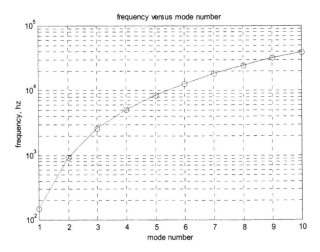

Figure 15.4: Resonant frequency versus mode number.

Figure 15.4 shows that modes 8, 9 and 10 have frequencies higher than the required 20 khz required by the problem, so our model should be adequate.

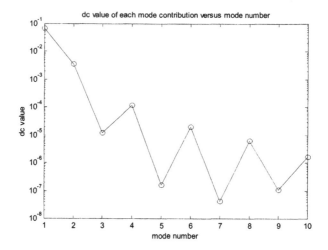

Figure 15.5: dc value of each mode contribution versus mode number.

Figure 15.5 shows the dc gain values for all the modes plotted versus mode number.   It is interesting that the low values for modes 3, 5, 7 and 9 correspond to small values of the midpoint node elements of the respective eigenvectors (see the bold highlighted entries in columns 3, 5, 7 and 9 in Table 15.3).   This means that the midpoint is nearly a "node" for those modes. Again, a "node" for a mode refers not to the number of the end point of the element but a location along the beam where the displacement is zero for a particular mode of vibration.

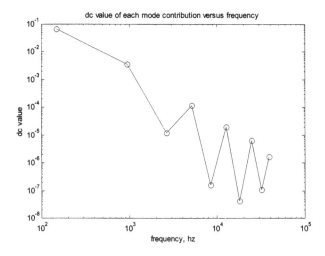

Figure 15.6: dc Value of each mode contribution versus resonant frequency.

Figure 15.6 shows dc gain versus frequency of the mode. Note that there is a general trend for lower gains as frequency increases. This is not always the case, as we shall see in Chapter 16.

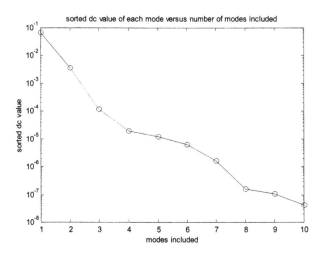

Figure 15.7: Sorted dc value of each mode versus number of modes included.

Figure 15.7 shows the sorted values for the dc gains, from largest to smallest. The list of mode numbers after sorting is given by "index_sort" below. The ordering can be seen in the dc value versus mode number plot in Figure 15.5.

```
index_sort =   1   2   4   6   3   8   10   5   9   7
```

15.6.4 Damping, Defining Reduced Frequencies and Modal Matrices

The section below asks for the damping value and whether to use the original ordering of modes or the modes sorted by dc gain. At this point, three different sets of modal matrices and eigenvalue vectors will be defined. The first set uses all the modes and frequencies and keeps them in their original, unsorted order. This set will be used to calculate frequency and step responses of the non-reduced model for comparison. The second set uses only the "num_modes_used" number of modes and keeps them in their original, unsorted order. This set will be used to see the effects of a simple truncation of higher frequency modes without sorting or ranking. The third set again uses the "num_modes_used" number of modes but includes only the modes with the highest dc gains. We will calculate frequency response and transient response results for both of the reduced cases and compare results with the "all modes included" case. The two reduced models are denoted with the "_nosort" and "_sort" suffixes throughout the code. We will see that because the dc gain values for this model generally decrease with frequency, the sorted and unsorted models will give almost the same results. The example in the next chapter, however, will not have this property.

```
        zeta = input('enter value for damping, .02 is 2% of critical (default) ... ');

        if (isempty(zeta))
            zeta = .02;
        end

%       all modes included model, use original order

        xnnew = xn(:,(1:num_modes_total));

        freqnew = freqvec((1:num_modes_total));

%       reduced, no sorting, just use the first num_modes_used modes in xnnew_nosort

        xnnew_nosort = xn(:,1:num_modes_used);

        freqnew_nosort = freqvec(1:num_modes_used);

%       reduced, sorting, use the first num_modes_used sorted modes in xnnew_sort

        xnnew_sort = xn(:,index_sort(1:num_modes_used));

        freqnew_sort = freqvec(index_sort(1:num_modes_used));
```

15.6.5 Setting up System Matrix "a"

The section below sets up three state space system "a" matrices. Since we know the form of the modal form state space equation from Chapter 10, it can be built automatically. The general form is given by (15.4). The system matrix is made up of eigenvalue and damping terms for each mode, and each mode is a 2x2 submatrix along the diagonal.

$$\dot{x} = Ax + Bu \qquad (15.4)$$

$$
\begin{bmatrix} \dot{x}_1 \\ \dot{x}_2 \\ \dot{x}_3 \\ \dot{x}_4 \\ \cdots \\ \cdots \end{bmatrix}
=
\begin{bmatrix}
0 & 1 & 0 & 0 & \cdots & \cdots \\
-\omega_1^2 & -2\zeta_1\omega_1 & 0 & 0 & \cdots & \cdots \\
0 & 0 & 0 & 1 & \cdots & \cdots \\
0 & 0 & -\omega_2^2 & -2\zeta_2\omega_2 & \cdots & \cdots \\
\cdots & \cdots & \cdots & \cdots & \cdots & \cdots \\
\cdots & \cdots & \cdots & \cdots & \cdots & \cdots
\end{bmatrix}
\begin{bmatrix} x_1 \\ x_2 \\ x_3 \\ x_4 \\ \cdots \\ \cdots \end{bmatrix}
+
\begin{bmatrix} 0 \\ F_{p1} \\ 0 \\ F_{p2} \\ \cdots \\ \cdots \end{bmatrix} u \qquad (15.5)
$$

The first system matrix, "a," is for the full, non-reduced system and includes all the modes in their original order. The second is "a_nosort" and has the reduced size with the original ordering of modes. The third is "a_sort" and has the reduced number of modes with dc gain ordering.

%	define variables for all modes included system matrix, a
	w = freqnew*2*pi; % frequencies in rad/sec
	w2 = w.^2;
	zw = 2*zeta*w;
%	define variables for reduced, nosorted system matrix, a_nosort
	w_nosort = freqnew_nosort*2*pi; % frequencies in rad/sec
	w2_nosort = w_nosort.^2;
	zw_nosort = 2*zeta*w_nosort;
%	define variables for reduced, sorted system matrix, a_sort
	w_sort = freqnew_sort*2*pi; % frequencies in rad/sec
	w2_sort = w_sort.^2;
	zw_sort = 2*zeta*w_sort;
%	define size of system matrix

```
        asize = 2*num_modes_total;

        asize_red = 2*num_modes_used;

        disp(' ');
        disp(' ');
        disp(['size of system matrix a is ',num2str(asize)]);
        disp(['size of reduced system matrix a is ',num2str(asize_red)]);

%       setup all modes included "a" matrix, system matrix

        a = zeros(asize);

        for  col = 2:2:asize

        row = col-1;

        a(row,col) = 1;

        end

        for  col = 1:2:asize

        row = col+1;

        a(row,col) = -w2((col+1)/2);

        end

        for  col = 2:2:asize

        row = col;

        a(row,col) = -zw(col/2);

        end

%       setup reduced, nosorted "a_nosort" matrix, system matrix

        a_nosort = zeros(asize_red);

        for  col = 2:2:asize_red

        row = col-1;

        a_nosort(row,col) = 1;

        end

        for  col = 1:2:asize_red

        row = col+1;

        a_nosort(row,col) = -w2_nosort((col+1)/2);
```

```
          end

          for  col = 2:2:asize_red

          row = col;

          a_nosort(row,col) = -zw_nosort(col/2);

          end

%         setup reduced, sorted "a_sort" matrix, system matrix

          a_sort = zeros(asize_red);

          for  col = 2:2:asize_red

          row = col-1;

          a_sort(row,col) = 1;

          end

          for  col = 1:2:asize_red

          row = col+1;

          a_sort(row,col) = -w2_sort((col+1)/2);

          end

          for  col = 2:2:asize_red

          row = col;

          a_sort(row,col) = -zw_sort(col/2);

          end
```

15.6.6 Setting up Input Matrix "b"

As with the system matrix above, here we will set up three different input matrices, "b," "b_nosort" and "b_sort." We begin with the force vector in physical coordinates, with "numdof" rows. The rows are all zeros except for the "mid_node_row," which has a value of 1.0 mN. The force vector in principal coordinates is obtained by premultiplying by the transpose of the modal matrix. The state space form of the force vector in principal coordinates is the "numdof x 1" force vector in principal coordinates padded with zeros to create the same number of rows as states.

```
%          setup input matrix b, state space forcing function in principal coordinates

%          f_physical is the vector of physical force
%          zeros at each output DOF and input force at the input DOF

           f_physical = zeros(numdof,1);     %          start out with zeros

           f_physical(mid_node_row) = 1.0;              %  input force at node 6, midpoint node

%          f_principal is the vector of forces in principal coordinates

           f_principal = xnnew'*f_physical;

%          b is the vector of forces in principal coordinates, state space form

           b = zeros(2*num_modes_total,1);

           for  cnt = 1:num_modes_total

                    b(2*cnt) = f_principal(cnt);

           end

%          f_principal_nosort is the vector of forces in principal coordinates

           f_principal_nosort = xnnew_nosort'*f_physical;

%          b_nosort is the vector of forces in principal coordinates, state space form

           b_nosort = zeros(2*num_modes_used,1);

           for  cnt = 1:num_modes_used

                    b_nosort(2*cnt) = f_principal_nosort(cnt);

           end

%          f_principal_sort is the vector of forces in principal coordinates

           f_principal_sort = xnnew_sort'*f_physical;

%          b_sort is the vector of forces in principal coordinates, state space form

           b_sort = zeros(2*num_modes_used,1);

           for  cnt = 1:num_modes_used

                    b_sort(2*cnt) = f_principal_sort(cnt);

           end
```

15.6.7 Setting up Output Matrix "c" and Direct Transmission Matrix "d"

The output matrices below, "c," "c_nosort" and "c_sort," are separated into displacement and velocity matrices, "cdisp" and "cvel," so that they can be premultiplied by the appropriate modal matrix to obtain vectors of displacements and velocities in physical coordinates. With the defined output displacement and velocity matrices, all displacement and velocity degrees of freedom in physical coordinates are available for plotting or further analysis. Since there is no direct feedthrough on this model, the "d" matrix is zero.

```
%      setup cdisp and cvel, padded xn matrices to give the displacement and velocity
%      vectors in physical coordinates
%      cdisp and cvel each have numdof rows and alternating columns consisting of columns
%      of xnnew and zeros to give total columns equal to the number of states

%      all modes included cdisp and cvel

       for col = 1:2:2*length(freqnew)

       for row = 1:numdof

       cdisp(row,col) = xnnew(row,ceil(col/2));

       cvel(row,col) = 0;

       end

       end

       for col = 2:2:2*length(freqnew)

       for row = 1:numdof

       cdisp(row,col) = 0;

            cvel(row,col) = xnnew(row,col/2);

       end

       end

%      reduced, nosorted cdisp and cvel

       for col = 1:2:2*length(freqnew_nosort)

       for row = 1:numdof

       cdisp_nosort(row,col) = xnnew_nosort(row,ceil(col/2));

       cvel_nosort(row,col) = 0;
```

```
            end

            end

        for  col = 2:2:2*length(freqnew_nosort)

        for  row = 1:numdof

        cdisp_nosort(row,col) = 0;

            cvel_nosort(row,col) = xnnew_nosort(row,col/2);

        end

        end

%       reduced, sorted cdisp and cvel

        for  col = 1:2:2*length(freqnew_sort)

        for  row = 1:numdof

        cdisp_sort(row,col) = xnnew_sort(row,ceil(col/2));

        cvel_sort(row,col) = 0;

        end

        end

        for  col = 2:2:2*length(freqnew_sort)

        for  row = 1:numdof

        cdisp_sort(row,col) = 0;

            cvel_sort(row,col) = xnnew_sort(row,col/2);

        end

        end

%       define output

        d = [0];  %
```

15.6.8 Frequency Range, "ss" Setup, Bode Calculations

The first part of this section defines the frequency range to be used for the frequency responses, logarithmically spaced frequency vectors in units of hz and rad/sec. Three "ss" state space systems are defined for the displacement of the tip of the beam, the non-reduced system, the "nosort" and the "sort." Since "cdisp" contains information about all the degrees of freedom, they are all available for output by defining the appropriate row. The "bode" command is used to calculate the magnitude and phase vectors over the defined frequency range, and the magnitudes are converted to db.

```
%        define frequency vector for frequency responses

         freqlo = 10;

         freqhi = 100000;

         flo=log10(freqlo) ;
         fhi=log10(freqhi) ;

         f=logspace(flo,fhi,200) ;
         frad=f*2*pi ;

%        take transfer functions, outputting the midpoint and tip node rows of the displacement
%        vector cdisp

%        define displacement state space system with the "ss" command

         sysdisptip = ss(a,b,cdisp(tip_node_row,:),d);

%        defined reduced systems using num_modes_used nosort modes

         sysdisptip_nosort = ss(a_nosort,b_nosort,cdisp_nosort(tip_node_row,:),d);

%        define reduced systems using num_modes_used sorted modes

         sysdisptip_sort = ss(a_sort,b_sort,cdisp_sort(tip_node_row,:),d);

%        use "bode" command to generate magnitude/phase vectors

         [magdisptip,phsdisptip]=bode(sysdisptip,frad) ;

         [magdisptip_nosort,phsdisptip_nosort]=bode(sysdisptip_nosort,frad) ;

         [magdisptip_sort,phsdisptip_sort]=bode(sysdisptip_sort,frad) ;

%        convert magnitude to db

         magdisptipdb = 20*log10(magdisptip);

         magdisptipdb_nosort = 20*log10(magdisptip_nosort);
```

```
           magdisptipdb_sort = 20*log10(magdisptip_sort);
```

15.6.9 Full Model - Plotting Frequency Response, Step Response

This section plots the frequency response for tip displacement due to a unit force at the beam midpoint. It then overlays the contribution of each individual mode to the overall response. Since the "a" matrix consists of 2x2 submatrices along the diagonal, all we have to do to get the contribution of each individual mode is to pull out successive 2x2 individual mode system matrices. Similarly, we take the appropriate rows and columns of "b" and "cdisp" for each mode. Because of the systematic form of the matrices, MATLAB can generate the individual mode matrices automatically. To facilitate comparison with the dc gain values calculated for all the modes (and used in their sorting), an "o" is plotted along the left-hand axis for each individual mode. Because the magnitude axis is in db units, the individual contributions cannot be combined graphically like with a linear magnitude axis as shown in Chapter 6. Nevertheless, using the overlaid plots to get a mental image of the combining modes is valuable.

For the unit force step response, a time vector, "t" and input vector "u" are defined for use with the MATLAB function "lsim."

```
%          start plotting

           if num_modes_used == num_modes_total

%          plot all modes included response

           semilogx(f,magdisptipdb(1,:),'k.-')
           title(['cantilever tip displacement for mid-length force, all ', ...
                   num2str(num_modes_used),' modes included'])
           xlabel('Frequency, hz')
           ylabel('Magnitude, db mm')
           grid off
           pause

           hold on

           max_modes_plot = num_modes_total;

           for pcnt = 1:max_modes_plot

                   index = 2*pcnt;

                   amode = a(index-1:index,index-1:index);

                   bmode = b(index-1:index);
```

```
                    cmode = cdisp(numdof,index-1:index);

                    dmode = [0];

                    sysdisptip_mode = ss(amode,bmode,cmode,dmode);

                    [magdisptip_mode,phsdisptip_mode]=bode(sysdisptip_mode,frad)  ;

                    magdisptip_modedb = 20*log10(magdisptip_mode);

                    semilogx(f,magdisptip_modedb(1,:),'k-')

        end

        dc_gain_freq = freqlo*ones(size(freqnew));

        semilogx(dc_gain_freq(1:num_modes_used),20*log10(dc_gain
                     (1:num_modes_used)),'ko:')

        pause

        hold off

%       now use lsim to calculate step response to a unit force

        ttotal = 0.1;

        t = linspace(0,ttotal,200);

        u = ones(size(t));

        [disptip,ts] = lsim(sysdisptip,u,t);

        plot(ts,disptip,'k-')
        title(['tip disp for mid-length step force, all ',num2str(num_modes_used), ...
                     ' modes included'])
        xlabel('time, sec')
        ylabel('displacement, mm')
        grid off
        pause
```

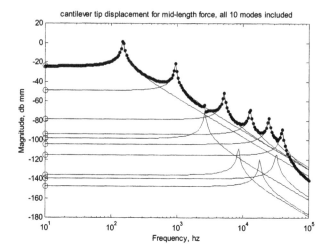

Figure 15.8: Cantilever tip displacement for mid-length force, all 10 modes included.

Figure 15.8 shows the overall frequency response with the overlaid sdof responses of all the individual modes for the 10-element model using all 10 available modes. The "o's" at the 10 hz frequency indicate the values of dc gain for each mode. Note that the fifth, seventh and ninth modes have such low gains that their resonant peaks are barely visible on the overall response. The third mode has a higher gain, as indicated by the small pole/zero combination between the second and fourth modes.

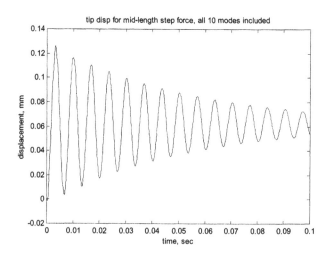

Figure 15.9: Cantilever tip displacement for mid-length force, all 10 modes included.

Figure 15.9 depicts the response of the beam tip due to a 1mN step force at the midpoint. We will be comparing the different modal truncation methods with this overall response.

15.6.10 Reduced Models – Plotting Frequency Response, Step Response

The following section of code does the same thing for the reduced unsorted and sorted models as the last section did for the full model. In all the plots, the full model results are overlaid with the reduced model results to show the differences. In the examples that follow, we will use four modes in the reduced models. The reader is encouraged to run the code using different numbers of reduced modes to see the effects on both frequency and time domain responses.

```
else

%       plot unsorted modal truncation

        semilogx(f,magdisptipdb(1,:),'k-',f,magdisptipdb_nosort(1,:),'k.-')
        title(['unsorted modal truncation:  cantilever tip displacement for mid- ...
                length force, first ',num2str(num_modes_used),' modes included'])
        legend('all modes','unsorted partial modes',3)

        dcgain_error_percent_nosort = 100*(magdisptip_nosort(1) ...
                        magdisptip(1))/magdisptip(1)

        xlabel('Frequency, hz')
        ylabel('Magnitude, db mm')
        grid off

        pause

        hold on

        max_modes_plot = num_modes_used;

        for  pcnt = 1:max_modes_plot

                index = 2*pcnt;

                amode = a_nosort(index-1:index,index-1:index);

                bmode = b_nosort(index-1:index);

                cmode = cdisp_nosort(numdof,index-1:index);

                dmode = [0];

                sysdisptip_mode = ss(amode,bmode,cmode,dmode);

                [magdisptip_mode,phsdisptip_mode]=bode(sysdisptip_mode,frad)  ;
```

```
                    magdisptip_modedb = 20*log10(magdisptip_mode);

                    semilogx(f,magdisptip_modedb(1,:),'k-')

        end

        dc_gain_freq_nosort = freqlo*ones(size(freqnew_nosort));

        semilogx(dc_gain_freq_nosort(1:num_modes_used),20*log10 ...
                    (dc_gain_nosort(1:num_modes_used)),'ko:')

        pause

        hold off

%       plot sorted modal truncation

        semilogx(f,magdisptipdb(1,:),'k-',f,magdisptipdb_sort(1,:),'k-')
        title(['sorted modal truncation:  cantilever tip displacement for mid-length force, ...
                    first ',num2str(num_modes_used),' modes included'])
        legend('all modes','sorted partial modes',3)

        dcgain_error_percent_sort = 100*(magdisptip_sort(1) - magdisptip(1))/magdisptip(1)

        xlabel('Frequency, hz')
        ylabel('Magnitude, db mm')
        grid off

        pause

        hold on

        max_modes_plot = num_modes_used;

        for  pcnt = 1:max_modes_plot

                    index = 2*pcnt;

                    amode = a_sort(index-1:index,index-1:index);

                    bmode = b_sort(index-1:index);

                    cmode = cdisp_sort(numdof,index-1:index);

                    dmode = [0];

                    sysdisptip_mode = ss(amode,bmode,cmode,dmode);

                    [magdisptip_mode,phsdisptip_mode]=bode(sysdisptip_mode,frad) ;

                    magdisptip_modedb = 20*log10(magdisptip_mode);

                    semilogx(f,magdisptip_modedb(1,:),'k-')
```

```
        end

        dc_gain_freq_sort = freqlo*ones(size(freqnew_nosort));

        semilogx(dc_gain_freq_sort(1:num_modes_used),20*log10 ...
                        (dc_gain_sort(1:num_modes_used)),'ko:')

        pause

        hold off

%       now use lsim to calculate step response to a unit force

        ttotal = 0.1;

        t = linspace(0,ttotal,200);

        u = ones(size(t));

        [disptip,ts] = lsim(sysdisptip,u,t);

        [disptip_nosort,ts_nosort] = lsim(sysdisptip_nosort,u,t);

        [disptip_sort,ts_sort] = lsim(sysdisptip_sort,u,t);

        plot(ts,disptip,'k-',ts_nosort,disptip_nosort,'k+-',ts_sort,disptip_sort,'k.-')
        title(['tip disp for mid-length step force, first ',num2str(num_modes_used) ...
                        ,' modes included'])
        legend('all modes','unsorted partial modes','sorted partial modes')
        xlabel('time, sec')
        ylabel('displacement, mm')
        grid off
        pause
```

15.6.11 Reduced Models – Plotted Results – Four Modes Used

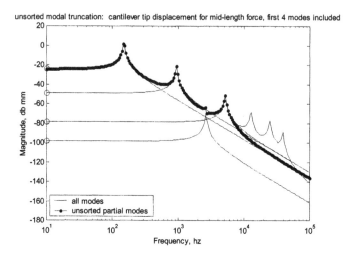

Figure 15.10: Cantilever tip displacement for mid-length force, first four modes included – unsorted modal truncation.

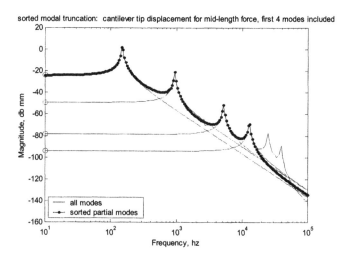

Figure 15.11: Cantilever tip displacement for mid-length force, first four modes included – sorted modal truncation.

Figure 15.10 depicts overall plus individual mode contributions for the four unsorted modes model. Note that the first four unsorted modes are used. The dc gain error relative to the full 10-mode model is +0.024% because the dc gain terms for the eliminated modes are not included. Note that the last three peaks in the "all modes" response are missed because the modes are not included.

Figure 15.11 shows overall plus individual mode contributions for the four sorted modes model. Note that this time the third mode is skipped and the fifth mode is used instead because it has a higher dc gain. The dc gain error relative to the full 10-mode model is –0.027%.

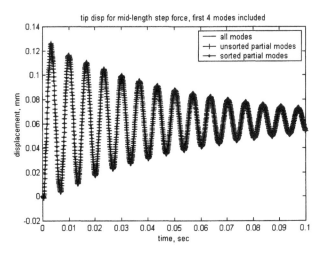

Figure 15.12: Comparison of step responses for all modes included and four modes included, unsorted and sorted.

Figure 15.12 shows step response for full, reduced unsorted and reduced sorted models. Because the dc gain for the two models is in error only by a fraction of a percent and because the eliminated modes are some 80db (four orders of magnitude) lower than the most significant first mode, there is no discernable difference in the responses of the full and two reduced models.

15.6.12 Modred Description

The MATLAB Control System Toolbox has a function, "modred" (MODel order REDuction), which can be used for reducing models while retaining the overall system dc gain. The "mdc" or "Matched DC" gain option for the function "modred" reduces defined states by setting the derivatives of the states to be eliminated to zero, then solving for the remaining states. The method essentially sets up the eliminated states to be "infinitely fast" and is analogous to Guyan reduction in that the low frequency effects of the eliminated states are included in the remaining states. The other option for "modred" is the "del" option, which simply eliminates the defined states, typically associated with the higher frequency modes.

The derivation of the "mdc" option follows. We start with the state space description of the system:

$$\dot{x} = Ax + Bu$$
$$y = Cx + Du$$

$$(15.6a,b)$$

Assume that we have a method of ordering the importance of the modes making up the **A**, **B** and **C** matrices, in our case using dc or peak gains. If we then rearrange and partition the matrices such that the states corresponding to the most important modes are separated from the less important modes, designating the important modes as x_r (reduced) and the unimportant modes to be eliminated as x_e, we get

$$\begin{bmatrix} \dot{x}_r \\ \dot{x}_e \end{bmatrix} = \begin{bmatrix} A_{rr} & A_{re} \\ A_{er} & A_{ee} \end{bmatrix} \begin{bmatrix} x_r \\ x_e \end{bmatrix} + \begin{bmatrix} B_r \\ B_e \end{bmatrix} u$$

$$y = \begin{bmatrix} C_r & C_e \end{bmatrix} \begin{bmatrix} x_r \\ x_e \end{bmatrix} + Du$$

$$(15.7a,b)$$

Expanding the matrices:

$$\dot{x}_r = A_{rr}x_r + A_{re}x_e + B_r u$$
$$\dot{x}_e = A_{er}x_r + A_{ee}x_e + B_e u$$

$$(15.8a,b)$$

Setting the $\dot{x}_e$ states equal to zero in (15.10) is analogous to setting (14.14) equal to zero in the Guyan reduction process. We are then, in effect, including the low frequency dc gain or static equilibrium characteristics of the eliminated modes in the reduced modes.

$$0 = A_{er}x_r + A_{ee}x_e + B_e u$$

$$(15.9)$$

Solving for x_e:

$$x_e = -A_{ee}^{-1}A_{er}x_r - A_{ee}^{-1}B_e u$$

$$(15.10)$$

Substituting back into the $\dot{x}_r$ equation and grouping terms:

$$\begin{aligned}
\dot{x}_r &= A_{rr}x_r + A_{re}\left(-A_{ee}^{-1}A_{er}x_r - A_{ee}^{-1}B_e u\right) + B_r u \\
&= \left(A_{rr} - A_{re}A_{ee}^{-1}A_{er}\right)x_r + \left(B_r - A_{re}A_{ee}^{-1}B_e\right)u
\end{aligned} \tag{15.11}$$

Substituting back into the expanded output equations:

$$\begin{aligned}
y &= C_r x_r + C_e x_e + Du \\
&= C_r x_r + C_e\left(-A_{ee}^{-1}A_{er}x_r - A_{ee}^{-1}B_e u\right) + Du \\
&= \left(C_r - C_e A_{ee}^{-1}A_{er}\right)x_r + \left(D - C_e A_{ee}^{-1}B_e\right)u
\end{aligned} \tag{15.12}$$

The new matrices for the reduced model become:

$$\begin{aligned}
A_{red} &= A_{rr} - A_{re}A_{ee}^{-1}A_{er} \\
B_{red} &= B_r - A_{re}A_{ee}^{-1}B_e \\
C_{red} &= C_r - C_e A_{ee}^{-1}A_{er} \\
D_{red} &= D - C_e A_{ee}^{-1}B_e
\end{aligned} \tag{15.13a,b,c,d}$$

The new state equations are:

$$\begin{aligned}
\dot{x}_{red} &= A_{red}x_{red} + B_{red}u \\
y_{red} &= C_{red}x_{red} + D_{red}u
\end{aligned} \tag{15.14a,b}$$

We will see (Figure 15.14) that the high frequency portion of the response when reducing using "modred" does not roll off quickly with frequency as we are used to seeing. Rather, it will be "flat" with frequency. The reason for the shape of the "modred" high frequency asymptote is in the D_{red} term in (15.13d). In many cases, the direct transmission term D is zero. When using "modred," however, even if D is zero, there is still the $-C_e A_{ee}^{-1}B_e$ portion of D_{red} to contend with. Repeating Figure 5.2 below, we can see the direct transmission term.

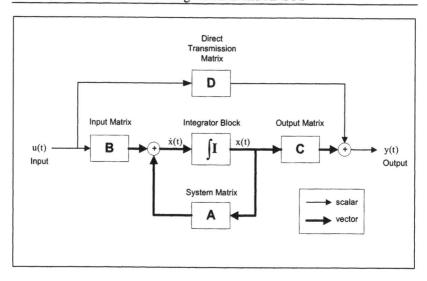

Figure 15.13: State space system block diagram.

At high frequencies, where the system matrix dynamics start to attenuate, the $-\mathbf{C}_e\mathbf{A}_{ee}^{-1}\mathbf{B}_e$ term of $\mathbf{D}_{red}$ starts to dominate the response – hence the "flat" high frequency response in Figure 15.15.

15.6.13 Defining Sorted or Unsorted Modes to be Used

The section of code below prompts the user to define whether the modes are to be sorted by dc gain or left in the original order for the "modred" operation. One argument of the "modred" command is to define the states to be eliminated. The states to be eliminated can be defined as a vector of arbitrary states or as a continuous partition of states. We will define them in the code below as a continuous block of states, from one index greater than the number of states to be kept to the total number of states. Therefore, if we sort by dc gain before using "modred," we would keep only the most important states. If we choose to use the unsorted states, we will be eliminating the higher frequency modes and keeping the lower frequency modes.

```
%        use modred to reduce, select whether to use sorted or unsorted modes for the reduction

         modred_sort = input('modred: enter "1" to use sorted modes for reduced runs, ...
                  "enter" to use unsorted ... ');

         if isempty(modred_sort)
                  modred_sort = 0
         end
```

```
if  modred_sort == 1              %  use sorted mode order

        xnnew = xn(:,index_sort(1:num_modes_total));

        freqnew = freqvec(index_sort(1:num_modes_total));

else                        %  use original mode order

        xnnew = xn(:,(1:num_modes_total));

        freqnew = freqvec((1:num_modes_total));

end
```

15.6.14 Defining System for Reduction

In this section we define a new set of "a," "b," "c" and "d" matrices which will be used with "modred."

```
%        define variables for all modes included system matrix, a

        w = freqnew*2*pi;              %            frequencies in rad/sec

        w2 = w.^2;

        zw = 2*zeta*w;

%        define size of system matrix

        asize = 2*num_modes_total;

%        setup all modes included "a" matrix, system matrix

        a = zeros(asize);

        for  col = 2:2:asize

        row = col-1;

        a(row,col) = 1;

        end

        for  col = 1:2:asize

        row = col+1;

        a(row,col) = -w2((col+1)/2);

        end
```

```
              for  col = 2:2:asize

              row = col;

              a(row,col) = -zw(col/2);

              end

%             setup input matrix b, state space forcing function in principal coordinates

%             f_physical is the vector of physical force
%             zeros at each output DOF and input force at the input DOF

              f_physical = zeros(numdof,1);              %          start out with zeros

              f_physical(mid_node_row) = 1.0;                     %  input force at node
                          6,midpoint node

%             f_principal is the vector of forces in principal coordinates

              f_principal = xnnew'*f_physical;

%             b is the vector of forces in principal coordinates, state space form

              b = zeros(2*num_modes_total,1);

              for  cnt = 1:num_modes_total

                      b(2*cnt) = f_principal(cnt);

              end

%             setup cdisp and cvel, padded xn matrices to give the displacement and velocity
%             vectors in physical coordinates
%             cdisp and cvel each have numdof rows and alternating columns consisting of columns
%             of xnnew and zeros to give total columns equal to the number of states

%             all modes included cdisp and cvel

              for  col = 1:2:2*length(freqnew)

              for  row = 1:numdof

              cdisp(row,col) = xnnew(row,ceil(col/2));

              cvel(row,col) = 0;

              end

              end

              for  col = 2:2:2*length(freqnew)

              for  row = 1:numdof
```

```
          cdisp(row,col) = 0;

              cvel(row,col) = xnnew(row,col/2);

      end

      end

%     define output

      d = [0];  %
```

15.6.15 Modred Calculations – "mdc" and "del"

This section defines a MATLAB state space, "ss," system using either the unsorted or sorted eigenvectors and eigenvalues from above, and then both the "mdc" and "del" options with "modred" to calculate two reduced systems. In order to be able to plot not only the overall frequency response from the reduced systems but also the individual mode contributions, we will use the "ssdata" function in MATLAB to define the reduced system matrices. In the next section we will use 2x2 submatrices of the reduced system matrix to define individual modal contributions. The "bode" command is then used to generate the magnitude/phase solution vectors, which are converted to db.

```
%     define state space system for reduction, ordered defined by modred_sort

      sysdisptip_red = ss(a,b,cdisp(tip_node_row,:),d);

%     define reduced matrices using matched dc gain method "mdc"

      states_elim = (2*num_modes_used+1):2*num_modes_total;

      sysdisptip_mdc = modred(sysdisptip_red,states_elim,'mdc');

      [adisptip_mdc,bdisptip_mdc,cdisptip_mdc,ddisptip_mdc] = ssdata(sysdisptip_mdc);

%     define reduced matrices by eliminating high frequency states, 'del

      sysdisptip_elim = modred(sysdisptip_red,states_elim,'del');

      [adisptip_elim,bdisptip_elim,cdisptip_elim,ddisptip_elim] = ssdata(sysdisptip_elim);

%     use "bode" command to generate magnitude/phase vectors for reduced systems

      [magdisptip_mdc,phsdisptip_mdc]=bode(sysdisptip_mdc,frad) ;

      [magdisptip_elim,phsdisptip_elim]=bode(sysdisptip_elim,frad) ;

%     convert magnitude to db

      magdisptip_mdcdb = 20*log10(magdisptip_mdc);
```

```
magdisptip_elimdb = 20*log10(magdisptip_elim);
```

15.6.16 Reduced Modred Models – Plotting Commands

This section plots the frequency responses with the individual mode contribution overlays for both the "mdc" and "del" options for "modred." The only difference between the code here and that of section 15.6.10 is that the cmode term goes from 1: instead of numdof: because we are using the results of the "modred" operation to define the reduced system matrix, which has only one row in cdisptip instead of numdof rows in cdisp. Once again, "lsim" is used to calculate the step response of the system.

```
%        plot modred using 'elim'

         semilogx(f,magdisptipdb(1,:),'k-',f,magdisptip_elimdb(1,:),'k.-')

         if  modred_sort == 1
                     title(['reduced elimination:  tip disp for mid-length step force, ...
                            first ',num2str(num_modes_used),' sorted modes included'])
         else
                     title(['reduced elimination:  tip disp for mid-length step force, ...
                            first ',num2str(num_modes_used),' unsorted modes included'])
         end

         legend('all modes','reduced elim',3)

         dcgain_error_percent_sort = 100*(magdisptip_elimdb(1) ...
                            - magdisptip(1))/magdisptip(1)

         xlabel('Frequency, hz')
         ylabel('Magnitude, db mm')
         grid off

         pause

         hold on

%        now plot the overlay of the tip displacement magnitude with each mode contribution

         max_modes_plot = num_modes_used;

         for  pcnt = 1:max_modes_plot

                     index = 2*pcnt;

                     amode = adisptip_elim(index-1:index,index-1:index);

                     bmode = bdisptip_elim(index-1:index);

                     cmode = cdisptip_elim(1,index-1:index);
```

```
                    dmode = [0];

                    sysdisptip_mode = ss(amode,bmode,cmode,dmode);

                    [magdisptip_mode,phsdisptip_mode]=bode(sysdisptip_mode,frad)  ;

                    magdisptip_modedb = 20*log10(magdisptip_mode);

                    semilogx(f,magdisptip_modedb(1,:),'k-')

        end

        dc_gain_freq_sort = freqlo*ones(size(freqnew_nosort));

        pause

        hold off

%       modred using 'mdc'

        semilogx(f,magdisptipdb(1,:),'k-',f,magdisptip_mdcdb(1,:),'k.-')

        if  modred_sort == 1
                    title(['reduced matched dc gain:  tip disp for mid-length step force, ...
                        first ',num2str(num_modes_used),' sorted modes included'])
        else
                    title(['reduced matched dc gain:  tip disp for mid-length step force, ...
                        first ',num2str(num_modes_used),' unsorted modes included'])
        end

        legend('all modes','reduced mdc',3)

        dcgain_error_percent_nosort = 100*(magdisptip_mdcdb(1) ...
                    - magdisptip(1))/magdisptip(1)

        xlabel('Frequency, hz')
        ylabel('Magnitude, db mm')
        grid off

        pause

        hold on

        max_modes_plot = num_modes_used;

        for  pcnt = 1:max_modes_plot

                    index = 2*pcnt;

                    amode = adisptip_mdc(index-1:index,index-1:index);

                    bmode = bdisptip_mdc(index-1:index);

                    cmode = cdisptip_mdc(1,index-1:index);
```

```
                    dmode = [0];

                    sysdisptip_mode = ss(amode,bmode,cmode,dmode);

                    [magdisptip_mode,phsdisptip_mode]=bode(sysdisptip_mode,frad)  ;

                    magdisptip_modedb = 20*log10(magdisptip_mode);

                    semilogx(f,magdisptip_modedb(1,:),'k-')

           end

           dc_gain_freq_nosort = freqlo*ones(size(freqnew_nosort));

           pause

           hold off

%          now use lsim to calculate step response to a unit force

           [disptip,ts] = lsim(sysdisptip,u,t);

           [disptip_elim,ts_elim] = lsim(sysdisptip_elim,u,t);

           [disptip_mdc,ts_mdc] = lsim(sysdisptip_mdc,u,t);

           plot(ts,disptip,'k-',ts_mdc,disptip_mdc,'k.-',ts_elim,disptip_elim,'k+-')

           if  modred_sort == 1
                    title(['modred cantilever tip disp for mid-length step force, ...
                       first ',num2str(num_modes_used),' sorted modes included'])
           else
                    title(['modred cantilever tip disp for mid-length step force ...
                       , first ',num2str(num_modes_used),' unsorted modes included'])
           end

           legend('all modes','reduced - mdc','reduced - elim')
           xlabel('time, sec')
           ylabel('displacement, mm')
           grid off
           pause

           end
```

15.6.17 Plotting Unsorted Modred Reduced Results – Eliminating High Frequency Modes

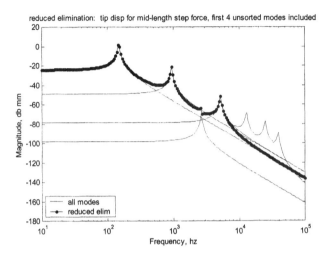

Figure 15.14: Cantilever tip displacement for mid-length force, first four modes included – unsorted modal truncation, modred "del" option.

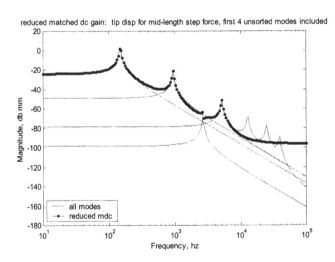

Figure 15.15: Cantilever tip displacement for mid-length force, first four modes included – unsorted modal truncation, modred "mdc" option.

Figure 15.14 shows overall frequency response with four overlaid individual mode contributions for the unsorted "del" "modred" option, with the six highest frequency modes eliminated. Note that at high frequencies the reduced curve attenuates with frequency similar to the "all modes" curve.

Figure 15.15 shows overall frequency response with four overlaid individual mode contributions for the unsorted "mdc" "modred" option, with the six highest frequency modes reduced. Note the rise in the high frequency portion of the magnitude curve as a result of the matrix reduction operations discussed at the end of Section 15.6.12. Depending on the purpose of the model, the high frequency discrepancy may or may not be important.

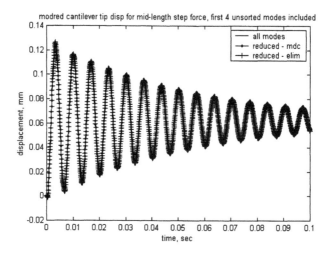

Figure 15.16: Comparison of step responses for all modes included and four modes included, "mdc" and "elim" "modred" options.

Figure 15.16 shows the overlay of step response for all mode model and "del" and "mdc" "modred" options. Note that there is no visible difference in the transient responses.

15.6.18 Plotting Sorted Modred Reduced Results – Eliminating Lower dc Gain Modes

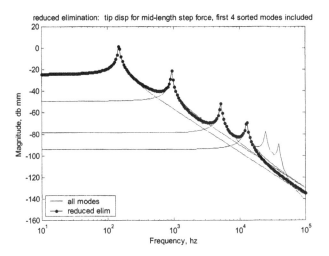

Figure 15.17: Cantilever tip displacement for mid-length force, first four sorted modes, modal truncation, "modred" "del" option.

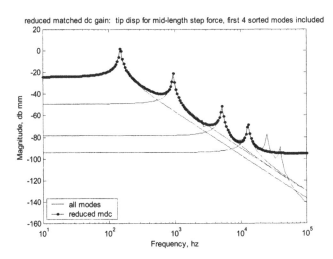

Figure 15.18: Cantilever tip displacement for mid-length force, first four sorted modes, "modred" "mdc" option.

Figure 15.17 shows overall frequency response with four overlaid individual mode contributions for the sorted "del" "modred" option, with the six lowest dc gain modes eliminated. Figure 15.18 shows overall frequency response with four overlaid individual mode contributions for the unsorted "mdc" "modred" option, with the six lowest dc gain modes reduced. Again, note the lack of high frequency attenuation with frequency for the "modred" reduction.

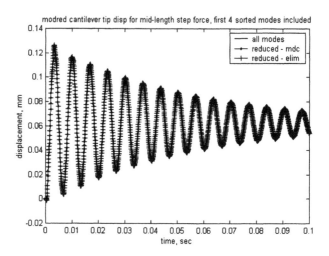

Figure 15.19: Comparison of step responses for all modes included and four sorted modes included, "mdc" and "elim" "modred" options.

Figure 15.19 depicts the overlay of step response for the all mode model and "del" and "mdc" "modred" options. Note that there is no visible difference in the transient responses.

15.6.19 Modred Summary

For this problem, where the dc gain of the response is dominated by the first several modes, there is not much difference between the sorted and unsorted responses. The "mdc" method minimizes low frequency errors by accounting for the dc gain of the unused modes but has high frequency behavior which deviates from the expected, and may not be desirable. The "del" method does not account for the dc gains of the unused modes, which can result in error in the low frequency portion of the frequency response. However, the "del" method has the advantage that it does not exhibit the unusual high frequency direct transmission matrix related behavior of the "mdc" method. If sorting of dc gain values is performed prior to the "del" operation, the system dc gain error may be acceptable while maintaining better high frequency performance.

15.7 ANSYS Code cantbeam_ss.inp Listing

The ANSYS code **cantbeam_ss.inp** solves for the eigenvalues and eigenvectors for a tip-loaded cantilever beam, with a sample output shown in Section 15.4. The user can define the number of elements to use for the cantilever and also choose whether to use the "Reduced" or "Block Lanczos" eigenvalue extraction method. The program then writes a frequency list out to a ".frq" file, outputs eigenvector listings to a ".eig" file and plots deformed/undeformed mode shapes to ".grp."

```
! cantbeam_ss.inp, 0.075 thick x 2 wide x 20mm long steel cant
! title automatically built based on number of elements and eigenvalue extraction method

/prep7

filename = 'cantbeam_ss'

! define number of elements to use

num_elem = 64

! define eigenvalue extraction method, 1 = reduced, 2 = block lanczos

eigext = 1

*if,eigext,eq,1, then
            nummodes = num_elem        ! only 1 displacement dof available for each element
*else
            nummodes = 2*num_elem      ! both disp and rotation dof's available for
                                       !   each  element
*endif

!          create the file name for storing data

!          first section of filename

aname = filename

!          second section of filename, number of elements

bname = num_elem

!          third section of filename, depends on eigenvalue extraction method

*if,eigext,ne,2, then
            cname = 'red'         ! reduced
*else
            cname = 'bl'          ! block Lanczos
*endif

! input the title, use %xxx% to substitute parameter name or parametric expression
```

```
aname_ti = 'cantbeam_ss - 0.075 thick x 2 wide x 20mm long steel cant'

/title,%aname_ti%, %bname%, %cname%

et,1,4                    ! element type for beam

! steel

ex,1,190e6                    ! mN/mm^2
dens,1,7.83e-6                ! kg/mm^3
nuxy,1,.293

! real value to define beam characteristics

r,1,0.15,0.05,0.00007031,0.075,0.2                    ! area, Izz, Iyy, TKz, TKy

! define plotting characteristics

/view,1,1,-1,1    ! iso view
/angle,1,-60      ! iso view
/pnum,mat,1       ! color by material
/num,1            ! numbers off
/type,1,0         ! hidden plot
/pbc,all,1        ! show all boundary conditions

csys,0                                        ! define global coordinate system

! nodes

n,1,0,0,0                                     ! left-hand node
n,num_elem+1,20,0,0           ! right-hand node

fill,1,num_elem+1             ! interior nodes

nall
nplo

! elements

type,1
mat,1
real,1
e,1,2
egen,num_elem,1,-1

! constrain left-hand end

nall
d,1,all,0                     ! constrain node 1, all dof's

! constrain all but uz and roty for all other nodes to allow only those dof's

nall
nsel,s,node,,2,num_elem+1
```

```
d,all,ux
d,all,uy
d,all,rotx
d,all,rotz

nall
eall
nplo
eplo

! ******************* eigenvalue run *****************

fini                    ! fini just in case not in begin

/solu                   ! enters the solution processor, needs to be here to do editing below

allsel                  ! default selects all items of specified entity type, typically nodes, elements

nsel,s,node,,2,num_elem+1
m,all,uz

*if,eigext,eq,1,then                    ! use reduced method

        antype,modal,new
        modopt,reduc,nummodes           ! method - reduced Householdert
        expass,off                      ! key = off, no expansion pass, key = on,
                                        !  do expansion
        mxpand,nummodes,,,no            ! nummodes to expand,freq beginning,freq
                                        !  ending,elcalc = yes - calculate stresses
        total,num_elem,1                ! total masters, 1 is exclude rotations

*elseif,eigext,eq,2                     ! use block lanczos

        antype,modal,new
        modopt,lanb,nummodes            ! no total required for block lanczos
                                        !  because calculates all eigenvalues
        expass,off
        mxpand,nummodes,,,no

*endif

allsel

solve                   ! starts the solution of one load step of a solution sequence, modal here

fini

! plot first mode

/post1

/format,,,,,10000

set,1,1
```

```
pldi,1

! *************** output frequencies **********************

save,%aname%%bname%%cname%,sav

/output,%aname%%bname%%cname%,frq                ! write out frequency list to ascii file .frq

set,list

/output,term                                     ! returns output to terminal

! ***************** output eigenvectors ************************

! define nodes for output:  forces applied or output displacements

nall

/output,%aname%%bname%%cname%,eig       ! write out frequency list to ascii file .eig

*do,i,1,nummodes
          set,,i
          /page,,,1000
          prdisp
*enddo

/output,term

! ***************** plot modes *****************

! pldi plots

/show,%aname%%bname%%cname%,grp,0                ! save mode shape plots to file .grp

allsel

/view,1,,-1,,                          ! side view for plotting
/angle,1,0
/auto

*do,i,1,nummodes
          set,1,i
   pldi,1
*enddo

/show,term
```

CHAPTER 16

GROUND ACCELERATION MATLAB
MODEL FROM ANSYS MODEL

16.1 Introduction

This chapter will continue to explore building MATLAB state space models
from ANSYS finite element results. We will use a different cantilever model,
where the cantilever has an additional tip mass and a tip spring all mounted on
a "shaker" base. This model will be a crude approximation of understanding
the effects of disk drive suspension resonances on undesired unloading of the
recording head during external vibration events. The problem shows how to
model ground acceleration forcing functions using ANSYS and MATLAB.
We will also see how to do sorting of modes in the presence of a rigid body
mode. In addition, there is a high frequency mode of the system with a large
dc gain, meaning that if unsorted modal truncation were used to decrease the
model size, the resulting model would have significant error.

16.2 Model Description

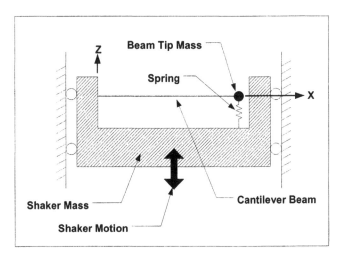

Figure 16.1: Ground displacement model for cantilever with tip mass and tip spring.

The figure above shows a schematic of the system to be analyzed. Once again,
the cantilever is a 2mm wide by 0.075mm thick by 20mm long steel beam. At
the tip, a lumped mass of 0.00002349 Kg is attached. The tip mass was
arbitrarily chosen to have the same mass as the beam. The spring attaching the

beam tip to the shaker has a stiffness of 1e6 mN/mm. The 0.05 Kg shaker mass was chosen to be approximately 1000 times the mass of the beam and tip mass combination, making the motions of the shaker insensitive to resonances of the beam. Thus, we can apply forces to the shaker and excite it to a known acceleration amplitude. This amplitude will then be transmitted to the base of the cantilever and the shaker attachment for the beam tip spring – effectively imparting a "ground acceleration" of any desired amplitude and shape to the flexible system. Of course, since the shaker body is not constrained, it will have large rigid body movements, but we are interested in the difference between the shaker motion and the motion of the tip, so we can ignore the rigid body motion.

In a disk drive, the cantilever would represent the "suspension," the small sheet metal device which supports the recording head, represented by the beam tip mass. The recording head is typically preloaded onto the disk with several grams of loading force by pre-bending and then displacing the suspension. This loading force is required to counteract the force generated by the air bearing when the disk is spinning, keeping the recording head a controlled distance from the disk and allowing efficient magnetic recording. During transportation of the disk drive it is subject to vibration and shock events in the z direction as indicated by the Shaker Motion arrow. Of course, vibration and shock occur in all directions, but the z direction is the most sensitive. In the z direction, the vibration or shock event may be large enough and have frequency content which will excite the suspension resonances, generating unloading forces at the head that could cause it to become momentarily unloaded. When unloaded, the slider will re-approach the disk and possibly damage the disk. Thus, understanding resonant characteristics of the suspension and the resulting tendency to unload the head is very important. Because the frequency content of typical vibration and shock events are less than several khz, having a good model of the resonant system up to roughly 10 khz is adequate.

16.3 Initial ANSYS Model Comparison –
Constrained-Tip and Spring-Tip Frequencies/Mode Shapes

The spring between the beam tip and the shaker is an artifice, created to allow measuring the forces between the beam tip and the shaker. If the spring had infinite stiffness, the tip would become simply supported. The stiffness of the spring used in the model was chosen to have the frequency of the mode involving the beam tip and the spring be very high relative to the first bending mode of the constrained-tip beam. This makes the tip simply supported at frequencies lower than the beam tip/spring mode and will allow a valid force measurement in the frequency range of the major beam bending modes.

There is always a compromise when using a spring artifice to replace a rigid boundary condition to enable calculating constraint forces. The compromise is that one would like a very stiff spring to make the model more accurate, however a very stiff spring would require more modes to be extracted because the frequency of the tip spring/tip mass mode would be higher. Thus, the eternal compromise with finite element models: between more accuracy (more elements) and a shorter time to solve the problem (fewer elements). The optimal model is always the smallest model which will give acceptable answers, no more, no less. This balance makes finite elements interesting!

In order to understand the effects of the tip spring on the resonances, we will use two ANSYS models. The first model will have the tip constrained in the z direction. The second model will be as described above, but with a tip spring connected to the shaker. The two models will be compared to ensure that the tip spring artifice does not significantly effect the major beam bending modes. The tip constrained model is **cantbeam_ss_tip_con.inp,** the spring-tip model is **cantbeam_ss_spring_shkr.inp**, which is listed at the end of the chapter. A comparison of resonant frequencies for the two models, each with 16-beam elements and using the Reduced method for eigenvalue extraction, is shown below:

Mode	Tip Constrained Freq, hz	Tip Spring Freq, hz	
1	0.0030932	0.0000	
2	654.37	654.36	
3	2120.2	2120.1	
4	4424.1	4423.3	
5	7567.0	7564.6	
6	11553.	11547.	
7	16392.	16378.	
8	22104.	22069.	
9	28730.	28590.	
10	36346.	32552.	**Note 32552 is tip/spring mode**
11	45079.	36547.	
12	55111.	45164.	
13	66628.	55171.	
14	79548.	66675.	
15	92830.	79583.	
16	0.10359E+06	92850.	

Table 16.1: Resonant frequencies for tip-constrained and spring-tip models.

The table above tells us that there is very good matching of resonant frequencies for the first 15 modes of the tip-constrained model and the tip spring model. The 92830 hz (15th) mode differs only 20 hz from the tip spring model 92850 hz mode. The difference between the two models is that the tip spring model has an additional mode at 32552 which is the tip spring/tip mass mode. Having good agreement between the two models up through 32552 hz means that we will get good results in the 0 to 10 khz range of interest. The ANSYS Display program can be used to plot the mode shapes of the two 16-element models by loading **cantbeam16red.grp** or **tipcon16red.grp** for the spring-tip or constrained-tip models, respectively. A MATLAB code, **cantbeam_shkr_modeshape.m**, can also be used to plot mode shapes for any of the spring-tip models, with selected modes plotted below for the 16-element model.

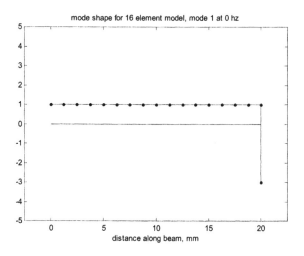

Figure 16.2: Rigid body mode, 0 hz.

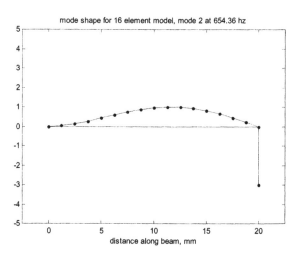

Figure 16.3: First bending mode, 654 hz.

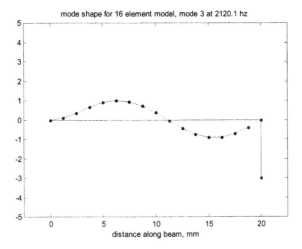

Figure 16.4: Second bending mode, 2120 hz.

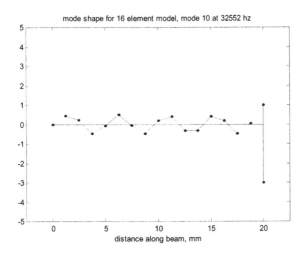

Figure 16.5: Beam tip / Spring mode at 32552 hz.

Note the deflection at the tip involving the spring for mode 10 for the 16-element model. Since we are interested in using the spring deflections to measure force exerted at the beam tip constraint, we will find that including the 10th mode is important because of its large dc gain value.

16.4   MATLAB State Space Model from ANSYS Eigenvalue Run – cantbeam_ss_shkr_modred.m

The MATLAB code used in this chapter is very similar to the code in Chapter 15. As such, some of the following descriptions will refer to the previous chapter.

The results shown and discussed in this chapter will be for the 16-element beam model; however, ANSYS data is available for 2-, 4-, 8-, 10-, 12-, 16-, 32- and 64-beam elements.

16.4.1 Input

This Section is similar to that in Section 15.6.1, with the same options available for choosing the number of elements to be analyzed. Eigenvalue/eigenvector results for all the models are available in the respective MATLAB .mat files and are called based on which menu item is picked.

```
%        cantbeam_ss_shkr_modred.m

         clear all;

         hold off;

         clf;

%        load the .mat file cantbeamXXred, containing evr - the modal matrix, freqvec -
%        the frequency vector and node_numbers - the vector of node numbers for the modal
%        matrix

         model = menu('choose which finite element model to use ... ', ....
                            '2 beam elements', ...
                            '4 beam elements', ...
                            '6 beam elements', ...
                            '8 beam elements', ...
                            '10 beam elements', ...
                            '12 beam elements', ...
                            '16 beam elements', ...
                            '32 beam elements', ...
                            '64 beam elements');

         if  model == 1
                   load cantbeam2red_shkr;
         elseif  model == 2
                   load cantbeam4red_shkr;
         elseif  model == 3
                   load cantbeam6red_shkr;
         elseif  model == 4
```

```
                    load cantbeam8red_shkr;
        elseif model == 5
                    load cantbeam10red_shkr;
        elseif model == 6
                    load cantbeam12red_shkr;
        elseif model == 7
                    load cantbeam16red_shkr;
        elseif model == 8
                    load cantbeam32red_shkr;
        elseif model == 9
                    load cantbeam64red_shkr;
        end
```

16.4.2 Shaker, Spring, Gram Force Definitions

The value of the beam tip spring stiffness is the same values as in the ANSYS code and is used to calculate the force between the beam tip and the shaker. The shaker mass value is the same value as in the ANSYS code and is used to define the force required in the MATLAB model to impart a desired acceleration level to the shaker. The force conversion from mN to gram force is defined as $1/9.807$.

```
kspring = 1000000;                      % mN/mm from ANSYS run

shaker_mass = 0.050;          % kg from ANSYS run

mn2gm_conversion = 0.101968; %  conversion factor from mn to gram-f, 1/9.807
```

16.4.3 Defining Degrees of Freedom and Number of Modes

This section of code is identical to that of Section 15.6.2.

```
%       define the number of degrees of freedom and number of modes from size of
%       modal matrix

        [numdof,num_modes_total] = size(evr);

%       define rows for shaker and tip nodes

        shaker_node_row = 1;

        tip_node_row = numdof;

        xn = evr;
```

16.4.4 Frequency Range, Sorting Modes by dc Gain and Plotting, Selecting Modes Used

As in Section 15.6.3, the next step in creating the model is to sort modes of vibration so that only the most important modes are kept. Repeating from Chapter 15 to obtain the frequency response at dc:

$$\frac{z_j}{F_k} = \sum_{i=1}^{m} \frac{z_{nji} z_{nki}}{\omega_i^2} \quad , \tag{16.1}$$

where the dc gain of for the ith mode is given by the expression:

$$i^{th} \text{ mode dc gain:} \quad \left(\frac{z_j}{F_k}\right) = \frac{z_{nji} z_{nki}}{\omega_i^2} \tag{16.2}$$

The difference between the code below and the code in Section 15.6.3 is that we have a rigid body, 0 hz, mode in this model and the previous cantilever did not. The problem is in dividing (16.1) by $\omega_i^2 = \omega_1^2 = 0$, which would give a dc gain of infinity for the rigid body mode. In order to get around this, we do not use zero for the rigid body frequency but instead use the frequency response lower bound frequency for calculating a "low frequency" gain. In this model the lower bound frequency is 100 hz. Another method of ranking would be to rank only the non rigid body modes, recognizing that the rigid body mode is always included.

Once again, dc gain will be used to rank the relative importance of modes. The dc gain calculation for each mode, "dc_value," is broken into two parts. The first part calculates the gain of the rigid body mode at the "freqlo" frequency while the second part calculates the dc gain of all the non rigid body modes.

The bulk of this section is similar to Section 15.6.3.

```
%   calculate the dc amplitude of the displacement of each mode by
%   multiplying the forcing function row of the eigenvector by the output row

        omega2 = (2*pi*freqvec)'.^2;    % convert to radians and square

%       define frequency range for frequency response

        freqlo = 100;

        freqhi = 100000;
```

```
flo=log10(freqlo) ;
fhi=log10(freqhi) ;

f=logspace(flo,fhi,200) ;
frad=f*2*pi ;

dc_gain = abs([xn(shaker_node_row,1)*xn(tip_node_row,1)/(frad(1)^2) ...
              (xn(shaker_node_row,2:num_modes_total) ...
           .*xn(tip_node_row,2:num_modes_total))./omega2(2:num_modes_total)]);

[dc_gain_sort,index_sort] = sort(dc_gain);

dc_gain_sort = fliplr(dc_gain_sort);

index_sort = fliplr(index_sort)

dc_gain_nosort = dc_gain;

index_orig = 1:num_modes_total;

semilogy(index_orig,freqvec,'k-');
title('frequency versus mode number')
xlabel('mode number')
ylabel('frequency, hz')
grid
disp('execution paused to display figure, "enter" to continue'); pause

semilogy(index_orig,dc_gain_nosort,'k-')
title('dc value of each mode contribution versus mode number')
xlabel('mode number')
ylabel('dc value')
grid
disp('execution paused to display figure, "enter" to continue'); pause

loglog([freqlo; freqvec(2:num_modes_total)],dc_gain_nosort,'k-')
title('dc value of each mode contribution versus frequency')
xlabel('frequency, hz')
ylabel('dc value')
grid
disp('execution paused to display figure, "enter" to continue'); pause

semilogy(index_orig,dc_gain_sort,'k-')
title('sorted dc value of each mode versus number of modes included')
xlabel('modes included')
ylabel('sorted dc value')
grid
disp('execution paused to display figure, "enter" to continue'); pause

num_modes_used = input(['enter how many modes to include, ...
            ',num2str(num_modes_total),' default, max ... ']);

if (isempty(num_modes_used))
        num_modes_used = num_modes_total;
end
```

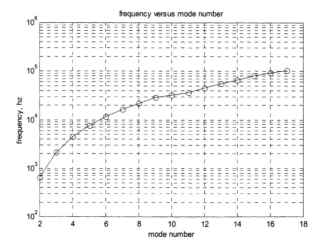

Figure 16.6: Resonant frequency versus mode number for 16-element model.

Figure 16.6 shows the resonant frequency versus mode number for the 16-element model, Reduced method of eigenvalue extraction, showing that modes six and higher have frequencies greater than the 10 khz frequency range of interest for this model. This would lead one to think that only the first six or eight modes would be required to define the force in the 0 to 10 khz frequency range, which is not the case as we shall see.

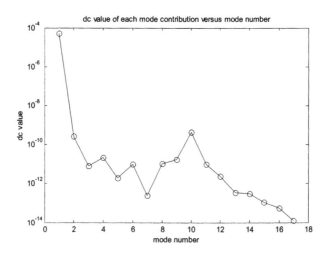

Figure 16.7: Low frequency and dc gains versus mode number.

Figure 16.7 shows the low frequency gain for the rigid body mode, mode 1, and the dc gains for all other modes, versus mode number. Note that the second most important mode (the second highest dc gain) is mode 10, and that it is even more important than the first bending mode of the cantilever.

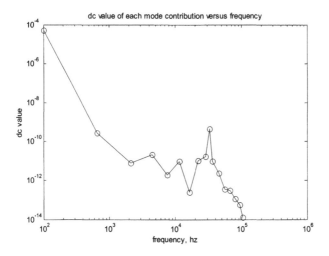

Figure 16.8: Low frequency and dc gain versus frequency.

Figure 16.8 shows the same data plotted against frequency instead of mode number. The tip mass / tip spring mode at 32552 hz is the mode with the high gain.

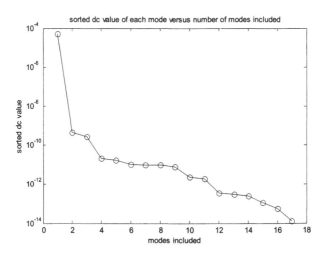

Figure 16.9: Sorted low frequency and dc gains versus number of modes.

In Figure 16.9 we can see the sorted values for the low frequency and dc gains, from largest to smallest. The list of sorted mode numbers is given in the table below. Once again, the 10th mode is the second most significant after the rigid body mode.

index_sort =   1   10   2   4   9   8   6   11   3   12   5   13   14   7   15   16   17

Table 16.2: Sorted low frequency and dc gain indices.

16.4.5 Damping, Defining Reduced Frequencies and Modal Matrices

This section is exactly like that in Section 15.6.4.

```
        zeta = input('enter value for damping, .02 is 2% of critical (default) ... ');

        if (isempty(zeta))
            zeta = .02;
        end

%       all modes included model, use original order

        xnnew = xn(:,(1:num_modes_total));

        freqnew = freqvec((1:num_modes_total));

%       reduced, no sorting, just use the first num_modes_used modes in xnnew_nosort

        xnnew_nosort = xn(:,1:num_modes_used);

        freqnew_nosort = freqvec(1:num_modes_used);

%       reduced, sorting, use the first num_modes_used sorted modes in xnnew_sort

        xnnew_sort = xn(:,index_sort(1:num_modes_used));

        freqnew_sort = freqvec(index_sort(1:num_modes_used));
```

16.4.6 Setting Up System Matrix "a"

This section is exactly like that in Section 15.6.5.

```
%       define variables for all modes included system matrix, a

        w = freqnew*2*pi;              %       frequencies in rad/sec

        w2 = w.^2;

        zw = 2*zeta*w;
```

```
%          define variables for reduced, nosorted system matrix, a_nosort

           w_nosort = freqnew_nosort*2*pi;          %          frequencies in rad/sec

           w2_nosort = w_nosort.^2;

           zw_nosort = 2*zeta*w_nosort;

%          define variables for reduced, sorted system matrix, a_sort

           w_sort = freqnew_sort*2*pi;              %          frequencies in rad/sec

           w2_sort = w_sort.^2;

           zw_sort = 2*zeta*w_sort;

%          define size of system matrix

           asize = 2*num_modes_total;

           asize_red = 2*num_modes_used;

           disp(' ');
           disp(' ');
           disp(['size of system matrix a is ',num2str(asize)]);
           disp(['size of reduced system matrix a is ',num2str(asize_red)]);

%          setup all modes included "a" matrix, system matrix

           a = zeros(asize);

           for  col = 2:2:asize

           row = col-1;

           a(row,col) = 1;

           end

           for  col = 1:2:asize

           row = col+1;

           a(row,col) = -w2((col+1)/2);

           end

           for  col = 2:2:asize

           row = col;

           a(row,col) = -zw(col/2);

           end
```

```
%        setup reduced, nosorted "a_nosort" matrix, system matrix

         a_nosort = zeros(asize_red);

         for  col = 2:2:asize_red

         row = col-1;

         a_nosort(row,col) = 1;

         end

         for  col = 1:2:asize_red

         row = col+1;

         a_nosort(row,col) = -w2_nosort((col+1)/2);

         end

         for  col = 2:2:asize_red

         row = col;

         a_nosort(row,col) = -zw_nosort(col/2);

         end

%        setup reduced, sorted "a_sort" matrix, system matrix

         a_sort = zeros(asize_red);

         for  col = 2:2:asize_red

         row = col-1;

         a_sort(row,col) = 1;

         end

         for  col = 1:2:asize_red

         row = col+1;

         a_sort(row,col) = -w2_sort((col+1)/2);

         end

         for  col = 2:2:asize_red

         row = col;

         a_sort(row,col) = -zw_sort(col/2);
```

```
                                    end
```

16.4.7 Setting Up Matrices "b," "c" and "d"

The only difference between this section and Sections 15.6.6 and 15.6.7 is in defining the force to be applied to the shaker to give 1g acceleration.

```
%          setup input matrix b, state space forcing function in principal coordinates

%          f_physical is the vector of physical force
%          zeros at each output DOF and input force at the input DOF

           f_physical = zeros(numdof,1);              %         start out with zeros

           f_physical(shaker_node_row) = 9807*shaker_mass*1.0;  % input force at shaker, 1g

%          now setup the principal force vector for the three cases, all modes, nosort, sort

%          f_principal is the vector of forces in principal coordinates

           f_principal = xnnew'*f_physical;

%          b is the vector of forces in principal coordinates, state space form

           b = zeros(2*num_modes_total,1);

           for cnt = 1:num_modes_total

                   b(2*cnt) = f_principal(cnt);

           end

%          f_principal_nosort is the vector of forces in principal coordinates

           f_principal_nosort = xnnew_nosort'*f_physical;

%          b_nosort is the vector of forces in principal coordinates, state space form

           b_nosort = zeros(2*num_modes_used,1);

           for cnt = 1:num_modes_used

                   b_nosort(2*cnt) = f_principal_nosort(cnt);

           end

%          f_principal_sort is the vector of forces in principal coordinates

           f_principal_sort = xnnew_sort'*f_physical;

%          b_sort is the vector of forces in principal coordinates, state space form
```

```
        b_sort = zeros(2*num_modes_used,1);

        for cnt = 1:num_modes_used

                b_sort(2*cnt) = f_principal_sort(cnt);

        end

%       setup cdisp and cvel, padded xn matrices to give the displacement and velocity
%       vectors in physical coordinates
%       cdisp and cvel each have numdof rows and alternating columns consisting of columns
%       of xnnew and zeros to give total columns equal to the number of states

%       all modes included cdisp and cvel

        for col = 1:2:2*length(freqnew)

        for row = 1:numdof

        cdisp(row,col) = xnnew(row,ceil(col/2));

        cvel(row,col) = 0;

        end

        end

        for col = 2:2:2*length(freqnew)

        for row = 1:numdof

        cdisp(row,col) = 0;

            cvel(row,col) = xnnew(row,col/2);

        end

        end

%       reduced, nosorted cdisp and cvel

        for col = 1:2:2*length(freqnew_nosort)

        for row = 1:numdof

        cdisp_nosort(row,col) = xnnew_nosort(row,ceil(col/2));

        cvel_nosort(row,col) = 0;

        end

        end

        for col = 2:2:2*length(freqnew_nosort)
```

```
                for row = 1:numdof

            cdisp_nosort(row,col) = 0;

                    cvel_nosort(row,col) = xnnew_nosort(row,col/2);

            end

            end

%           reduced, sorted cdisp and cvel

            for col = 1:2:2*length(freqnew_sort)

            for row = 1:numdof

            cdisp_sort(row,col) = xnnew_sort(row,ceil(col/2));

            cvel_sort(row,col) = 0;

            end

            end

            for col = 2:2:2*length(freqnew_sort)

            for row = 1:numdof

            cdisp_sort(row,col) = 0;

                    cvel_sort(row,col) = xnnew_sort(row,col/2);

            end

            end

%           define output

            d = [0]; %
```

16.4.8 "ss" Setup, Bode Calculations

This section differs from that of Section 15.6.8 in that the frequency range definition that exists in 15.6.8 was moved earlier in this code to allow the use of "freqlo" to calculate the low frequency gain of the rigid body mode. Also, the "ss" model below for "sysforce" directly calculates the force in the spring by subtracting the displacement of the shaker from that beam tip and multiplying the difference by the spring stiffness and the mN to gram force conversion. The output then indicates the variation of force between the beam tip and the shaker, or for the disk drive the variation in force which is preloading the recording head to the disk. If the variation in force exceeds the preload force, the head will tend to unload.

```
%        define tip force state space system with the "ss" command

         sysforce = ss(a,b,mn2gm_conversion*kspring*(cdisp(tip_node_row,:)- ...
                      cdisp(shaker_node_row,:)),d);

%        define reduced system using nosort modes

         sysforce_nosort = ss(a_nosort,b_nosort,mn2gm_conversion*kspring* ...
                      (cdisp_nosort(tip_node_row,:)-cdisp_nosort(shaker_node_row,:)),d);

%        define reduced system using sorted modes

         sysforce_sort = ss(a_sort,b_sort,mn2gm_conversion*kspring* ...
                      (cdisp_sort(tip_node_row,:)-cdisp_sort(shaker_node_row,:)),d);

%        use "bode" command to generate magnitude/phase vectors

         [magforce,phsforce] = bode(sysforce,frad);

         [magforce_nosort,phsforce_nosort] = bode(sysforce_nosort,frad);

         [magforce_sort,phsforce_sort] = bode(sysforce_sort,frad);
```

16.4.9 Full Model – Plotting Frequency Response, Shock Response

The code in this section is similar to that in Section 15.6.9, where the overall frequency response and its individual mode contributions are plotted. The "lsim" command is used to calculate the response to a half-sine shock pulse.

```
%        start plotting

         if num_modes_used == num_modes_total

%        plot all modes included response

         loglog(f,magforce(1,:),'k.-')
         title(['cantilever tip force for mid-length force, all ',num2str(num_modes_used), ...
                  ' modes included'])
         xlabel('Frequency, hz')
         ylabel('Force, gm')
         grid on
         disp('execution paused to display figure, "enter" to continue'); pause

         hold on

         max_modes_plot = num_modes_used;

         for pcnt = 1:max_modes_plot
```

```
                    index = 2*pcnt;

                    amode = a_nosort(index-1:index,index-1:index);

                    bmode = b_nosort(index-1:index);

                    cmode_shaker = cdisp_nosort(1,index-1:index);

                    cmode_tip = cdisp_nosort(numdof,index-1:index);

                    dmode = [0];

                    sysforce_mode = ss(amode,bmode,mn2gm_conversion*kspring* ...
                            (cmode_tip - cmode_shaker),dmode);

                    [magforce_mode,phsforce_mode]=bode(sysforce_mode,frad)  ;

                    loglog(f,magforce_mode(1,:),'k-')

          end

          disp('execution paused to display figure, "enter" to continue'); pause

          hold off

%         now use lsim to calculate force due to a 0.002 sec half-sine 100g shock pulse

          ttotal = 0.03;

          shock_amplitude = 100;

          pulse_width = input('enter half-sine shock pulse width, sec, default is 0.002 ... ');

          if isempty(pulse_width)
                    pulse_width = 0.002;
          end

          t = linspace(0,ttotal,1000);

          dt = t(2) - t(1);

          for  cnt = 1:length(t)

                    if  t(cnt) < pulse_width

                            u(cnt) = shock_amplitude*sin(2*pi*(1/(2*pulse_width))*t(cnt));

                    else

                    u(cnt) = 0;

          end

          end
```

```
plot(t,u,'k-')
title('acceleration of shaker mass')
xlabel('time, sec')
ylabel('acceleration, g')
grid on
disp('execution paused to display figure, "enter" to continue'); pause

[force,ts] = lsim(sysforce,u,t);

plot(ts,force,'k-')
title(['cantilever tip force for ',num2str(shock_amplitude),'g, ',num2str(pulse_width) ...
                       ,' sec input, all ',num2str(num_modes_used),' modes included'])
xlabel('time, sec')
ylabel('Force, gm')
grid on
disp('execution paused to display figure, "enter" to continue'); pause

peak_force = max(abs(force))
```

Plots for the 16-beam element model are shown below.

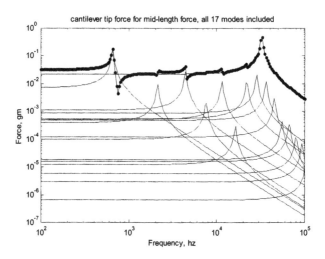

Figure 16.10: Overall frequency response with overlaid individual mode contributions.

Figure 16.10 shows the overall frequency response with overlaid individual mode contributions for all 16 modes. Note the significant dc gain of the 32 khz beam tip/spring mode, which is higher than even the first bending mode dc gain. One can imagine how the overall response would be changed if the 32 khz mode were not included. Without the dc gain of the mode, the overall dc gain would be significantly in error.

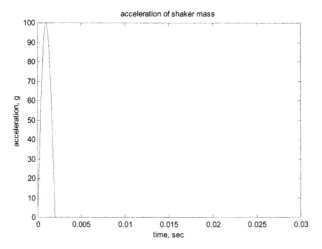

Figure 16.11: Acceleration versus time for the 100g, 2msec shock pulse applied to the system.

Figure 16.11 shows the acceleration versus time profile that is applied to the shaker body.

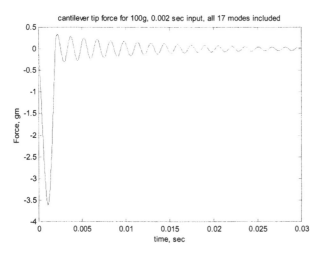

Figure 16.12: Force in the spring versus time, reflecting the change in preload force applied to the head.

For the shock pulse in Figure 16.11, the force in the spring versus time is shown in Figure 16.12. If the preload force were 3 gm, the head would be in danger of unloading from the disk since the peak variation in preload force is 3.6 gm.

16.4.10 Reduced Models – Plotting Frequency Response, Shock Response

This section is similar to Section 15.6.10, setting up frequency response and half-sine shock response for sorted and unsorted modes.

```
          else

%         unsorted modal truncation

          loglog(f,magforce(1,:),'k-',f,magforce_nosort(1,:),'k.-')
          title(['unsorted modal truncation:  cantilever tip force for mid-length force, ...
                    first ',num2str(num_modes_used),' modes included'])
          legend('all modes','unsorted partial modes',3)

          dcgain_error_percent_nosort = 100*(magforce_nosort(1) - magforce(1))/magforce(1)

          xlabel('Frequency, hz')
          ylabel('Force, gm')
          grid on

          disp('execution paused to display figure, "enter" to continue'); pause

          hold on

          max_modes_plot = num_modes_used;

          for  pcnt = 1:max_modes_plot

                    index = 2*pcnt;

                    amode = a_nosort(index-1:index,index-1:index);

                    bmode = b_nosort(index-1:index);

                    cmode_shaker = cdisp_nosort(1,index-1:index);

                    cmode_tip = cdisp_nosort(numdof,index-1:index);

                    dmode = [0];

                    sysforce_mode = ss(amode,bmode,mn2gm_conversion*kspring* ...
                              (cmode_tip - cmode_shaker),dmode);

                    [magforce_mode,phsforce_mode]=bode(sysforce_mode,frad) ;

                    loglog(f,magforce_mode(1,:),'k-')

          end

          disp('execution paused to display figure, "enter" to continue'); pause
```

```
          hold off

%         sorted modal truncation

          loglog(f,magforce(1,:),'k-',f,magforce_sort(1,:),'k.-')
          title(['sorted modal truncation:  cantilever tip force for mid-length force, ...
                     first ',num2str(num_modes_used),' modes included'])
          legend('all modes','sorted partial modes',3)

          dcgain_error_percent_sort = 100*(magforce_sort(1) - magforce(1))/magforce(1)

          xlabel('Frequency, hz')
          ylabel('Force, gm')
          grid on

          disp('execution paused to display figure, "enter" to continue'); pause

          hold on

%         now plot the overlay of the tip force magnitude with each mode contribution

          max_modes_plot = num_modes_used;

          for  pcnt = 1:max_modes_plot

                  index = 2*pcnt;

                  amode = a_nosort(index-1:index,index-1:index);

                  bmode = b_nosort(index-1:index);

                  cmode_shaker = cdisp_nosort(1,index-1:index);

                  cmode_tip = cdisp_nosort(numdof,index-1:index);

                  dmode = [0];

                  sysforce_mode = ss(amode,bmode,mn2gm_conversion*kspring* ...
                          (cmode_tip - cmode_shaker),dmode);

                  [magforce_mode,phsforce_mode]=bode(sysforce_mode,frad)  ;

                  loglog(f,magforce_mode(1,:),'k-')

          end

          disp('execution paused to display figure, "enter" to continue'); pause

          hold off

%         now use lsim to calculate force due to a 0.002 sec half-sine 100g shock pulse

          ttotal = 0.03;
```

```
shock_amplitude = 100;
pulse_width = input('enter half-sine shock pulse width, sec, default is 0.002 ... ');

if isempty(pulse_width)
        pulse_width = 0.002;
end

t = linspace(0,ttotal,1000);

dt = t(2) - t(1);

for cnt = 1:length(t)

if t(cnt) < pulse_width

        u(cnt) = shock_amplitude*sin(2*pi*(1/(2*pulse_width))*t(cnt));

else

        u(cnt) = 0;

        end

end

plot(t,u,'k-')
title('acceleration of shaker mass')
xlabel('time, sec')
ylabel('acceleration, g')
grid on
disp('execution paused to display figure, "enter" to continue'); pause

[force,ts] = lsim(sysforce,u,t);

[force_nosort,ts_nosort] = lsim(sysforce_nosort,u,t);

[force_sort,ts_sort] = lsim(sysforce_sort,u,t);

plot(ts,force,'k-',ts_nosort,force_nosort,'k+:',ts_sort,force_sort,'k.-')
title(['cantilever tip force for ',num2str(shock_amplitude),'g, ',num2str(pulse_width) ...
        ,' sec input, ',num2str(num_modes_used),' modes included'])
legend('all modes','unsorted partial modes','sorted partial modes',4)
xlabel('time, sec')
ylabel('Force, gm')
grid on
disp('execution paused to display figure, "enter" to continue'); pause

max_force = max(abs(force));

max_force_nosort = max(abs(force_nosort));
max_force_sort = max(abs(force_sort));

error_nosort_percent = 100*(max_force_nosort - max_force)/max_force
error_sort_percent = 100*(max_force_sort - max_force)/max_force
```

16.4.11 Reduced Models – Plotted Results, Four Modes Used

Note that in all the frequency response plots that follow, the title will indicate that "four" modes are included, the four being the rigid body mode at 0 hz and the first three either sorted or unsorted resonances.   Because we are subtracting the displacement of the tip from the displacement of the shaker to find the force in the spring, the rigid body mode is effectively subtracted out, allowing us to see the detailed motion of the beam/mass relative to the shaker. This is why the rigid body mode does not show up as one of the four individual modes used.

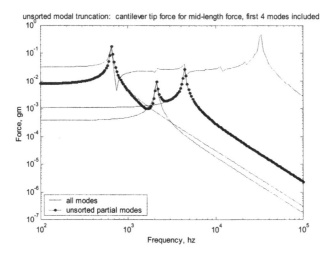

Figure 16.13: Overall plus individual mode contributions for the four unsorted mode model.

In Figure 16.13 the first four unsorted modes are used, so the 32 khz beam tip mode is not included and the overall response is poor. Both the dc gain and high frequency behavior are badly in error. The dc gain error is 75%.

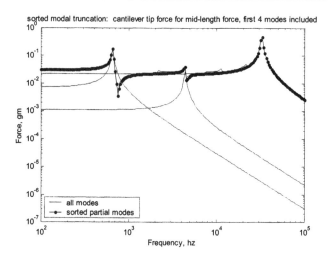

Figure 16.14: Overall plus individual mode contributions for the four sorted mode model.

In Figure 16.14 the 32 khz beam tip mode is one of the included modes. Both the overall dc gain and high frequency behavior are quite good matches with the "all modes included" model with only four modes included. The dc gain error is −6.2%.

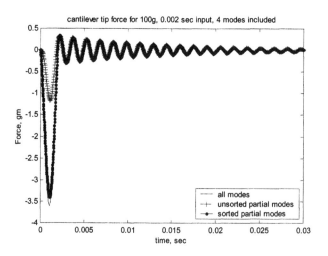

Figure 16.15: Half-sine shock pulse response for full, reduced unsorted and reduced sorted models.

Figure 16.15 shows the how the dc gain error in the frequency domain for the unsorted model shows up as a significant error in peak response in the time domain, 67%. The error in the sorted peak response is only 5.6%.

16.4.12 Modred – Setting up, "mdc" and "del" Reduction, Bode Calculations

In this section the user is prompted for whether to use the sorted or original mode order, then the corresponding system matrices are defined.   The "modred" command is used with both the "mdc" and "del" options to define two reduced systems.   The "bode" command is used to calculate frequency responses.

```
%        use modred to reduce, select whether to use sorted or unsorted modes for the reduction

         modred_sort = input('modred: enter "1" to use sorted modes for reduced runs, ...
              "enter" to use unsorted ... ');

         if isempty(modred_sort)
                modred_sort = 0
         end

         if modred_sort == 1            % use sorted mode order

                xnnew = xn(:,index_sort(1:num_modes_total));

                freqnew = freqvec(index_sort(1:num_modes_total));

         else                            % use original mode order

                xnnew = xn(:,(1:num_modes_total));

                freqnew = freqvec((1:num_modes_total));

         end

%        define variables for all modes included system matrix, a

         w = freqnew*2*pi;              %          frequencies in rad/sec

         w2 = w.^2;

         zw = 2*zeta*w;

%        setup all modes included "a" matrix, system matrix

         a = zeros(asize);

         for col = 2:2:asize

         row = col-1;
```

```
           a(row,col) = 1;

           end

           for  col = 1:2:asize

           row = col+1;

           a(row,col) = -w2((col+1)/2);

           end

           for  col = 2:2:asize

           row = col;

           a(row,col) = -zw(col/2);

           end

%          setup input matrix b, state space forcing function in principal coordinates

%          f_physical is the vector of physical force
%          zeros at each output DOF and input force at the input DOF

           f_physical = zeros(numdof,1);              % start out with zeros

           f_physical(shaker_node_row) = 9807*shaker_mass*1.0;  % input force at shaker, 1g

%          now setup the principal force vector for the three cases, all modes, nosort, sort

%          f_principal is the vector of forces in principal coordinates

           f_principal = xnnew'*f_physical;

%          b is the vector of forces in principal coordinates, state space form

           b = zeros(2*num_modes_total,1);

           for  cnt = 1:num_modes_total

                     b(2*cnt) = f_principal(cnt);

           end

%          setup cdisp and cvel, padded xn matrices to give the displacement and velocity
%          vectors in physical coordinates
%          cdisp and cvel each have numdof rows and alternating columns consisting of columns
%          of xnnew and zeros to give total columns equal to the number of states

%          all modes included cdisp and cvel

           for  col = 1:2:2*length(freqnew)

           for  row = 1:numdof
```

```
          cdisp(row,col) = xnnew(row,ceil(col/2));

          cvel(row,col) = 0;

          end

          end

          for  col = 2:2:2*length(freqnew)

          for  row = 1:numdof

          cdisp(row,col) = 0;

              cvel(row,col) = xnnew(row,col/2);

          end

          end
%         define output

          d = [0];  %

%         define state space system for reduction, ordered defined by modred_sort

          sysforce_red = ss(a,b,mn2gm_conversion*kspring*(cdisp(tip_node_row,:)- ...
                          cdisp(shaker_node_row,:)),d);

%         define reduced matrices using matched dc gain method "mdc"

          states_elim = (2*num_modes_used+1):2*num_modes_total;

          sysforce_mdc = modred(sysforce_red,states_elim,'mdc');

          [aforce_mdc,bforce_mdc,cforce_mdc,dforce_mdc] = ssdata(sysforce_mdc);

%         define reduced matrices by eliminating high frequency states, 'del'

          sysforce_elim = modred(sysforce_red,states_elim,'del');

          [aforce_elim,bforce_elim,cforce_elim,dforce_elim] = ssdata(sysforce_elim);

%         use "bode" command to generate magnitude/phase vectors for reduced systems

          [magforce_mdc,phsforce_mdc]=bode(sysforce_mdc,frad) ;

          [magforce_elim,phsforce_elim]=bode(sysforce_elim,frad) ;

%         convert magnitude to db

          magforce_mdcdb = 20*log10(magforce_mdc);

          magforce_elimdb = 20*log10(magforce_elim);
```

16.4.13 Reduced Modred Models – Plotting Commands

Both the "del" and "mdc" reduced systems are plotted and compared with the original, non-reduced system. The individual mode contributions to the two reduced responses are also plotted.

```
%        start plotting

%        modred using 'elim'

         loglog(f,magforce(1,:),'k-',f,magforce_elim(1,:),'k.-')

         if  modred_sort == 1
                 title(['reduced elimination: cantilever tip force for mid-length force, ...
                     first ',num2str(num_modes_used),' sorted modes included'])
                         dcgain_error_percent_elim_sort = 100*(magforce_elim(1) ...
                         - magforce(1))/magforce(1)
         else
                 title(['reduced elimination: cantilever tip force for mid-length force, ...
                         first ',num2str(num_modes_used),' unsorted modes included'])
                 dcgain_error_percent_elim_nosort = 100*(magforce_elim(1) ...
                 - magforce(1))/magforce(1)
         end

         legend('all modes','reduced elimination',3)

         xlabel('Frequency, hz')
         ylabel('Force, gm')
         grid on

         disp('execution paused to display figure, "enter" to continue'); pause

         hold on

         max_modes_plot = num_modes_used;

         for  pcnt = 1:max_modes_plot

                 index = 2*pcnt;

                 amode = aforce_elim(index-1:index,index-1:index);

                 bmode = bforce_elim(index-1:index);

                 cmode = cforce_elim(1,index-1:index);

                 dmode = [0];

                 sysforce_mode = ss(amode,bmode,cmode,dmode);

                 [magforce_mode,phsforce_mode]=bode(sysforce_mode,frad) ;

                 loglog(f,magforce_mode(1,:),'k-')
```

```
        end

        disp('execution paused to display figure, "enter" to continue'); pause

        hold off

%       modred using 'mdc'

        loglog(f,magforce(1,:),'k-',f,magforce_mdc(1,:),'k.-')

        if  modred_sort == 1
                    title(['reduced matched dc gain: cantilever tip force for mid-length ...
                        force, first ',num2str(num_modes_used),' sorted modes included'])
                            dcgain_error_percent_mdc_sort = 100*(magforce_mdc(1) ...
                        - magforce(1))/magforce(1)
        else
                    title(['reduced matched dc gain:  cantilever tip force for mid-length  ...
                        f orce, first ',num2str(num_modes_used),' unsorted modes included'])
                        dcgain_error_percent_mdc_nosort = 100*(magforce_mdc(1) ...
                        - magforce(1))/magforce(1)
        end

        legend('all modes','reduced mdc',3)


        xlabel('Frequency, hz')
        ylabel('Force, gm')
        grid on

        disp('execution paused to display figure, "enter" to continue'); pause

        hold on

        max_modes_plot = num_modes_used;

        for  pcnt = 1:max_modes_plot

                    index = 2*pcnt;

                    amode = aforce_mdc(index-1:index,index-1:index);

                    bmode = bforce_mdc(index-1:index);

                    cmode = cforce_mdc(1,index-1:index);

                    dmode = [0];

                    sysforce_mode = ss(amode,bmode,cmode,dmode);

                    [magforce_mode,phsforce_mode]=bode(sysforce_mode,frad)  ;

                    loglog(f,magforce_mode(1,:),'k-')

        end
```

```
            disp('execution paused to display figure, "enter" to continue'); pause

            hold off

%           now use lsim to calculate force due to a 0.002 sec half-sine 100g shock pulse

            [force_mdc,ts_mdc] = lsim(sysforce_mdc,u,t);

            [force_elim,ts_elim] = lsim(sysforce_elim,u,t);

            plot(ts,force,'k-',ts_mdc,force_mdc,'k.-',ts_elim,force_elim,'k+-')

            if  modred_sort == 1
                    title(['modred cantilever tip force for ',num2str(shock_amplitude),'g, ...
                        ',num2str(pulse_width) ,' sec input, ',num2str(num_modes_used), ...
                        ' sorted modes included'])
            else
                    title(['modred cantilever tip force for ',num2str(shock_amplitude),'g, ...
                        ',num2str(pulse_width) ,' sec input, ',num2str(num_modes_used), ...
                        ' unsorted modes included'])
            end

            legend('all modes','reduced - mdc','reduced - elim',4)
            xlabel('time, sec')
            ylabel('Force, gm')
            grid on
            disp('execution paused to display figure, "enter" to continue'); pause

            max_force_mdc = max(abs(force_mdc));
            max_force_elim = max(abs(force_elim));

            peak_error_mdc_percent = 100*(max_force_mdc - max_force)/max_force
            peak_error_elim_percent = 100*(max_force_elim - max_force)/max_force

            end
```

16.4.14 Plotting Unsorted Modred Reduced Results – Eliminating High Frequency Modes

This section looks at how well "modred" performs when unsorted modes are used. We will see that the "del" option using the first four unsorted modes does a poor job of matching the original response while the "mdc" option using the same four unsorted modes does a good job of matching the lower frequency range of the response while missing the tenth mode resonance. The overall transient response of the system is matched well by the "mdc" option while the "del" option has significant error.

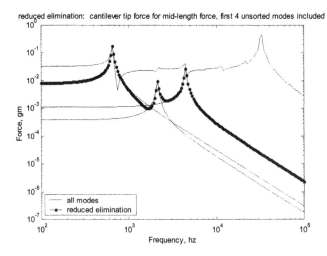

Figure 16.16: Overall frequency response with overload individual mode contributions for unsorted "del" modred option, with the 12 highest frequency modes eliminated.

Figure 16.16 displays the same response as the "unsorted" plot in Figure 16.13 because the "del" option in modred and our simple modal truncation method are equivalent. The dc gain is in error by 75%.

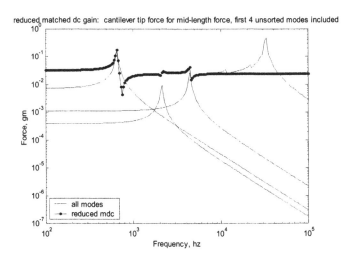

Figure 16.17: Overall frequency response with overlaid individual mode contributions for unsorted "mdc" modred option, with the 12 highest frequency modes reduced.

In Figure 16.17, the dc error is very small, 0.0008%. Even though the 32 khz mode is not included, the gain in the portion from 1 to 20 khz is close to the full model gain.

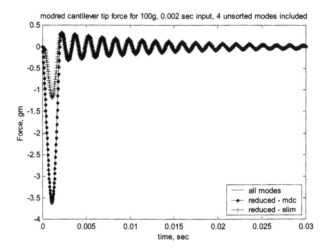

Figure 16.18: Half-sine shock pulse response for full, reduced unsorted "mdc" and reduced unsorted "del" models.

Figure 16.18 shows that the effect of the dc gain error in the frequency domain for the unsorted model shows up as a significant error in peak response in the time domain, 67%. The error in the unsorted peak response is only 0.09% for the "mdc" reduction.

16.4.15 Plotting Sorted Modred Reduced Results –
Eliminating Lower dc Gain Modes

This Section repeats the analysis of the previous Section but the sorted modes are used, retaining the higher dc gain modes. Since the important tenth mode is included in the retained sorted modes, we would expect that the reduced responses would match the original, all modes included response.

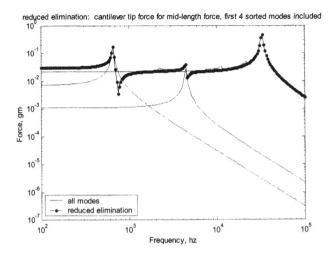

Figure 16.19: Overall frequency response with overload individual mode contributions for sorted "del" modred option, with the 12 lowest dc gain modes eliminated.

Figure 16.19 shows the same response as the "sorted" plot in Figure 16.14 because the "del" option in modred and our simple sorted modal truncation methods are equivalent. The dc gain is in error by 6.2%.

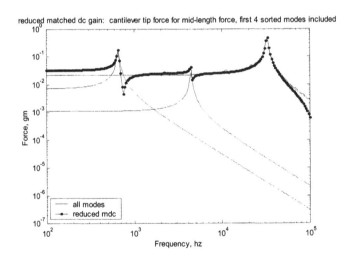

Figure 16.20: Overall frequency response with overload individual mode contributions for sorted "mdc" modred option, with the 12 lowest dc gain modes eliminated.

Note the high frequency discrepancy in Figure 16.20, related to using the "mdc" modred option. For this problem, which is dominated by the low frequency (<10khz) response and the dc gain of the 32 khz mode, the high frequency response is not important. The dc gain is in error by only 0.0025%.

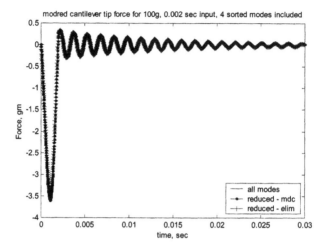

Figure 16.21: Half-sine shock pulse response for full, reduced unsorted and reduced sorted models.

The errors in peak response are 5.6% for the "del" method and 0.0773% for the "mdc" method.

16.4.16 Model Reduction Summary

Reduction Method Used	Dc gain error, percent	Peak error, percent	Comments
Nosort	75.45	67	This case should show the worst error because the 32 khz beam tip/spring mode is not included in the lowest four frequency modes.
Nosort, elim	75.45	67	The modred "del" option is the same as the "nosort" case because it just eliminates (truncates) the twelve highest frequency modes.
Sort	6.19	5.61	Sorting for dc gain with four modes includes the 32 khz mode, so the dc gain error is reduced. However, it still contains errors because the dc gain terms from the 12 unused modes are not included.
Sort, elim	6.19	5.61	The modred "del" option is the same as the "sort" case because it just eliminates (truncates) the twelve lowest dc gain modes.
Nosort, mdc	0.0007	0.0913	The modred "mdc" option, even though it does not use the 32 khz mode, takes its dc gain into effect, resulting in the small dc gain error. Because the frequency content of the shock pulse is low (~250 hz), the low frequency portion of the overall transfer function dominates the accuracy of the shock response.
Sort, mdc	0.0025	0.0773	Sorting the modes before reducing does not have a significant effect on the dc gain because the "mdc" operations take into account the dc gain effects of the unused modes.

Table 16.3: Summary of model reduction methods used, ranked from highest to lowest errors, with comments about each method.

Table 16.3 shows that using the modred "mdc," ("matched dc gain") method is the preferred method for this problem to obtain accurate results. For results that have accuracy in the 5 to 6% range, sorting by dc gain and then removing the lower dc gain modes is another available approach. It is clear that arbitrarily truncating high frequency modes can lead to significant errors because a single, important mode is neglected. Another source of error would occur if the ANSYS model had not included enough elements (modes) to take into account the beam tip mode or if a selected range of eigenvalues had not included the mode.

In summary, every model reduction problem provides new challenges and needs to be analyzed before making a decision about which reduction method is most appropriate.

16.5 ANSYS Code cantbeam_ss_spring_shkr.inp Listing

The ANSYS code in this section is similar to the code **cantbeam_ss.inp** in Section 15.7 with the exception that a tip spring and "shaker" mass are added.

```
! cantbeam_ss_spring_shkr.inp, 0.075 thick x 2 wide x 20mm long steel cant with tip
! mass and spring on shaker, shaker mass at cantilever base and coupled to spring ground
! title automatically built based on number of elements and eigenvalue extraction method

/prep7

filename = 'cantbeam_ss_spring_shkr'

! define number of elements to use

num_elem = 10

! define eigenvalue extraction method, 1 = reduced, 2 = block lanczos

eigext = 2

*if,eigext,eq,1, then
            nummodes = num_elem+1                ! only 1 displacement dof available for
each element
*else
            nummodes = 2*(num_elem+1)   ! both disp and rotation dof's available for each
                                        !  element
*endif

!        create the file name for storing data

!        first section of filename

aname = 'cantbeam'

!        second section of filename, number of elements

bname = num_elem

!        third section of filename, depends on eigenvalue extraction method
```

```
*if,eigext,ne,2, then
            cname = 'red'          ! reduced
*else
            cname = 'bl'           ! block Lanczos
*endif

! input the title, use %xxx% to substitute parameter name or parametric expression

aname_ti = 'cantbeam'

/title,%aname_ti%, %bname%, %cname%, spring tip

et,1,4                 ! element type for beam
et,2,14                                ! element type for spring
et,3,21                                ! element type for mass

! steel

ex,1,190e6                     ! mN/mm^2
dens,1,7.83e-6                 ! kg/mm^3
nuxy,1,0.293

! real value to define beam characteristics

r,1,0.15,0.05,0.00007031,0.075,0.2          ! beam properties: area, Izz, Iyy, TKz, TKy
r,2,1000000                                 ! spring stiffness, mN/mm
r,3,0.00002349,0.00002349,0.00002349        ! mass at tip, Kg
r,4,0.050,0.050,0.050                       ! shaker mass, Kg, approximately 1000 times mass

! define plotting characteristics

/view,1,1,-1,1     ! iso view
/angle,1,-60       ! iso view
/pnum,mat,1        ! color by material
/num,1             ! numbers off
/type,1,0          ! hidden plot
/pbc,all,1         ! show all boundary conditions

csys,0                                 ! define global coordinate system

! nodes

n,1,0,0,0                                      ! left-hand node
n,num_elem+1,20,0,0            ! right-hand node

fill,1,num_elem+1              ! interior nodes

n,num_elem+2,20,0,-3          ! spring connection node

nall
nplo

! elements

!  beam
```

```
type,1
mat,1
real,1
e,1,2
egen,num_elem,1,-1

! spring at tip

type,2
real,2
e,num_elem+1,num_elem+2

! mass at tip

type,3
real,3
e,num_elem+1

! shaker mass

type,3
real,4
e,1

! couple mass and spring end

nall
d,1,ux,0                              ! constrain all except uz for node 1
d,1,uy,0
d,1,rotx,0
d,1,roty,0
d,1,rotz,0

d,num_elem+2,ux,0          ! constrain all except uz for spring end node
d,num_elem+2,uy,0
d,num_elem+2,rotx,0
d,num_elem+2,roty,0
d,num_elem+2,rotz,0

!          d,1,uz,0

cp,1,uz,1,num_elem+2          ! uz couple shaker mass and spring end node

! constrain all but uz and roty for all other nodes to allow only those dof's

nall
nsel,s,node,,2,num_elem+1
d,all,ux
d,all,uy
d,all,rotx
d,all,rotz

nall
eall
```

```
nplo
eplo

! ****************** eigenvalue run ******************

fini                    ! fini just in case not in begin

/solu                   ! enters the solution processor, needs to be here to do editing below

allsel                  ! default selects all items of specified entity type, typically nodes, elements

nsel,s,node,,2,num_elem+1
m,all,uz

*if,eigext,eq,1,then                        ! use reduced method

        antype,modal,new
        modopt,reduc,nummodes        ! method - reduced Householder, nummodes –
                                     ! no to extract
        expass,off           ! key = off, no expansion pass, key = on, do expansion
        mxpand,nummodes,,,no         ! nummodes to expand,freq beginning,freq
                                     ! ending,elcalc = yes - calculate stresses
        total,nummodes,1             ! total masters, 1 is exclude rotations

*elseif,eigext,eq,2                          ! use block lanczos

        antype,modal,new
        modopt,lanb,nummodes         ! no total required for block lanczos because
                                     !   calculates all eigenvalues
        expass,off
        mxpand,nummodes,,,no

*endif

allsel

solve                   ! starts the solution of one load step of a solution sequence, modal here

fini

! plot first mode

/post1

/format,,,,,10000

set,1,1

pldi,1

! ****************** output frequencies ********************

save,%aname%%bname%%cname%,sav

/output,%aname%%bname%%cname%,frq                ! write out frequency list to ascii file .frq
```

```
set,list

/output,term                          ! returns output to terminal

! ********************* output eigenvectors *******************

! define nodes for output:  forces applied or output displacements

nsel,s,node,,1,num_elem+1

/output,%aname%%bname%%cname%,eig      ! write out frequency list to ascii file .eig

*do,i,1,nummodes
         set,,i
         /page,,,1000
         prdisp
*enddo

/output,term

! ****************** plot modes ******************

! pldi plots

/show,%aname%%bname%%cname%,grp,0              ! save mode shape plots to file .grp

allsel

/view,1,,-1,,                         ! side view for plotting
/angle,1,0
/auto

*do,i,1,nummodes
         set,1,i
   pldi,1
*enddo

/show,term
```

CHAPTER 17

SISO DISK DRIVE ACTUATOR MODEL

17.1 Introduction

This chapter will use an ANSYS model of a complete disk drive actuator/suspension system to expand on the methods and examples of the last two chapters.

While simple in appearance, a disk drive actuator/suspension system must fulfill a number of exacting requirements. The suspension system is required to provide a stiff connection between the actuator and the head in the seeking/track-following direction, while providing a compliant system in a direction perpendicular to the plane of the disk. This allows the air bearing supported head to comply to the shape and vibration of the disk. The actuator is designed with low mass to allow fast seeking. It must have resonant characteristics which provide small residual vibration following a seek from one track to another. Since the entire disk drive is subject to various shock and vibration events, the actuator dynamics must aid in preventing the head from unloading from the disk during the event.

The actuator/suspension system used as the example for this and the next chapter is a single disk actuator, with two arms and two suspensions. It is purposely designed with poor resonance characteristics (different thickness arms, coil positioned off the mass center of the system, etc.) in order to provide a richer resonance picture for analysis.

We will assume that the servo system used with the actuator is a sampled system with a 20khz sample rate, meaning that the Nyquist frequency is 10khz. We need to understand all the modes of vibration of the system up to at least 20khz because the sampled system will alias frequencies that are higher than 10khz back into the 0 to 10khz range.

We will find that **the dynamics of this ANSYS model with approximately 21000 degrees of freedom can be described well using between 8 and 20 modes of vibration** (16 to 40 states), depending on what measure of "goodness" is used. If we are interested in impulse response, we will see in the next chapter that using only eight modes results in a system with approximately a 5% error. For a good fit in the frequency domain through 10 khz only 8 modes are required, while a good fit through 20 khz requires 20 modes. In a well-designed actuator (this example is poorly designed as mentioned earlier) fewer than 20 modes are required since symmetry will couple in fewer modes.

This actuator/suspension model is a good example of what the book is all about: generating low order models of complicated systems, in this case a model which is approximately 1000 times smaller than the original model.

Once the ANSYS model results are available, a MATLAB model will be created. Then we will analyze several methods of reducing the size of the model. In the previous chapters, we used dc gains of the individual modes of vibration to rank the most important modes to keep. If we use uniform damping (the same zeta value for all modes) we will reach the same ranking conclusion using either dc gain or peak gain. However, if we use non-uniform damping, peak gain ranking is required. The MATLAB code will prompt for whether uniform or non-uniform damping is being used and will choose the appropriate ranking, dc gain or peak gain. The next chapter will introduce another, more elegant method of ranking modes to be eliminated, balanced reduction.

17.2 Actuator Description

Figure 17.1 shows top and cross-sectioned side views of the actuator used for the analysis. The global XYZ coordinate system for the model is indicated.

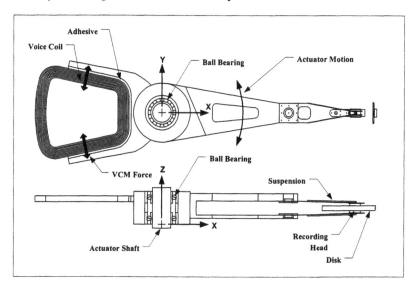

Figure 17.1: Drawing of actuator/suspension system.

The shaft is constrained in all directions, providing a fixed reference about which the actuator rotates on two axially preloaded ball bearings.   This actuator is purposely designed to have poor dynamic characteristics, as seen in the side view.   The coil, to which the Voice Coil Motor (VCM) forces are applied, is not centered between the two bearings and the two arms are of unequal thickness.   Both the coil force mispositioning and the unequal arm thickness inertial effects will tend to excite rotations about the x axis.

The coil is bonded to the aluminum actuator body. During operation, current passes through the coil windings. The current interacts with the magnetic field from pairs of magnets above and below the straight legs of the coil (not shown), creating forces on the straight legs.   The direction of the force is dependent on the direction of the current in the coil, clockwise or counterclockwise. The motion of the actuator due to the coil force is indicated by "Actuator Motion."

The suspensions are designed to provide a preload of several grams force onto the disk surface. During operation the preload is counterbalanced by the air bearing lifting force, controlling the flying height spacing between the head and disk to less than several microinches. During shipment, the preload tends to hold the head down on the disk surface in the event of shock and vibration events, preventing potential damage caused by the head lifting off and striking the disk.

17.3 ANSYS Suspension Model Description

Before analyzing the complete actuator/suspension system, we will analyze only the suspension system.   Understanding the dynamics of sensitive components of larger assemblies as components can add considerable insight to interpretation of the dynamics of the overall system.

The suspension portion of the actuator/suspension model is shown in Figures 17.2 and 17.3. The complete suspension is depicted in Figure 17.2, and the "flexure" portion of the suspension is shown in Figure 17.3.

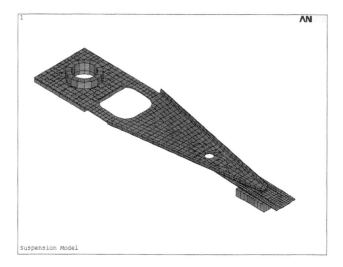

Figure 17.2: Suspension model.

The recording head (slider) is bonded to the center section of the flexure. The "dimple" at the center of the slider tongue provides a point contact about which the slider can rotate in the pitch and roll directions. The tip of the dimple and the contact point on the underside of the loadbeam are constrained to move together in translation. The flexure body is laser welded to the loadbeam (the triangular section), which is itself laser welded to the swage plate at the left-hand end.

The boundary conditions for the suspension model are: the swage plate is constrained in the x and z directions and the four slider corners are constrained in the z direction. A large mass is attached at the swage plate to allow for y direction ground acceleration forcing function. Because there is no constraint in the y direction there will be a zero-frequency, rigid body mode in that direction.

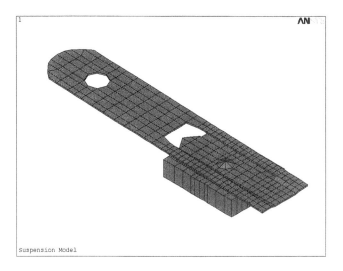

Figure 17.3: Flexure and recording head (slider) portion of suspension. Note the "dimple" at the center of the slider, a point about which the slider rotates to comply with the disk topology.

The model is built with the ability to easily change the critical flatness and forming parameters because the dynamics of the suspension are so dependent on the geometry. Small (0.025 mm, 0.001 inch) defects in critical forming and flatness parameters can drastically change the resonance characteristics,

The suspension model is made completely of eight-node brick elements. Laser welds and bonded joints are simulated by "merging" the nodes being welded or bonded, essentially creating a rigid joint at that connection.

The ANSYS suspension-only model, **srun.inp**, is included in the available downloads but will not be discussed. Running the model with different values for the three input parameters "zht," "bump" and "offset" will show the extreme sensitivity of the first torsion mode (described below) to these parameters.

17.4 ANSYS Suspension Model Results

The suspension has six modes of vibration in the 0 to 10 khz frequency range. The ANSYS frequency response plot for the suspension is shown in Figure 17.4. The six modes in the 0 to 10 khz will be plotted and described below.

17.4.1 Frequency Response

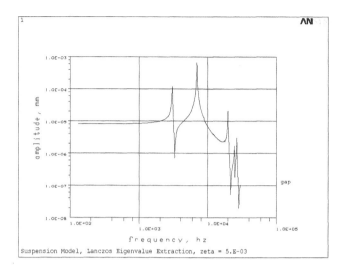

Figure 17.4: Suspension frequency response for a y direction forcing function.

17.4.2 Mode Shape Plots

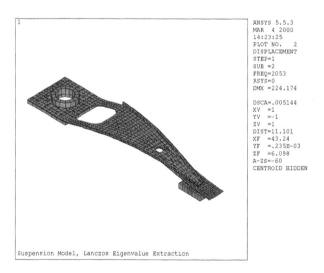

Figure 17.5: Mode 2, 2053 hz, first bending mode.

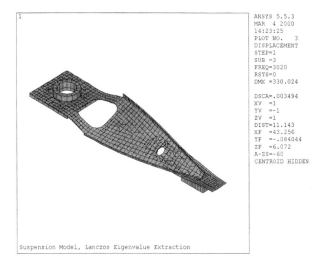

```
1                                           ANSYS 5.5.3
                                            MAR  4 2000
                                            14:23:25
                                            PLOT NO.   3
                                            DISPLACEMENT
                                            STEP=1
                                            SUB =3
                                            FREQ=3020
                                            RSYS=0
                                            DMX =330.024

                                            DSCA=.003494
                                            XV  =1
                                            YV  =-1
                                            ZV  =1
                                            DIST=11.143
                                            XF  =43.256
                                            YF  =-.084044
                                            ZF  =6.072
                                            A-ZS=-60
                                            CENTROID HIDDEN




Suspension Model, Lanczos Eigenvalue Extraction
```

Figure 17.6: Mode 3, 3020 hz, first torsion mode.

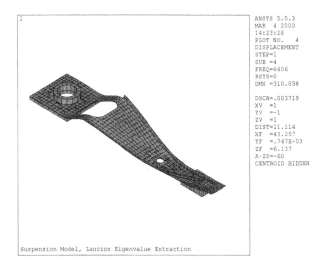

```
1                                           ANSYS 5.5.3
                                            MAR  4 2000
                                            14:23:26
                                            PLOT NO.   4
                                            DISPLACEMENT
                                            STEP=1
                                            SUB =4
                                            FREQ=6406
                                            RSYS=0
                                            DMX =310.058

                                            DSCA=.003719
                                            XV  =1
                                            YV  =-1
                                            ZV  =1
                                            DIST=11.114
                                            XF  =43.257
                                            YF  =.747E-03
                                            ZF  =6.137
                                            A-ZS=-60
                                            CENTROID HIDDEN




Suspension Model, Lanczos Eigenvalue Extraction
```

Figure 17.7: Mode 4, 6406 hz, second bending mode.

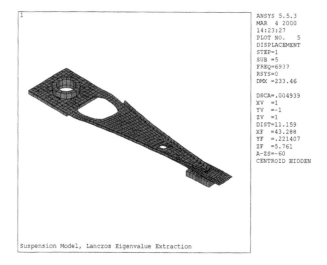

Figure 17.8: Mode 5, 6937 hz, sway or lateral mode.

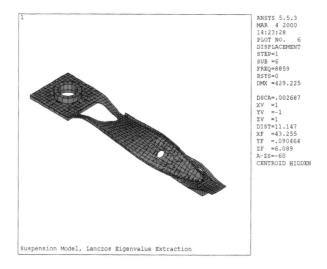

Figure 17.9: Mode 6, 8859 hz, second torsion mode.

The suspension frequency response plot and mode shape plots complement each other and help to develop a visual, intuitive understanding of modal coupling. The only modes that have y direction motion of the slider relative to the swage plate are the first torsion and sway modes as can be seen in the frequency response plot of Figure 17.4. All the other modes have motions which are orthogonal to the motion of interest. The first bending mode is the most obvious example. Since its motion in only in the z direction, it cannot be excited by a y direction forcing function, and thus, does not couple into the frequency response.

17.5 ANSYS Actuator/Suspension Model Description

The complete actuator/suspension model is shown in Figure 17.10. It also is made of eight-node brick elements except for the inclusion of spring elements which are used to simulate the ball bearings' individual ball stiffnesses.

The shaft and inner radii of the two ball bearing inner rings are fully constrained. The four corners of each of the sliders are constrained for zero motion in the z direction, essentially creating an infinitely stiff air bearing.

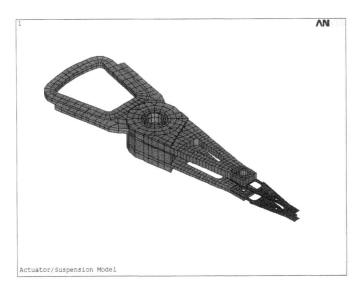

Figure 17.10: Complete actuator/suspension model.

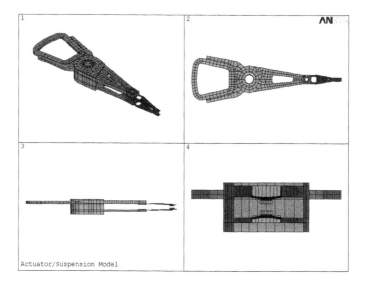

Figure 17.11: Actuator / suspension model, four views.

The primary motion of the actuator is rotation about the pivot bearing, therefore the final model has the coordinate system transformed from a Cartesian x,y,z coordinate system to a Cylindrical, r, θ and z system, with the two origins coincident.

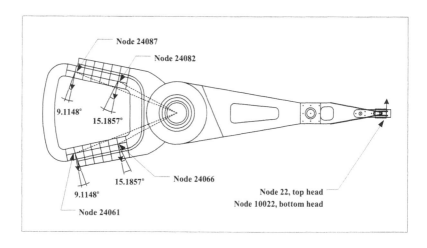

Figure 17.12: Nodes used for reduced MATLAB model. Shown with partial finite element mesh at coil.

For reduced models we only require eigenvector information for degrees of freedom where forces are applied and where displacements are required. Figure 17.12 shows the nodes used for the reduced MATLAB model. The four nodes 24061, 24066, 24082 and 24087 are located in the center of the coil in the z direction and are used for simulating the VCM force. The forces created by the interactions between the current in the straight legs of the coil and the magnetic field are perpendicular to the straight leg sections. Since the coordinate system is cylindrical, the forces are decomposed into radial and circumferential components as shown in Figure 17.12. Nodes 22 and 10022 are the nodes for the top and bottom heads (heads 1 and 0), respectively. The arrows at the nodes indicate the direction of forces, and the angles show the directions of the force, measured from the circumferential direction. The components in the radial and circumferential directions are taken using the angles.

The model uses only the circumferential motion of the heads, which, if divided by the radius from the pivot to the head, will give output in radians.

The actuator/suspension ANSYS code, **arun.inp,** is too large to be listed here but is available for downloading.

17.6 ANSYS Actuator/Suspension Model Results

A recommended sequence for analyzing dynamic finite element models is:

1) Plot resonant frequencies versus mode numbers to get a feel for the frequency range. See if there are any significant jumps in frequency between modes which can indicate the system transitioning from one type of characteristic motion to another. For example, a sequence of bending modes transitioning into a sequence of torsional modes.

2) Plot frequency responses to define which modes couple into the response.

3) Plot and animate the mode shapes that contribute to the response, identifying modes that couple into motions in directions of interest and those that do not. Visually get a sense of how the geometry of the structure affects the modes.

4) Run parameter studies to understand the sensitivity of critical modes to design variables: dimensions, tolerances, material properties, etc.

17.6.1 Eigenvalues, Frequency Responses

The actuator/suspension model was run using the Block Lanczos method to extract the first 50 eigenvalues and eigenvectors. The plot of frequency versus number of modes is shown in Figure 17.13. The first mode, the rigid body mode, was calculated to be 0.0101 hz, with the first oscillatory mode frequency at 785 hz.

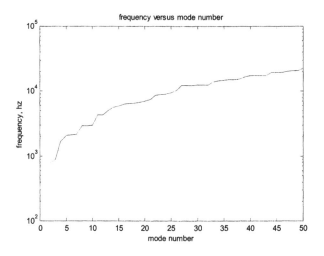

Figure 17.13: Frequencies versus mode number.

Mode 50 is at 22350 hz, which is slightly higher than our objective of including all the modes through 20 khz.

Frequency responses for the displacements of heads 0 and 1 (bottom and top heads) for coil input force can be seen in Figures 17.14 and 17.15. Mode shape plots, with undeformed and deformed shapes, are then shown for the modes which are evident in the frequency response plots. In addition, some typical modes that do not couple into the frequency response are shown.

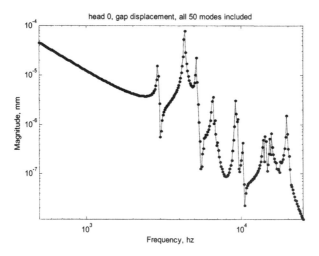

Figure 17.14: Frequency response for head 0 for coil input.

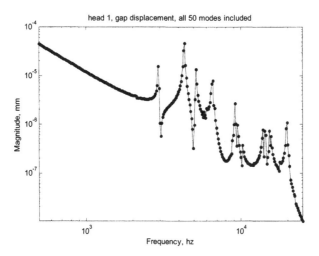

Figure 17.15: Frequency response for head 1 for coil input.

17.6.2 Mode Shape Plots

In this section we will plot overlaid undeformed and deformed modes shapes for selected modes, which will then be described and discussed in the next section.

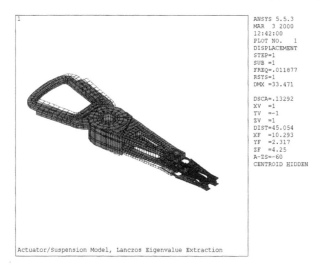

Figure 17.16: Mode 1 undeformed/deformed mode shape plot, 0.012 hz rigid body rotation.

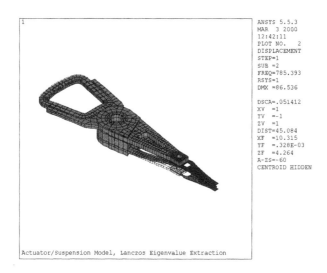

Figure 17.17: Mode 2 mode shape plot, 785 hz. Bending of bottom arm.

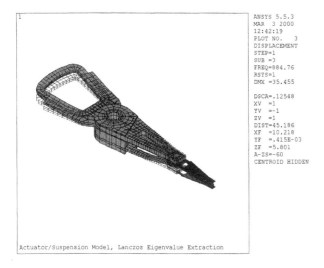

Figure 17.18:   Mode 3 mode shape plot, 885 hz, coil and bottom arm bending.

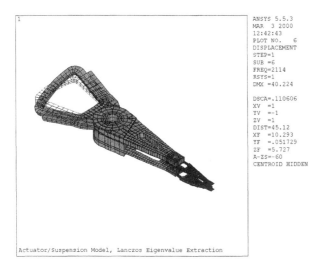

Figure 17.19:   Mode 6 mode shape plot, 2114 hz, coil torsion.

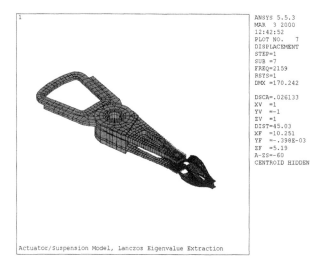

Figure 17.20: Mode 7 mode shape plot, 2159 hz, suspension bending modes.

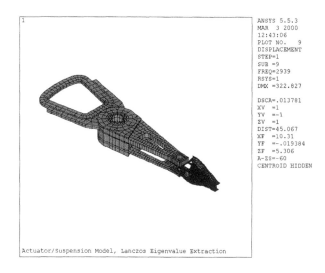

Figure 17.21: Mode 9 mode shape plot, 2939 hz, suspension torsion mode.

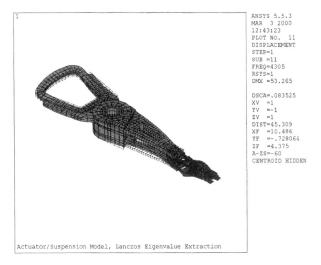

```
1                                                    ANSYS 5.5.3
                                                     MAR  3 2000
                                                     12:43:23
                                                     PLOT NO.  11
                                                     DISPLACEMENT
                                                     STEP=1
                                                     SUB =11
                                                     FREQ=4305
                                                     RSYS=1
                                                     DMX =53.265

                                                     DSCA=.083525
                                                     XV  =1
                                                     YV  =-1
                                                     ZV  =1
                                                     DIST=45.309
                                                     XF  =10.486
                                                     YF  =-.728064
                                                     ZF  =4.375
                                                     A-ZS=-60
                                                     CENTROID HIDDEN

Actuator/Suspension Model, Lanczos Eigenvalue Extraction
```

Figure 17.22: Mode 11 mode shape plot, 4305 hz, system mode.

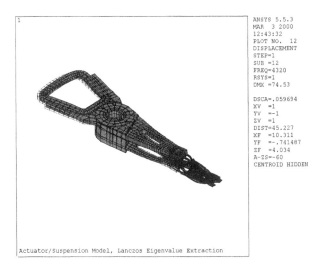

```
1                                                    ANSYS 5.5.3
                                                     MAR  3 2000
                                                     12:43:32
                                                     PLOT NO.  12
                                                     DISPLACEMENT
                                                     STEP=1
                                                     SUB =12
                                                     FREQ=4320
                                                     RSYS=1
                                                     DMX =74.53

                                                     DSCA=.059694
                                                     XV  =1
                                                     YV  =-1
                                                     ZV  =1
                                                     DIST=45.227
                                                     XF  =10.311
                                                     YF  =-.741487
                                                     ZF  =4.034
                                                     A-ZS=-60
                                                     CENTROID HIDDEN

Actuator/Suspension Model, Lanczos Eigenvalue Extraction
```

Figure 17.23: Mode 12 mode shape plot, 4320 hz, radial mode.

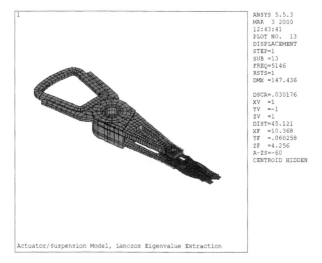

Figure 17.24: Mode 13 mode shape plot, 5146 hz.

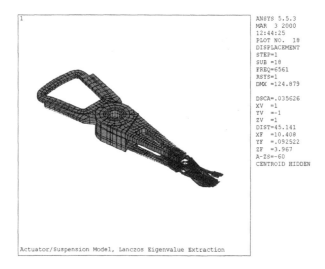

Figure 17.25: Mode 18 mode shape plot, 6561 hz.

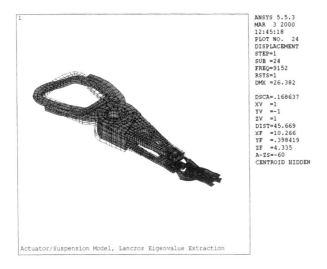

Figure 17.26: Mode 24 mode shape plot, 9152 hz.

17.6.3 Mode Shape Discussion

We will now correlate the two frequency response plots, Figures 17.14 and 17.15, with the mode shape plots above to start getting an intuitive feel for which modes couple into the response plots and which modes do not.

Mode 1, the rigid body mode, shows up as the 40db/decade low frequency slopes on both frequency responses, head 0 and head 1.

Modes 2 and 3, at 785 and 884 hz, are representative of modes that do not couple because of the direction of the motion.   Both modes involve only bending motions of arms and/or coil in the x-z plane. Since the motions are perpendicular (orthogonal) to the direction of force and to the direction of the head in the circumferential direction, the modes should not couple into the frequency response plots. Therefore we see no resonance peaks at these two frequencies.

Mode 6 at 2114 hz is a coil/actuator torsion mode that shows up as the small pole/zero pair in the head 1 frequency response.

Mode 7 at 2159 hz is a suspension bending mode that does not couple into the response.

Mode 9 at 2939 hz is a suspension torsion mode that interacts with the rigid body mode to create the significant pole/zero pair at 2939 hz.

Modes 11 and 12 at 4305 hz and 4320 hz are the major system modes with significant y direction motion of the coil, bearings, arms and suspensions. These are the two modes associated with the highest resonant peak in the frequency response. What appears to be a single peak is actually two peaks.

Mode 13 at 5146 hz is a mode which involves torsion of the coil and actuator body about the x axis with the suspensions moving torsionally and laterally.

Mode 18 at 6561 hz is a suspension sway mode, where the suspension-only mode at 6937 hz (Figure 17.8) is reduced to 6561 hz because it is attached to the flexible actuator.

Mode 24 at 9152 hz is a highly deformed actuator mode, in which the actuator hub moves significantly about the ball bearing, the coil deforms and suspensions and arms deflect.

17.6.4 ANSYS Output Example Listing

A partial listing of the eigenvector output (actrl.eig) for modes 1, 2, 11 and 12 is shown below. These four modes were chosen for listing and discussion because they illustrate some key points about interpreting ANSYS eigenvector output. The important information in each of the eigenvector sections is highlighted in bold type. The "SUBSTEP" is the mode number, and "FREQ" is the eigenvalue in hz. Since the output is in cylindrical coordinates, UX, UY and UZ refer to radial, circumferential and z axis coordinates, respectively. Since all the elements attached to the six nodes listed are eight-node brick elements, with only translational degrees of freedom, all the rotation eigenvector values are zero. The six nodes listed correspond to the two heads, 22 and 10022 and the four coil forcing function nodes, 24061, 24066, 24082 and 24087. See Figure 17.12 for node locations. We need both radial (UX) and circumferential (UY) directions because the forces applied by the VCM to the coil are perpendicular to the straight legs of the coil, and have both radial and circumferential components.

```
PRINT DOF  NODAL SOLUTION PER NODE

***** POST1 NODAL DEGREE OF FREEDOM LISTING *****

LOAD STEP=   1 SUBSTEP=   1
FREQ= 0.11877E-01  LOAD CASE=  0

THE FOLLOWING DEGREE OF FREEDOM RESULTS ARE IN COORDINATE SYSTEM
1

NODE     UX        UY       UZ        ROTX     ROTY    ROTZ
  22 0.30718E-06 32.772  0.85804E-12 0.0000   0.0000  0.0000
```

```
10022 0.30759E-06 32.772   -0.49994E-10 0.0000    0.0000    0.0000
24061 0.11969E-06 16.968   -0.17668E-08 0.0000    0.0000    0.0000
24066 0.77415E-07 10.274   -0.15751E-08 0.0000    0.0000    0.0000
24082 0.68508E-07 10.274   -0.15395E-08 0.0000    0.0000    0.0000
24087 0.10089E-06 16.968   -0.16990E-08 0.0000    0.0000    0.0000
```

MAXIMUM ABSOLUTE VALUES
NODE   10022   22   24061   0   0   0
VALUE   0.30759E-06 32.772   -0.17668E-08 0.0000   0.0000   0.0000

*ENDDO INDEX= I

***** POST1 NODAL DEGREE OF FREEDOM LISTING *****

LOAD STEP= 1 **SUBSTEP=   2**
FREQ= 785.39   LOAD CASE= 0

THE FOLLOWING DEGREE OF FREEDOM RESULTS ARE IN COORDINATE SYSTEM
1

NODE	UX	UY	UZ	ROTX	ROTY	ROTZ
22	-0.25631	-0.19637E-01	0.15936E-04	0.0000	0.0000	0.0000
10022	0.92764	-0.10736	0 .29519E-02	0.0000	0.0000	0.0000
24061	0.18573	-0.67085E-01	-5.7724	0.0000	0.0000	0.0000
24066	0.17688	-0.88331E-01	-2.1255	0.0000	0.0000	0.0000
24082	0.17616	0.95885E-01	-2.1213	0.0000	0.0000	0.0000
24087	0.18506	0.79278E-01	-5.7661	0.0000	0.0000	0.0000

MAXIMUM ABSOLUTE VALUES
NODE   10022   10022   24061   0   0   0
VALUE   0.92764   -0.10736   -5.7724   0.0000   0.0000   0.0000

***** POST1 NODAL DEGREE OF FREEDOM LISTING *****

LOAD STEP= 1 **SUBSTEP= 11**
FREQ= 4305.3   LOAD CASE= 0

THE FOLLOWING DEGREE OF FREEDOM RESULTS ARE IN COORDINATE SYSTEM
1

NODE	UX	UY	UZ	ROTX	ROTY	ROTZ
22	-4.4488	27.588	-0.66528E-04	0.0000	0.0000	0.0000
10022	3.9832	41.657	0.44809E-01	0.0000	0.0000	0.0000
24061	-0.43605	-10.023	-8.7664	0.0000	0.0000	0.0000
24066	0.35112	-3.5631	-11.532	0.0000	0.0000	0.0000
24082	3.9625	-1.1137	-14.210	0.0000	0.0000	0.0000
24087	5.0136	-7.8562	-6.0297	0.0000	0.0000	0.0000

MAXIMUM ABSOLUTE VALUES
NODE   24087   10022   24082   0   0   0
VALUE   5.0136   41.657   -14.210   0.0000   0.0000   0.0000

***** POST1 NODAL DEGREE OF FREEDOM LISTING *****

```
LOAD STEP=  1 SUBSTEP=  12
FREQ=  4320.1    LOAD CASE=  0

THE FOLLOWING DEGREE OF FREEDOM RESULTS ARE IN COORDINATE SYSTEM
1

  NODE     UX        UY        UZ        ROTX      ROTY      ROTZ
    22    4.3947    36.811  -0.25761E-02  0.0000    0.0000    0.0000
 10022   -0.88223   62.097   0.34209E-01  0.0000    0.0000    0.0000
 24061   -5.3622   -11.584   3.9397       0.0000    0.0000    0.0000
 24066   -3.9590   -2.2258   10.513       0.0000    0.0000    0.0000
 24082    0.81662  -4.0070   7.7931       0.0000    0.0000    0.0000
 24087    2.0281   -13.160   6.6813       0.0000    0.0000    0.0000

MAXIMUM ABSOLUTE VALUES
NODE   24061    10022    24066      0         0         0
VALUE  -5.3622   62.097   10.513   0.0000    0.0000    0.0000
```

We will now discuss the eigenvector listings above in light of the frequency response and mode shape plots reviewed earlier. Once again, we will make the connection between modes that contribute to frequency responses and those that do not.

Mode 1 shows that all the UX and UZ entries are essentially zero, which is appropriate for a rigid body mode where the actuator is rotating about the shaft, with only circumferential, UY, displacements. The relative amplitudes of each UY entry are related by their radial distances from the shaft. The frequency calculated is not exactly zero because of rounding and slight geometric errors which create small stiffnesses in rotation about the shaft.

Mode 2 is the first oscillatory mode, the arm bending mode. A mode which involves only UZ motion will have no cross-coupling in the y direction since the actuator system is symmetrical about the x axis. In a typical disk drive, the actuator is not perfectly symmetrical, and modes whose motions are primarily in the vertical direction will couple in the y direction. All of the UY entries for this mode are very small relative to the UZ entries, indicating that the contribution of this mode to the y direction motion of the head should be small.

Modes 12 and 13 are the major system modes, those modes with the highest amplitude motion on the frequency response plot. The entries in the UY column are significant relative to the entries for mode 2 and are of the same order of magnitude as those in mode 1. This indicates that this mode is relatively important for our desired frequency response.

The eigenvalues and UX and UY eigenvector entries are stripped out of the actrl.eig file and stored in the MATLAB .mat file actrl_eig.mat (Appendix 1). Now we are ready to read the ANSYS results into MATLAB and start developing the reduced model.

17.7 MATLAB Model, MATLAB Code act8.m Listing and Results

17.7.1 Code Description

The code starts by reading in the ANSYS model eigenvalue and eigenvector results for all 50 modes from actrl_eig.mat. The VCM force components in the radial and circumferential directions are then defined using the angles shown in Figure 17.12.

The user is prompted to specify whether the same zeta value is to be used for all modes (uniform damping), or whether each mode can have different values, non-uniform damping. If uniform damping is specified, the user is prompted to enter a value for zeta, a vector of uniform damping values is created and dc gains are calculated. If non-uniform damping is chosen, a damping vector is read in from **zetain.m** and peak gains are calculated. The appropriate gains are then sorted and plotted, indicating the most important modes to retain. Typically uniform damping is taken in the range of 0.005 (0.5% of critical damping) to 0.02 (2% of critical damping). If experimental data is available, the damping values for each mode in **zetain.m** can be matched to its experimentally determined value.

Once the user defines the number of modes to be retained, two state space systems are automatically built. The first includes all 50 modes and the second includes the sorted, reduced number of modes. The 50-mode response is plotted for either head 0 or head 1 with individual mode contributions overlaid.

Since the servo system postulated for the actuator has a 20 khz sample frequency, the Nyquist frequency is half that, or 10 khz. This means that resonances higher in frequency than the Nyquist frequency will be aliased back to the 0 to 10 khz range. The user is prompted for the sample frequency to be used (default 20 khz). The MATLAB "c2d" command is used to create a discrete model of the original continuous system. A discrete frequency response, with upper limit of the Nyquist frequency, is created and plotted, overlaying the original continuous frequency response. If the sample rate is high enough, this overlay allows one to see that it will not alias critical modes of vibration. Experimentally, the only information available from a discrete servo system frequency response is up to the Nyquist frequency. Measurements which are independent of the servo system (such as from an external laser measurement system) are required to identify modes higher than the Nyquist frequency. An example of using a very low sampling frequency with this actuator system will be shown.

Frequency responses are calculated using the reduced, sorted modes, truncating the less important modes and using the "modred" "mdc" option. Truncating is the same as using the "del" option on the MATLAB "modred" command.

17.7.2 Input, dof Definition

The first section of code reads in the eigenvalue/eigenvector data from **actrl_eig.mat** and defines explicitly the degrees of freedom used. The original ANSYS model has approximately 21000 degrees of freedom. By defining only the degrees of freedom required for the desired frequency response, we can reduce the number of degrees required for the MATLAB model to 12: the radial and circumferential components of the two head nodes and the four coil forcing function nodes.

```
%        act8.m

         clear all;

         hold off;

         clf;

%        load the Block Lanczos .mat file actrl_eig.mat, containing evr - the modal matrix,
%        freqvec - the frequency vector and node_numbers - the vector of node numbers
%        for the modal matrix

%        the output for the ANSYS run is the following dof's

% dof node       dir     where
%  1      22     ux - radial, top head gap
%  2   10022     ux - radial, bottom head gap
%  3   24061     ux - radial, coil
%  4   24066     ux - radial, coil
%  5   24082     ux - radial, coil
%  6   24087     ux - radial, coil
%  7      22     uy - circumferential, top head gap
%  8   10022     uy - circumferential, bottom head gap
%  9   24061     uy - circumferential, coil
% 10   24066     uy - circumferential, coil
% 11   24082     uy - circumferential, coil
% 12   24087     uy - circumferential, coil

         load actrl_eig;
```

```
[numdof,num_modes_total] = size(evr);

freqvec(1) = 0;        % set frequency of rigid body mode to zero

xn = evr;
```

17.7.3 Forcing Function Definition, dc Gain Calculation

A vector of the squares of the eigenvalues, in rad/sec units, for use in the gain calculations is generated. Like the dc gain calculation with a rigid body mode discussed in the last chapter, we will again calculate the low frequency gain of the rigid body mode using the lowest frequency defined in the frequency response calculation.

The forcing function components for the four coil nodes are defined, again using Figure 17.12 as the reference. A unity force is applied at the coil, and evenly distributed among the four nodes. The force at each coil node is decomposed into its components in the radial and circumferential (x and y) directions. The coil forces in physical coordinates are then defined for each coil node and where the ux and uy force entries for the head nodes, dof 1, 2, 7 and 8 are all zero.

A discussion of what is meant by "Single Input Single Output" (SISO) is appropriate here. This model is a "SI" or Single Input model because the same force is applied to all four coil nodes, requiring only a single column vector for the input matrix "b." The fact that forces are applied to multiple nodes has no significance relative to the "SI" definition.

In Chapter 15, (15.2) and (15.3), we found that the dc gain and peak gain of for the ith mode are given by the expressions:

$$\frac{z_{ji}}{F_{ki}} = \frac{z_{nji} z_{nki}}{\omega_i^2} , \qquad (17.1)$$

$$\frac{z_{ji}}{F_{ki}} = \frac{-j}{2\zeta_i} (\text{dc gain}) \qquad (17.2)$$

where $z_{nji} z_{nki}$, the residue, is the product of the jth (output) row and kth (force applied) row terms of the ith eigenvector divided by the square of the eigenvalue for the ith mode and ζ_i is the damping for the ith mode. For all the models so far in the book, forces have been applied at a single node and displacements have been taken at a single node, making the above definitions clear. Here we are applying the same force to four coil nodes, so we will define a composite forcing function which will consist of the force applied to each node times the eigenvector value for that node, f_physical'*xn. The dimensions of this operation are (1 x ndof) x (ndof x nmodes) = (1 x nmodes), so we have a **composite** force vector for each mode.

This composite force vector is then multiplied element by element by the rows of the eigenvector matrix corresponding to the uy direction displacements of the two heads.

We will calculate and plot the gains for both head 0 and head 1 but will only calculate frequency response results for one or the other (user defined). Thus there is no ambiguity about whether to rank modes based on the gains of head 0 or head 1, only the one chosen for frequency response calculations is used for ranking.

```
%   calculate the dc amplitude of the displacement of each mode by
%   multiplying the composite forcing function by the output row

        omega2 = (2*pi*freqvec)'.^2;    % convert to radians and square

%       define frequency range for frequency response

        freqlo = 501;

        freqhi = 25000;

        flo=log10(freqlo) ;
        fhi=log10(freqhi) ;

        f=logspace(flo,fhi,300) ;
        frad=f*2*pi ;

%       define radial and circumferential forces applied at four coil force nodes
%       "x" is radial, "y" is circumferential, total force is unity

        n24061fx = 0.25*sin(9.1148*pi/180);
        n24061fy = 0.25*cos(9.1148*pi/180);

        n24066fx = 0.25*sin(15.1657*pi/180);
        n24066fy = 0.25*cos(15.1657*pi/180);

        n24082fx = -0.25*sin(15.1657*pi/180);
        n24082fy = 0.25*cos(15.1657*pi/180);

        n24087fx = -0.25*sin(9.1148*pi/180);
        n24087fy = 0.25*cos(9.1148*pi/180);

%       f_physical is the vector of physical force
%       zeros at each output dof and input force at the input dof
```

```
                f_physical = [    0
                                  0
                             n24061fx
                             n24066fx
                             n24082fx
                             n24087fx
                                  0
                                  0
                             n24061fy
                             n24066fy
                             n24082fy
                             n24087fy  ];
```

%   define composite forcing function, force applied to each node times
         eigenvector value
%   for that node

```
         force = f_physical'*xn;
```

%   choose which head to use for frequency responses

```
         head = input('enter "0" default for head 0 or "1" for head 1 ... ');

         if  isempty(head)
                    head = 0;
         end
```

%   prompt for uniform or variable zeta

```
         zeta_type = input('enter "1" to read in damping vector (zetain.m) ...
                            or "enter" for uniform damping ... ');

         if  (isempty(zeta_type))

                  zeta_type = 0;

                  zeta_uniform = input('enter value for uniform damping, ...
                                       .005 is 0.5% of critical (default) ... ');

                  if (isempty(zeta_uniform))
                            zeta_uniform = 0.005;
                  end

                  zeta_unsort = zeta_uniform*ones(num_modes_total,1);

                  gainstr = 'dc gain';

         else

                  zetain;      % read in zeta_unsort damping vector from zetain.m file

                  gainstr = 'peak gain';

         end
```

```
        if  length(zeta_unsort) ~= num_modes_total

                error(['error - zetain vector has ',num2str(length(zeta_unsort)), ...
                        ' entries instead of ',num2str(num_modes_total)]);

        end

%       calculate dc gains if uniform damping, peak gains if non-uniform

        if  zeta_type == 0              % dc gain

                gain_h0 = abs([force(1)*xn(8,1)/frad(1) ...
                        force(2:num_modes_total).*xn(8,2:num_modes_total) ...
                        ./omega2(2:num_modes_total)]);

                gain_h1 = abs([force(1)*xn(7,1)/frad(1) ...
                        force(2:num_modes_total).*xn(7,2:num_modes_total) ...
                        ./omega2(2:num_modes_total)]);

        elseif  zeta_type == 1          % peak gain

                gain_h0 = abs([force(1)*xn(8,1)/frad(1) ...
                        force(2:num_modes_total).*xn(8,2:num_modes_total) ...
                        ./((2*zeta_unsort(2:num_modes_total))'.*omega2(2:num_modes_total))]);

                gain_h1 = abs([force(1)*xn(7,1)/frad(1) ...
                        force(2:num_modes_total).*xn(7,2:num_modes_total) ...
                        ./((2*zeta_unsort(2:num_modes_total))'.*omega2(2:num_modes_total))]);

        end

%       sort gains, keeping track of original and new indices so can rearrange
%       eigenvalues and eigenvectors

        [gain_h0_sort,index_h0_sort] = sort(gain_h0);

        [gain_h1_sort,index_h1_sort] = sort(gain_h1);

        gain_h0_sort = fliplr(gain_h0_sort);                    % max to min

        gain_h1_sort = fliplr(gain_h1_sort);                    % max to min

        index_h0_sort = fliplr(index_h0_sort)                   % max to min indices

        index_h1_sort = fliplr(index_h1_sort)                   % max to min indices

        index_orig = 1:num_modes_total;

        if  head == 0

                index_sort = index_h0_sort;

                headstr = 'head 0';
```

```
                    index_out = 2;

        elseif head == 1

                    index_sort = index_h1_sort;

                    headstr = 'head 1';

                    index_out = 1;

        end

%       plot results

        semilogy(index_orig(2:num_modes_total),freqvec(2:num_modes_total),'k-');
        title('frequency versus mode number')
        xlabel('mode number')
        ylabel('frequency, hz')
        grid off
        disp('execution paused to display figure, "enter" to continue'); pause

        semilogy(index_orig,gain_h0,'k-',index_orig,gain_h1,'k.-')
        title('dc value of each mode contribution versus mode number')
        xlabel('mode number')
        ylabel('dc value')
        legend('head 0','head 1')
        grid off
        disp('execution paused to display figure, "enter" to continue'); pause

        loglog(freqvec(2:num_modes_total),gain_h0(2:num_modes_total),'k-', ...
                    freqvec(2:num_modes_total),gain_h1(2:num_modes_total),'k.-')
        title('dc value of each mode contribution versus frequency')
        xlabel('frequency, hz')
        ylabel('dc value')
        legend('head 0','head 1')
        axis([500 25000 -inf 1e-4])
        grid off
        disp('execution paused to display figure, "enter" to continue'); pause

        semilogy(index_orig,gain_h0_sort,'k-',index_orig,gain_h1_sort,'k.-')
        title('sorted dc value of each mode versus number of modes included')
        xlabel('modes included')
        ylabel('sorted dc value')
        legend('head 0','head 1')
        grid off

%       choose number of modes to use based on ranking of dc gain values

        num_modes_used = input(['enter how many modes (including rigid body) ...
                to include, 'num2str(num_modes_total),' max, 8 default ... ']);

        if (isempty(num_modes_used))
                    num_modes_used = 8;
        end
```

```
num_states_used = 2*num_modes_used;
```

17.7.4 Ranking Results

Here, we will begin by reviewing the frequency versus mode number plot to get a feel for the frequency range of the model.

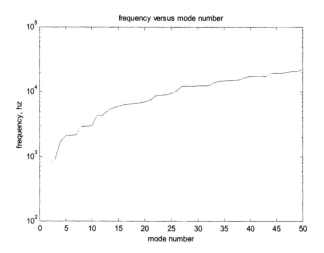

Figure 17.27: Frequency versus mode number.

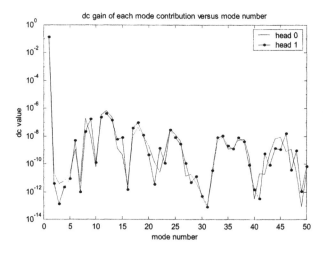

Figure 17.28: dc gain versus mode number, uniform damping zeta 0.005 (0.5% of critical damping) for all modes.

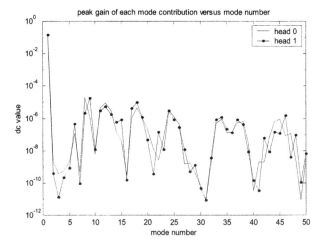

Figure 17.29: Peak gain versus mode number, non-uniform damping, zeta = 0.04 (4% of critical damping) for modes 11, 12 and 13.

The dc and peak gain plots for both head 0 and head 1 are shown above. Note the relative heights of the dc and peak gains for modes 11, 12 and 13. In the peak gain plot, those three gains are lower than the two gains immediately to the left. Conversely, in the dc gain plot the three modes are the highest gains with the exception of the rigid body mode.

The same two plots versus frequency, instead of mode number:

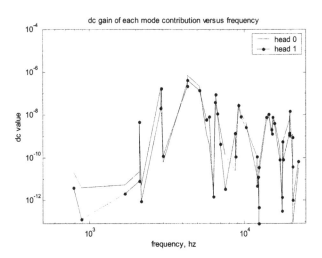

Figure 17.30: dc gain versus frequency.

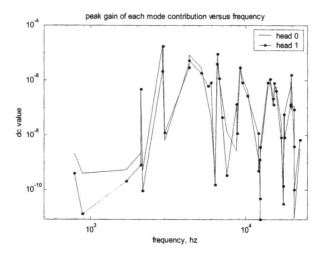

Figure 17.31: Peak gain versus frequency.

The gain plots versus mode number include the rigid body mode low frequency gain, while the gain plots versus frequency do not include the rigid body mode.

Figure 17.32 shows the modes ranked from most to least significant for the uniform damping (dc gain) case and includes the low frequency (500 hz) dc gain of the rigid body mode.

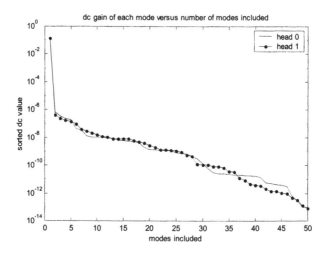

Figure 17.32: Sorted dc gain versus number of modes included.

Relative to the 500 hz low frequency gain of the rigid body mode, the next most significant mode is lower by almost six orders of magnitude. Note that both head 0 and head 1 have similar magnitude curves, although the ordering of individual ranked modes are different. Furthermore, after the drop in dc gain from the rigid body mode to the second mode, there are no other significant drops. Gain is changing gradually, so there is no clear demarcation indicating the number of modes needed to be included. Picking the number of modes to use will be quite subjective, with each additional mode improving the model only slightly.

17.7.5 Building State Space Matrices

To prepare for building the system matrices, two sets of eigenvalue vectors and eigenvector matrices are defined. The first set is the original, unsorted eigenvalues and eigenvectors. The second set consists of the rearranged eigenvalues, eigenvectors and the damping vector, sorted by dc or peak gain. Using the same techniques defined in earlier chapters, the a, b and c matrices are formed.

```
%       define eigenvalues and eigenvectors for unsorted and sorted modes

%       all modes included model, use original order

        xnnew = xn(:,(1:num_modes_total));

        freqnew = freqvec((1:num_modes_total));

        zeta = zeta_unsort;

%       all modes included, sorted

        xnnew_sort = xn(:,index_sort(1:num_modes_total));

        freqnew_sort = freqvec(index_sort(1:num_modes_total));

        zeta_sort = zeta_unsort(index_sort(1:num_modes_total));

%       define variables for all modes included system matrix, a

        w = freqnew*2*pi;              %  frequencies in rad/sec

        w2 = w.^2;

        zw = 2*zeta_unsort.*w;

%       define variables for all modes included sorted system matrix, a_sort

        w_sort = freqnew_sort*2*pi;    %  frequencies in rad/sec
```

```
        w2_sort = w_sort.^2;

        zw_sort = 2*zeta_sort.*w_sort;

%       define size of system matrix

        asize = 2*num_modes_total;

        disp(' ');
        disp(' ');
        disp(['size of system matrix a is ',num2str(asize)]);

%       setup system matrix for all modes included model

        a = zeros(asize);

        for  col = 2:2:asize

        row = col-1;

        a(row,col) = 1;

        end

        for  col = 1:2:asize

        row = col+1;

        a(row,col) = -w2((col+1)/2);

        end

        for  col = 2:2:asize

        row = col;

        a(row,col) = -zw(col/2);

        end

%       setup system matrix for sorted all modes included model

        a_sort = zeros(asize);

        for  col = 2:2:asize

        row = col-1;

        a_sort(row,col) = 1;

        end

        for  col = 1:2:asize
```

```
        row = col+1;

        a_sort(row,col) = -w2_sort((col+1)/2);

        end

        for  col = 2:2:asize

        row = col;

        a_sort(row,col) = -zw_sort(col/2);

        end

%       setup input matrix b, state space forcing function in principal coordinates

%       now setup the principal force vector for the three cases, all modes, sort

%       f_principal is the vector of forces in principal coordinates

        f_principal = xnnew'*f_physical;

%       b is the vector of forces in principal coordinates, state space form

        b = zeros(2*num_modes_total,1);

        for cnt = 1:num_modes_total

               b(2*cnt) = f_principal(cnt);

        end

%       f_principal_sort is the vector of forces in principal coordinates

        f_principal_sort = xnnew_sort'*f_physical;

%       b_sort is the vector of forces in principal coordinates, state space form

        b_sort = zeros(2*num_modes_total,1);

        for  cnt = 1:num_modes_used

               b_sort(2*cnt) = f_principal_sort(cnt);

        end

%       setup cdisp and cvel, padded xn matrices to give the displacement and velocity
%       vectors in physical coordinates
%       cdisp and cvel each have numdof rows and alternating columns
%       consisting of columns  of xnnew and zeros to give total columns equal
%       to the number of states

%       all modes included cdisp and cvel

        for  col = 1:2:2*length(freqnew)
```

```
                        for  row = 1:numdof

                                c_disp(row,col) = xnnew(row,ceil(col/2));

                                cvel(row,col) = 0;

                end

        end

        for  col = 2:2:2*length(freqnew)

                        for  row = 1:numdof

                                c_disp(row,col) = 0;

                                cvel(row,col) = xnnew(row,col/2);

                end

        end

%       all modes included sorted cdisp and cvel

        for  col = 1:2:2*length(freqnew_sort)

                        for  row = 1:numdof

                                cdisp_sort(row,col) = xnnew_sort(row,ceil(col/2));

                                cvel_sort(row,col) = 0;

                end

        end

        for  col = 2:2:2*length(freqnew_sort)

                        for  row = 1:numdof

                                cdisp_sort(row,col) = 0;

                                cvel_sort(row,col) = xnnew_sort(row,col/2);

                end

        end

%       define output

        d = [0];  %
```

17.7.6 Define State Space Systems, Original and Reduced

Now that the original and sorted state space matrices are available, we can use the "ss" command to define the systems for analysis. The following systems are set up:

1) unsorted model with all modes included

2) sorted model with all modes included

3) sorted, truncated reduced model using the sorted model from 2) above (same as the "modred" "del" option)

4) sorted, "modred" "mdc" option reduction using the sorted model from 2) above

The bode command is used to define magnitude and phase vectors for (1), (3) and (4) above.

In order to see the effects of different servo sample rates on aliasing of high frequency modes, the user is prompted to enter a sample frequency, which defaults to 20 khz. Examples of several sample rates are shown below. A discussion of aliasing is outside the scope of the book but several references are recommended (Franklin 1994 and Franklin 1998).

```
%        define state space systems with the "ss" command, outputs are the
%        two gap displacements

%        define unsorted all modes included system

         sys = ss(a,b,c_disp(7:8,:),d);

%        define sorted all modes included system

         sys_sort = ss(a_sort,b_sort,cdisp_sort(7:8,:),d);

%        define sorted reduced system

         a_sort_red = a_sort(1:num_states_used,1:num_states_used);

         b_sort_red = b_sort(1:num_states_used);

         cdisp_sort_red = cdisp_sort(7:8,1:num_states_used);

         sys_sort_red = ss(a_sort_red,b_sort_red,cdisp_sort_red,d);

%        define modred "mdc" reduced system, modred "del" option same as sorted reduced
                above
```

```
            states_del = (2*num_modes_used+1):2*num_modes_total;

            sys_mdc = modred(sys_sort,states_del,'mdc');

            sys_mdc_nosort = modred(sys,[17:100],'mdc');
%           use "bode" command to generate magnitude/phase vectors

            [mag,phs] = bode(sys,frad);

            [mag_sort_red,phs_sort_red] = bode(sys_sort_red,frad);

            [mag_mdc,phs_mdc]=bode(sys_mdc,frad) ;

            [mag_mdc_nosort,phs_mdc_nosort]=bode(sys_mdc_nosort,frad) ;
%           convert magnitude to db

            magdb = 20*log10(mag);

            mag_sort_reddb = 20*log10(mag_sort_red);

            mag_mdcdb = 20*log10(mag_mdc);
%           check on discretized system aliasing

            sample_freq = input('enter sample frequency, khz, default 20 khz ... ');

            if isempty(sample_freq)

                    sample_freq = 20;

            end

            nyquist_freq = sample_freq/2;

            disp(['Nyquist frequency is ',num2str(nyquist_freq),' khz']);

            ts = 1/(1000*sample_freq);

            freqdlo = 500;

            freqdhi = 1000*nyquist_freq;        % only take frequency response to nyquist_freq

            fdlo=log10(freqdlo) ;
            fdhi=log10(freqdhi) ;

            fd=logspace(fdlo,fdhi,400) ;
            fdrad=fd*2*pi ;

            sysd = c2d(sys,ts);

            [magd,phsd] = bode(sysd,fdrad);
```

```
magddb = 20*log10(magd);
```

17.7.7 Plotting of Results

The code section below plots the frequency response for the model including all 50 modes and overlaying the individual mode contributions. The sampled frequency response is also plotted, with an overlay of the original 50-mode model response for comparison.

The two reduced models are then plotted, including the individual mode contributions.

The workspace in saved in **act8_data.mat** for use in the **balreal.m** code in Chapter 18.

```
%        start plotting

%        plot all modes included response

         loglog(f,mag(index_out,:),'k.-')
         title([headstr ', gap displacement, all ',num2str(num_modes_total),' modes included'])
         xlabel('Frequency, hz')
         ylabel('Magnitude, mm')
         axis([500 25000 -inf 1e-4])
         grid off
         disp('execution paused to display figure, "enter" to continue'); pause

         hold on

         max_modes_plot = num_modes_total;

         for  pcnt = 1:max_modes_plot

                  index = 2*pcnt;

                  amode = a(index-1:index,index-1:index);

                  bmode = b(index-1:index);

                  cmode = c_disp(7:8,index-1:index);

                  dmode = [0];

                  sys_mode = ss(amode,bmode,cmode,dmode);

                  [mag_mode,phs_mode]=bode(sys_mode,frad)  ;

                  mag_modedb = 20*log10(mag_mode);
```

```
                    loglog(f,mag_mode(index_out,:),'k-')

        end

        axis([500 25000 -inf 1e-4])

        disp('execution paused to display figure, "enter" to continue'); pause

        hold off

        loglog(f,mag(index_out,:),'k-',fd,magd(index_out,:),'k.-')
        title([headstr ', gap displacement, all ',num2str(num_modes_total), ...
                ' modes included, Nyquist frequency ',num2str(nyquist_freq),' hz'])
        xlabel('Frequency, hz')
        ylabel('Magnitude, mm')
        legend('continuous','discrete')
        axis([500 25000 1e-8 1e-4])
        grid off

        disp('execution paused to display figure, "enter" to continue'); pause

        if  num_modes_used < num_modes_total     % calculate and plot reduced models

%        sorted modal truncation

        loglog(f,mag(index_out,:),'k-',f,mag_sort_red(index_out,:),'k.-')
        title([headstr ', sorted modal truncation:  gap displacement, first ', ...
                num2str(num_modes_used),' modes included'])
        legend('all modes','sorted partial modes',3)
        xlabel('Frequency, hz')
        ylabel('Magnitude, mm')
        axis([500 25000 1e-8 1e-4])
        grid off

        disp('execution paused to display figure, "enter" to continue'); pause

        hold on

        for  pcnt = 1:max_modes_plot

                index = 2*pcnt;

                amode = a_sort(index-1:index,index-1:index);

                bmode = b_sort(index-1:index);

                cmode = cdisp_sort(7:8,index-1:index);

                dmode = [0];

                sys_mode = ss(amode,bmode,cmode,dmode);

                [mag_mode,phs_mode]=bode(sys_mode,frad)  ;
```

```
                     loglog(f,mag_mode(index_out,:),'k-')

        end

        axis([500 25000 -inf 1e-4])

        disp('execution paused to display figure, "enter" to continue'); pause

        hold off

%       modred using 'mdc'

        loglog(f,mag(index_out,:),'k-',f,mag_mdc(index_out,:),'k.-')
        title([headstr ', reduced matched dc gain:  gap displacement, first ', ...
                   num2str(num_modes_used),' sorted modes included'])
        legend('all modes','reduced mdc',3)
        xlabel('Frequency, hz')
        ylabel('Magnitude, mm')
        axis([500 25000 1e-8 1e-4])
        grid off

        disp('execution paused to display figure, "enter" to continue'); pause

        hold on

        for  pcnt = 1:max_modes_plot

                 index = 2*pcnt;

                 amode = a_sort(index-1:index,index-1:index);

                 bmode = b_sort(index-1:index);

                 cmode = cdisp_sort(7:8,index-1:index);

                 dmode = [0];

                 sys_mode = ss(amode,bmode,cmode,dmode);

                 [mag_mode,phs_mode]=bode(sys_mode,frad)  ;

                 loglog(f,mag_mode(index_out,:),'k-')

        end

        axis([500 25000 -inf 1e-4])

        disp('execution paused to display figure, "enter" to continue'); pause

        hold off

%       modred using 'mdc' with unsorted modes

        loglog(f,mag(index_out,:),'k-',f,mag_mdc_nosort(index_out,:),'k.-')
        title([headstr ', reduced unsorted matched dc gain:  gap displacement, first ', ...
```

```
                        num2str(num_modes_used),' sorted modes included'])
        legend('all modes','reduced mdc',3)
        xlabel('Frequency, hz')
        ylabel('Magnitude, mm')
        axis([500 25000 1e-8 1e-4])
        grid off

        disp('execution paused to display figure, "enter" to continue'); pause

        hold on

        for  pcnt = 1:num_modes_used

                index = 2*pcnt;

                amode = a(index-1:index,index-1:index);

                bmode = b(index-1:index);

                cmode = c_disp(7:8,index-1:index);

                dmode = [0];

                sys_mode = ss(amode,bmode,cmode,dmode);

                [mag_mode,phs_mode]=bode(sys_mode,frad)  ;

                loglog(f,mag_mode(index_out,:),'k-')

        end

        axis([500 25000 -inf 1e-4])

        disp('execution paused to display figure, "enter" to continue'); pause

        hold off

        end

%       save the workspace for use in balred.m

        save act8_data
```

Plots using the code above are discussed in the following sections.

17.8 Uniform and Non-Uniform Damping Comparison

The four figures below show a comparison between the uniform and non-uniform damping cases. The first two depict uniform damping, while the second two show non-uniform damping, with higher damping for modes 11, 12 and 13.

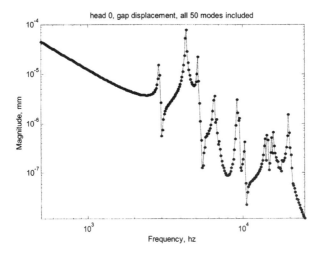

Figure 17.33: Head 0 frequency response, all 50 modes included, uniform damping with zeta = 0.005.

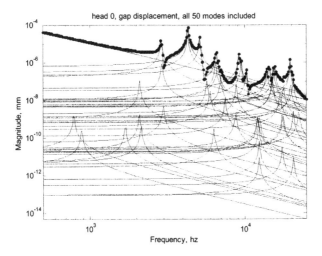

Figure 17.34: Head 0 frequency response, overlay of individual mode contributions, 50 modes included, uniform damping with zeta = 0.005.

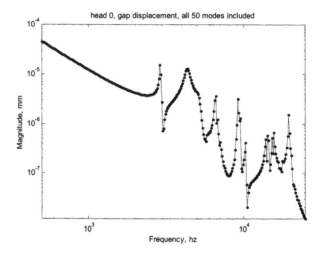

Figure 17.35: Head 0 frequency response, all 50 modes included, non-uniform damping with zeta = 0.005 for all modes except modes 11, 12 and 13, which have zeta = 0.04.

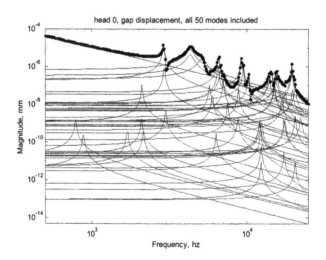

Figure 17.36: Head 0 frequency response, overlay of individual mode contributions, 50 modes included, non-uniform damping with zeta = 0.005 for all modes except modes 11, 12 and 13, which have zeta = 0.04.

Note the lower gain of the three modes in the 4 to 5.5 khz range for the non-uniform damping case.

17.9 Sample Rate and Aliasing Effects

In the two figures below we can see the effects of aliasing for two different servo system sample rates.

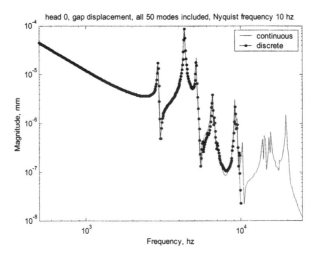

Figure 17.37: Discrete system frequency response overlaid on continuous system, sample rate 20 khz, Nyquist frequency 10 khz.

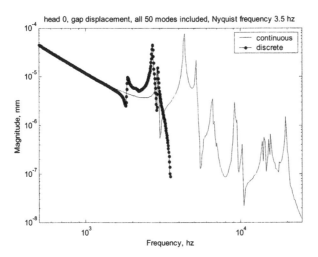

Figure 17.38: Discrete system frequency response overlaid on continuous system, sample rate 7 khz, Nyquist frequency 3.5 khz, showing aliasing effects.

The discrete system frequency response in Figure 17.37, which has a sample frequency of 20 khz, shows only small differences from the original continuous system response.   The discrete system response stops at the Nyquist frequency, 10 khz.

Unlike Figure 17.37, Figure 17.38, which has a much lower sample rate of 7 khz, shows a significant difference from the original continuous system. If one uses the sampled system to experimentally measure the frequency response, it can only measure the response in the 0-Nyquist frequency range. If the discrete system shown in Figure 17.33 were measured, there would be no way to know that the peak at 2.68 khz is not an actual mechanical resonance at 2.68 khz but is the system mode at 4.32 khz which is aliased. As mentioned earlier, only a measurement using a separate system, such as a laser measurement system, will reveal the actual mechanical system response.

17.10 Reduced Truncation and Matched dc Gain Results

This section compares sorted reduced truncation and sorted match dc gain (mdc) methods, both using eight modes.

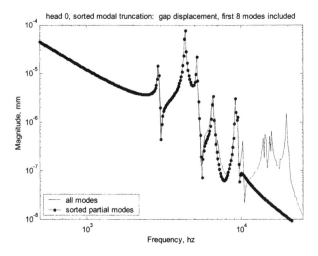

Figure 17.39: Reduced sorted modal truncation frequency response, eight modes included.

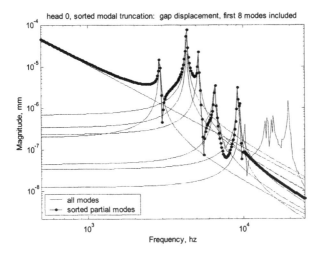

Figure 17.40: Reduced sorted modal truncation frequency response, eight modes included, showing overlay of eight individual modes.

The reduced sorted truncated system shown in Figures 17.37 and 17.38 matches the original 50-mode system frequency response quite well in the 0 to 10 khz range, but misses four modes between 10 and 20 khz.

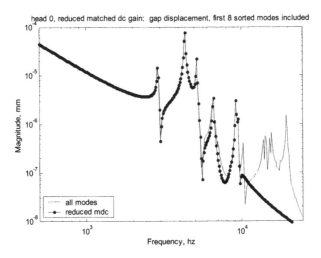

Figure 17.41: Reduced "modred" matched dc gain frequency response, eight modes included.

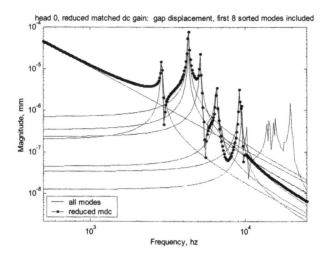

Figure 17.42: Reduced "modred" matched dc gain frequency response, eight modes included, showing overlay of eight individual modes.

The reduced "modred" matched dc (mdc) gain frequency response is virtually identical to the reduced sorted modal truncation response because the modes were sorted prior to using the matched method and the modes which were eliminated have low dc gain relative to the rigid body gain. Also, since the eliminated modes have such a small contribution to the overall response, the "flat" high frequency portion of the curve (highlighted in Figures 15.15 and 16.17) is not seen. To be sure that this was the case, the "modred" matched dc gain reduction was run on the system with unsorted modes, using the first eight modes. The results are shown below and show that the "flat" high frequency portion of the frequency response has returned.

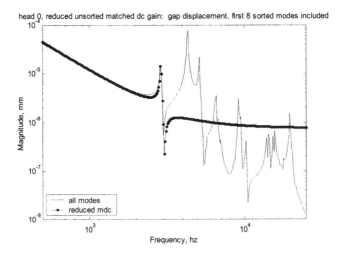

Figure 17.43: Unsorted Reduced "modred" matched dc gain frequency response, first eight unsorted modes included.

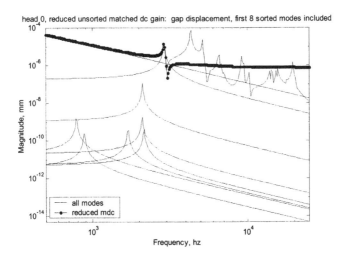

Figure 17.43: Unsorted Reduced "modred" matched dc gain frequency response, first eight unsorted modes included, showing overlay of eight individual modes.

Only eight modes were used for the reduced frequency responses in this chapter. In Chapter 18 we will compare responses for different number of reduced modes to get a sense for how many modes are required to define the pertinent dynamics.

CHAPTER 18

BALANCED REDUCTION

18.1 Introduction

In this chapter another method of reducing models, "balanced reduction," will be introduced. We will compare it with the dc and peak gain ranking methods using the disk drive actuator/suspension model from the last chapter.

We have developed a strong mental picture of ranking individual modes using dc and peak gains. Furthermore, we have developed the ranking method intuitively by graphically showing how the individual modes combine to create the overall frequency response.

The concepts of controllability and observability, commonly referenced in the control community, can be used to rank modes but there is some ambiguity involved. In general, the controllability of a given mode is not related to its observability, and vice versa. The balanced reduction technique simultaneously takes into account both controllability and observability in its ranking and overcomes the uncertainty involved in using either controllability or observability alone.

We will see that for the SISO actuator model introduced in the previous chapter the balanced method provides slightly better impulse response results than the dc gain method, for models with the same number of retained modes/states. For frequency response, the balanced method fits one additional mode over that of the dc gain method, in cases where the same number of reduced modes are used for both methods.

One issue with balanced reduction is that we lose the ability to directly identify individual modes in the reduced system model. After balanced reduction one needs to examine the system matrix to identify which modes are included, while the dc and peak gain ranking techniques retain the identities of the individual modes.

Unlike SISO models, which can be easily ranked using simple dc and peak gain techniques, MIMO models will require the balanced reduction method because it easily handles the problem of ranking multiple inputs and outputs. In the next chapter we will examine a MIMO example, a disk drive actuator with a second stage of actuation in addition to the voice coil motor.

Gawronski [1996, 1998] are two excellent advanced level texts that cover balanced reduction and balanced control of structures for those interested in examining the subject more deeply.

18.2 Reviewing dc Gain Ranking, MATLAB Code balred.m

So far we have used dc or peak gains of the individual modes to rank the importance of including each mode in the reduced system. Repeating (17.1) and (17.2), the dc gain and peak gain expressions:

$$\frac{z_{ji}}{F_{ki}} = \frac{z_{nji}z_{nki}}{\omega_i^2} , \tag{18.1}$$

$$\frac{z_{ji}}{F_{ki}} = \frac{-j}{2\zeta_i} (dc\ gain) , \tag{18.2}$$

where $z_{nji}z_{nki}$ is the product of the jth (output) row and kth (force applied) row terms of the ith eigenvector divided by the square of the eigenvalue for the ith mode.

For any mode, if the degree of freedom associated with the applied force has a zero value, then the force applied at that degree of freedom cannot excite that mode, so the dc and peak gains will also be zero. If the mode cannot be excited, then it has no effect on the frequency response and can be eliminated. Similarly, if the degree of freedom associated with the output has a zero value, then no matter how much force is applied to that mode, there will be no output. The dc and peak gains are zero, and the mode can be eliminated because it also will have no effect on the frequency response.

Loosely speaking, a mode which cannot be excited by the applied force is uncontrollable and a mode which has no output in the desired direction is unobservable. Conversely, modes which have "large" values for the forcing function degree of freedom are said to be "controllable" and modes with "large" values for the output degree of freedom are said to be "observable."

The code below, the input section from **balred.m**, reads in the stored output from **act8.m** (Chapter 17), stored in **act8_data.mat**. It then calculates and plots the input and output contributors to the dc gain, z_{nki}/ω_i and z_{nji}/ω_i and the resulting dc gain. This is the first time we have separated the input and output contributors to the dc gain term; in the past we have dealt only with the dc gain itself. The reason we are highlighting the two contributors is to bridge to understanding of the new concepts of controllability and observability.

```
%        balred.m   balanced modred reduction of actuator/suspension model

         clear all;

         hold off;

         clf;

         load act8_data;

%        plot dc gain and two contibutors, force and xn, versus mode

         index_states = 1:num_modes_total-1;

         omega1 = 2*pi*freqvec';    % convert to radians

         semilogy(index_orig(2:num_modes_total)-1,gain_h0(2:num_modes_total),'k.-', ...
                  index_orig(2:num_modes_total)-1,abs(force(2:num_modes_total)./ ...
                  omega1(2:num_modes_total)),'k-', ...
                  index_orig(2:num_modes_total)-1, ...
                  abs(xn(8,2:num_modes_total)./omega1(2:num_modes_total)),'ko-')
         title([headstr ' dc gain, force and xn values versus mode number'])
         xlabel('mode number')
         ylabel('dc value')
         legend('dc gain','force','xn',3)
         grid off

         disp('execution paused to display figure, "enter" to continue'); pause
```

Figure 18.1 shows the force and output (xn) components which when multiplied create the dc gain for each mode.

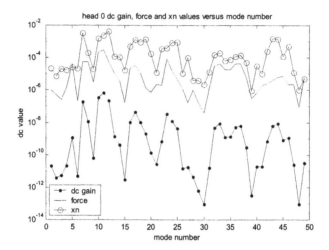

Figure 18.1: Force, output and dc gain for each mode.

It is evident from the curves of force and xn in Figure 18.1 that none of the modes has values for the input or output that go to zero, but that there is a three to four order of magnitude span for both the force and xn values. This three to four order of magnitude span for the force and xn vectors, when multiplied, results in an approximate seven order of magnitude span for the dc gain. We have used this span in dc gain values in previous chapters to rank the relative importance of modes, identifying modes for elimination.

18.3 Controllability, Observability

The intuitive descriptions of controllability and observability given above can be stated precisely using standard state space notation. See Chen [1999], Zhou [1996, 1998], Kailath [1980] and Bay [1999] for derivations and more detail.

For a state space system described by

$$\dot{x} = Ax + Bu$$
$$y = Cx$$

(18.3a,b)

the following definitions of controllability hold:

1) If there is an input "u" that can move the system from some arbitrary state x_1 to another arbitrary state x_2 in a finite time then the system is controllable.

2) A controllability matrix C can be formed as:

$$C = [\mathbf{B} \ \ \mathbf{AB} \ \ \mathbf{A^2B} \ ... \ \mathbf{A^{n-1}B}] \quad\quad\quad (18.4)$$

If C has full (row) rank n, the system is controllable. The controllability matrix gives no insight into the relative controllability of the different modes, it shows only whether the entire system is controllable or not. If one mode of the system is not controllable, the system is not controllable.

3) Another definition of controllability involves the controllability gramian, $\mathbf{W_c}$, the solution to the Lyapunov equation:

$$\mathbf{AW_c} + \mathbf{W_c A^T} + \mathbf{BB^T} = 0 \quad\quad\quad (18.5)$$

defined as:

$$\mathbf{W_c} = \int_0^\infty e^{\mathbf{A}\tau} \mathbf{BB^T} e^{\mathbf{A^T}\tau} \, d\tau \quad\quad\quad (18.6)$$

If the solution $\mathbf{W_c}(t)$ is non-singular (determinant is non-zero), then the system is controllable.

Diagonal elements of the controllability gramian give information about the relative controllability of the different modes and can be used in a manner similar to our use of dc gains to rank the relative controllability of individual modes.

Gramians exists only for systems that have all their poles strictly to the left of the "$j\omega$" axis. The actuator/suspension system we are analyzing has two rigid body mode poles at the origin, so we will have to analyze only the oscillatory portion of the system. We will do this by partitioning the modal form state matrices into the rigid body mode and the non-rigid body oscillatory modes. Then the definitions of controllability will be applied to only the oscillatory partition.

A similar set of definitions can be made for observability:

1) If the initial state x_0 of a system can be inferred from knowledge of the input u and the output y over a finite time $(0,t)$ then the system is said to be observable.

2) An observability matrix O can be formed as:

$$O = \begin{bmatrix} C \\ CA \\ ... \\ CA^{n-1} \end{bmatrix}$$

(18.7)

If O has full (column) rank n, the system is observable.

3)   Another definition of observability involves the observability gramian, W_0, the solution to the Lyapunov equation:

$$A^T W_0 + W_0 A + C^T C = 0$$

(18.8)

defined as:

$$W_0 = \int_0^\infty e^{A^T \tau} C^T C e^{A \tau} \, d\tau$$

(18.9)

If the solution $W_0(t)$ is non-singular (determinant is non-zero) then the system is observable.

The diagonal elements of the observability gramian give information about the relative observability of the different modes and can be used in a manner similar to using dc gains to rank the relative observability of modes.

Because we know the form of the **A**, **B** and **C** matrices for the state space modal form, we are able to substitute those matrices into the Lyapunov equations above and derive closed form controllability and observability gramians (Gawronski 1998). It is interesting to see how the closed form gramian expressions compare with the force and xn components of the dc and peak gains. We saw earlier that the dc gain can be looked at as a product of a "force" and an "output," xn.

$$\frac{z_{ji}}{F_{ki}} = \frac{z_{nji}z_{nki}}{\omega_i^2} = \left(\frac{z_{nji}}{\omega_i}\right)\left(\frac{z_{nki}}{\omega_i}\right) = (\text{output})(\text{force}),\qquad(18.10)$$

Similarly for the peak gain at resonance:

$$\left|\frac{z_{ji}}{F_{ki}}\right| = \frac{(\text{dc gain})}{2\zeta_i} = \frac{1}{2\zeta_i}\left(\frac{z_{nji}}{\omega_i}\right)\left(\frac{z_{nki}}{\omega_i}\right) = \left(\frac{z_{nji}}{\sqrt{2\zeta_i}\,\omega_i}\right)\left(\frac{z_{nki}}{\sqrt{2\zeta_i}\,\omega_i}\right)\quad(18.11)$$

Gawronski shows that the closed loop expression for the largest diagonal term in the 2x2 controllability gramian for mode "i" is given by:

$$w_{ci} = \frac{\|B_i\|_2^2}{4\zeta_i\omega_i},\qquad(18.12)$$

where the $\|\ \|_2$ notation represents the Euclidean norm, the square root of the sum of the squares of the elements of a vector.

The largest diagonal term in the 2x2 observability gramian for mode "i" is given by:

$$w_{oi} = \frac{\|C_i\|_2^2}{4\zeta_i\omega_i}\qquad(18.13)$$

The smaller of the two diagonal terms for both the controllability and observability gramians is derived from the larger term by dividing by the square of the eigenvalue for that respective mode.

The **B** and **C** matrices for mode "i" with input at dof "k" and displacement output at dof "j" are as follows:

$$B_i = \begin{bmatrix} 0 \\ F_k z_{nki} \end{bmatrix}\qquad(18.14)$$

$$C_i = \begin{bmatrix} z_{nji} & 0 \end{bmatrix}\qquad(18.15)$$

Substituting into the two equations above for the closed loop gramians:

$$W_{ci} = \frac{\|\mathbf{B}_i\|_2^2}{4\zeta_i\omega_i} = \frac{\left\|\begin{bmatrix} 0 \\ F_k z_{nki} \end{bmatrix}\right\|_2^2}{4\zeta_i\omega_i} = \frac{F_k^2 z_{nki}^2}{4\zeta_i\omega_i} \tag{18.16}$$

$$W_{oi} = \frac{\|\mathbf{C}_i\|_2^2}{4\zeta_i\omega_i} = \frac{\left\|\begin{bmatrix} z_{nji} & 0 \end{bmatrix}\right\|_2^2}{4\zeta_i\omega_i} = \frac{z_{nji}^2}{4\zeta_i\omega_i} \tag{18.17}$$

Comparing the peak gain terms and the gramian terms:

Force component of dc gain: $\dfrac{z_{nki}}{\sqrt{2\zeta_i\omega_i}}$ (18.18)

Controllability diagonal: $\dfrac{z_{nki}^2}{4\zeta_i\omega_i}$ (18.19)

Output component of dc gain: $\dfrac{z_{nji}}{\sqrt{2\zeta_i\omega_i}}$ (18.20)

Observability diagonal: $\dfrac{z_{nji}^2}{4\zeta_i\omega_i}$ (18.21)

When we have ranked using peak gains, we have used the expression:

$$\text{peak gain} = \frac{z_{nji}\, z_{nki}}{2\zeta_i\omega_i^2} \tag{18.22}$$

If we had used the controllability and observability gramian terms for each mode to rank, we would have ranked based on

$$\frac{z_{nki}^2\, z_{nji}^2}{16\zeta_i^2\omega_i^2} \tag{18.23}$$

In the controllability and observability gramian ranking of modes, we deal with the product of the squares of the eigenvector components while peak gain uses the product without squaring. Both rankings divide by the square of the eigenvalue and there is a difference in the two multipliers "2" and "16" as well as the squaring of the damping term.

18.4 Controllability, Observability Gramians

The following code section starts by defining a system which consists of the oscillatory modes of the system, excluding the first, rigid body mode. As mentioned above, gramians exist only for strictly stable systems, where all the poles strictly to the left of the " jω " axis. The two rigid body poles at the origin need to be eliminated from the system to be able to calculate gramians. In the modal form of the equations, where the modes are uncoupled, we can partition the system into rigid body and oscillatory modes. We can then calculate a reduced oscillatory system based on reducing the oscillatory modes. The full system is then ready to be re-assembled by augmenting the rigid body mode with the reduced oscillatory modes.

The controllability and observability gramians are calculated, plotted with their amplitudes on the z axis and then the diagonal entries are plotted. The position and velocity state terms are identified in each of the gramians and plotted separately.

```
%         define oscillatory system from unsorted model from act8.m, which only
%         has one output, either head 0 or head 1 so that when use balreal, will only
%         be taking into account a siso system, not the outputs of both heads 0 and 1

%         in act8.m, used output matrix with two rows so both head 0 and head 1 were available

          a_syso = a(3:asize,3:asize);        % ao is a for oscillatory system

          b_syso = b(3:asize);

          c_syso = c_disp(index_out+6,3:asize);

          syso = ss(a_syso,b_syso,c_syso,d);

%         define controllability and observability gramians for oscillatory system, syso

          wc = gram(syso,'c');

          wo = gram(syso,'o');

          [row_syso,col_syso] = size(a_syso);

          statevec = 1:row_syso;

%         calculate closed form gramians

%         define frequencies for oscillatory states

          omega1 = 2*pi*freqvec';    %  convert to radians
```

```
           ctr = 0;

           for  cnt = 1:num_modes_total

                   ctr = ctr + 2;

                   omega12(ctr-1) = omega1(cnt);

                   omega12(ctr) = omega1(cnt);

                   zeta_unsort12(ctr-1) = zeta_unsort(cnt);

                   zeta_unsort12(ctr) = zeta_unsort(cnt);

           end

%          the notation below is "wc" or "wo" for controllability or observability gramians,
%          "cf" for closed-form, and "1" or "2" for maximum and minimum values for a mode

           wccf1 = (b_syso.*b_syso)./(4*zeta_unsort12(3:2*num_modes_total)' ...
                    .*omega12(3:2*num_modes_total)');    % maximum terms

           wccf12 = wccf1(2:2:row_syso);            % pick out velocity terms

           wccf2 = (b_syso.*b_syso)./(4*zeta_unsort12(3:2*num_modes_total)' ...
                    .*omega12(3:2*num_modes_total)'.^3);      % minimum terms

           wccf22 = wccf2(2:2:row_syso);            % pick out displacement terms

           wocf1 = (c_syso.*c_syso)./(4*zeta_unsort12(3:2*num_modes_total) ...
                    .*omega12(3:2*num_modes_total));      % maximum terms

           wocf12 = wocf1(1:2:row_syso);            % pick out displacement terms

           wocf2 = (c_syso.*c_syso)./(4*zeta_unsort12(3:2*num_modes_total) ...
                    .*omega12(3:2*num_modes_total).^3);       % minimum terms

           wocf22 = wocf2(1:2:row_syso);            % pick out velocity terms

%          plot controllability and observability gramians

           meshz(wc);
           view(60,30);
           title([headstr ', controllability gramian for oscillatory system'])
           xlabel('state')
           ylabel('state')
           grid on

           disp('execution paused to display figure, "enter" to continue'); pause

           meshz(wo);
           view(60,30);
           title([headstr ', observability gramian for oscillatory system'])
           xlabel('state')
           ylabel('state')
```

```
                grid on

                disp('execution paused to display figure, "enter" to continue'); pause

%               pull out diagonal elements

                wc_diag = diag(wc);

                wo_diag = diag(wo);

                modevec = 2*(1:num_modes_total-1);

%               plot diagonal terms of controllability and observability gramians, calculated with
%               gram function and closed form

                semilogy(statevec,wc_diag,'k.-',statevec(2:2:row_syso),wccf12,'ko', ...
                            statevec(1:2:row_syso),wccf22,'ko')
                title([headstr ', controllability gramian diagonal terms'])
                xlabel('states')
                ylabel('diagonal')
                legend('calculated with gram','closed form',3)
                grid off

                disp('execution paused to display figure, "enter" to continue'); pause

                semilogy(statevec,wo_diag,'k.-',statevec(1:2:row_syso),wocf12,'ko', ...
                            statevec(2:2:row_syso),wocf22,'ko')
                title([headstr ', observability gramian diagonal terms'])
                xlabel('states')
                ylabel('diagonal')
                legend('calculated with gram','closed form',3)
                grid off

                disp('execution paused to display figure, "enter" to continue'); pause

%               position and velocity states plotted separately

                semilogy(statevec(1:2:row_syso),wc_diag(1:2:row_syso),'k.-', ...
                            statevec(2:2:row_syso),wc_diag(2:2:row_syso),'k-', ...
                            statevec(2:2:row_syso),wccf12,'ko', ...
                            statevec(1:2:row_syso),wccf22,'ko')
                title([headstr ', controllability gramian diagonal terms'])
                xlabel('states')
                ylabel('diagonal')
                legend('position states','velocity states','closed form','closed form',3)
                grid off

                disp('execution paused to display figure, "enter" to continue'); pause

                semilogy(statevec(1:2:row_syso),wo_diag(1:2:row_syso),'k.-', ...
                            statevec(2:2:row_syso),wo_diag(2:2:row_syso),'k-', ...
                            statevec(1:2:row_syso),wocf12,'ko', ...
                            statevec(2:2:row_syso),wocf22,'ko')
                title([headstr ', observability gramian diagonal terms'])
                xlabel('states')
```

```
ylabel('diagonal')
legend('position states','velocity states','closed form','closed form',3)
grid off

disp('execution paused to display figure, "enter" to continue'); pause

semilogy(index_states,wc_diag(2:2:row_syso),'k.-', ...
                        index_states,wo_diag(1:2:row_syso),'ko-')
title([headstr ', head 0 controllability and observability state gramians'])
xlabel('mode number')
ylabel('gramian')
legend('controllability velocity state','observability position state',3)
grid off

disp('execution paused to display figure, "enter" to continue'); pause
```

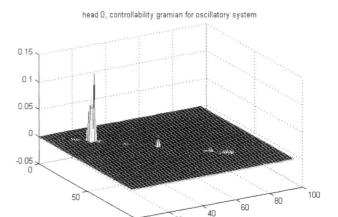

Figure 18.2: Controllability gramian values.

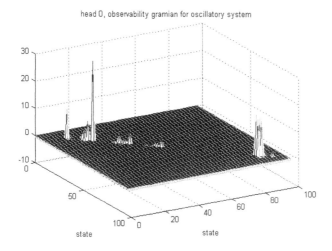

head 0, observability gramian for oscillatory system

Figure 18.3: Observability gramian values.

Figures 18.2 and 18.3 plot the controllability and observability gramian values on a linear z axis scale versus location in the matrix. As noted in Gawronski [1998], for systems described in modal coordinates (with small damping, small ζ values) the gramians are diagonally dominant, meaning that the off diagonal elements are small with respect to the diagonal elements. The largest controllability terms lie along the diagonal in approximately the state 20 to 22 positions, which are the 10th and 11th oscillatory modes. With the rigid body mode included, these become the 11th and 12th modes of the full system, which we identified in the previous chapter as the two system modes in the 4 khz range and identified with the dc gain as the modes with the highest values. Note that there are not any large entries in the higher state numbers for the controllability gramian. The observability gramian plot, however, shows some very high frequency states ($\sim$80 to 100) that have circumferential motion at head 0. Intuitively, the relatively heavy coil is not going to have many modes with circumferential motion at high frequencies, while the stiff, low mass suspension will have a number of high frequency modes with circumferential motion.

The diagonal entries of both gramians are plotted versus state in Figures 18.4 and 18.5, where the odd-numbered states are position states and the even-numbered states are velocity states. Values from the "gram" function and the closed form solution (18.16) (18.17) are shown.

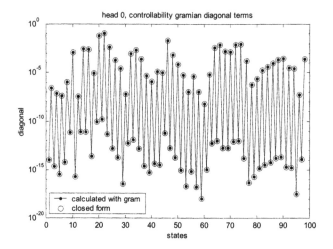

Figure 18.4: Controllability gramian diagonal terms.

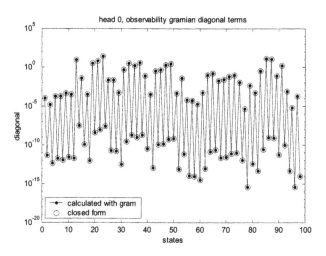

Figure 18.5: Observability gramian diagonal terms.

Figures 18.6 and 18.7 show the position and velocity terms of each gramian diagonal plotted separately. The position state and velocity state curves are offset by the square of the eigenvalue of each mode.

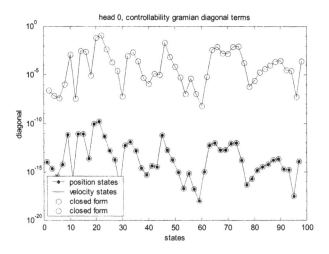

Figure 18.6: Controllability gramian diagonal position and velocity state terms.

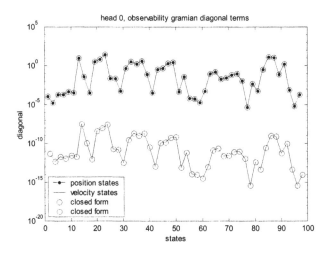

Figure 18.7: Observability gramian diagonal position and velocity state terms.

18.5 Ranking Using Controllability/Observability

Figure 18.8 shows the controllability gramian velocity state and the observability gramian position state (chosen such that the two curves have similar magnitudes for visual comparison). We could use the controllability curve to rank the states for controllability and eliminate those states with low controllability. Alternately, we could use the observability curve to rank the states for observability and then eliminate states with low observability. The problem with this approach is that the joint controllability/observability is not taken into account. There is no problem if a state chosen for elimination has a small controllability value and simultaneously a small observability value. However, if as in modes 43 and 44 (states 85 to 88) in Figure 18.8, the controllability value is small but the observability is relatively high, do we eliminate the mode or not? This is the source of ambiguity in ranking using **only** controllability or **only** observability gramians.

With the dc and peak gain ranking methods referenced earlier we used the product of the input and output (controllability measure and observability measure), jointly taking into account a measure of the controllability and observability of each mode.

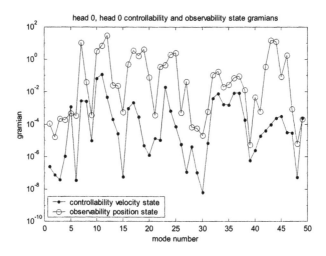

Figure 18.8: Controllability gramian velocity state and observability gramian position state diagonal terms.

18.6 Balanced Reduction

Balanced reduction was introduced in the control community by Moore [1981]. The algorithm used in the MATLAB balancing function "balreal" is taken from Laub [1987].

The algorithm creates a system with identical diagonal controllability and observability gramians. Since the two gramians are equal, **either** the diagonal or controllability gramian can be used to rank states for elimination and the ambiguity of using either only controllability or only observability is removed.

For the system "sys" defined by the following equations:

$$\dot{x} = Ax + Bu$$
$$y = Cx + Du$$
(18.24a,b)

the syntax for the MATLAB "balreal" function is:

$$[sysb,g,T,Ti] = balreal(sys),$$
(18.25)

where "sysb" is the new, balanced system and "g" is the diagonal of the joint gramian. "T" is the transformation matrix that is used to create "sysb." "Ti" is the inverse of "T."

The diagonal terms of the joint gramian, g, are squares of the Hankel singular values of the system. The Hankel matrix is the product of the controllability and observability gramians. Hankel singular values are the squares of the eigenvalues of the Hankel matrix. See Gawronski [1998] for a MATLAB script "bal_op_loop.m" that uses Singular Value Decomposition to calculate the Hankel singular values.

T is the state transformation matrix that is used along with its inverse, T^{-1}, to create "sysb" from "sys" using:

$$\dot{x}_b = TAT^{-1}x_b + TBu$$
$$y = CT^{-1}x_b + Du$$
(18.26a,b)

The gramians are also transformed by T to identical diagonal form:

$$W_{bo} = W_{bc} = diag(g)$$
(18.27)

Because the controllability and observability gramians are identical, there is no ambiguity in deciding whether the most controllable or the most observable states should be chosen. The states to be kept are the states with the largest diagonal terms.

The code below uses "balreal" to calculate the balanced system, "sysob," and plots the resulting gramians.

```
%       use balreal to rank oscillatory states and modred to reduce for comparison

        [sysob,g,T,Ti] = balreal(syso);
```

```
%          define controllability and observability gramians for balanced
%          oscillatory system, sysob

           wcb = gram(sysob,'c');

           wob = gram(sysob,'o');

           wcb_diag = diag(wcb);

           wob_diag = diag(wob);

           modevec = 2*(1:num_modes_total-1);

%          plot balanced controllability and observability gramians

           meshz(wcb);
           view(60,30);
           title([headstr ', oscillatory system balanced controllability gramian'])
           xlabel('state')
           ylabel('state')
           grid on

           disp('execution paused to display figure, "enter" to continue'); pause

           meshz(wob);
           view(60,30);
           title([headstr ', oscillatory system balanced observability gramian'])
           xlabel('state')
           ylabel('state')
           grid on

           disp('execution paused to display figure, "enter" to continue'); pause

%          plot diagonal terms of balanced controllability and observability gramians

           semilogy(statevec,wcb_diag,'k.-',statevec,wob_diag,'ko-')
           title([headstr ', balanced system controllability and observability gramian ...
                          diagonal terms'])
           xlabel('states')
           ylabel('diagonal')
           legend('controllability','observability',3)
           grid off

           disp('execution paused to display figure, "enter" to continue'); pause
```

Figures 18.9 and 18.10 plot terms of the controllability and observability gramian matrices for the balanced system, with the values plotted along the z axis.   Comparing them to the original, unbalanced, controllability and observability gramian plots in Figures 18.2 and 18.3, we see that the balanced plots are identical and strictly diagonal.

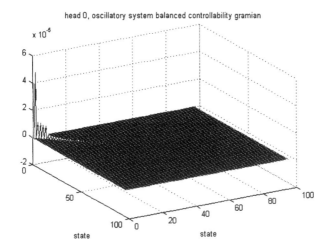

Figure 18.9: Balanced controllability gramian.

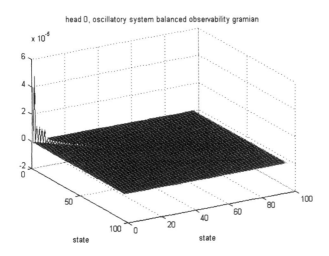

Figure 18.10: Balanced observability gramian.

Plotting diagonal terms of the controllability and observability gramians versus states, Figure 18.11, shows that the two curves overlay one another and that they are ranked from large to small by virtue of the balancing operations.

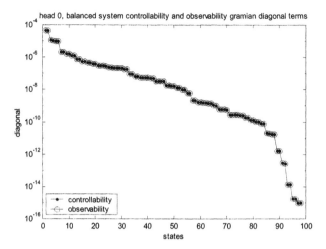

Figure 18.11: Balanced system controllability and observability gramian diagonal terms.

We are now in a position to use the balanced system gramian (either controllability or observability) to decide which states are relatively less important and can be eliminated. Since the states in the balanced system are organized from most to least significant, the MATLAB function "modred" can be used with either the "del" or "mdc" option to eliminate the states with the lowest joint controllability/observability, the higher numbered states in the balanced system.

18.7 Balanced and dc Gain Ranking Frequency Response Comparison

The code in this section starts by plotting the Hankel singular values and the sorted dc gain of the oscillatory modes to see their similarities. The modred function is then used to reduce the system to the number of modes chosen in the last **act8.m** run, using both the "del" and "mdc" options. The complete system is then rebuilt by augmenting the reduced oscillatory system with the rigid body mode. Finally, the code plots frequency responses and compares the results of dc gain ranking from **act8.m** and balanced ranking from **balred.m**.

```
%       plot sorted diagonal values and dc gain

        [row_syso,col_syso] = size(a_syso);

        semilogy(statevec,g,'k.-',2*index_orig((2:num_modes_total)-1), ...
                gain_h0_sort(2:num_modes_total),'k-')
        title([headstr ', sorted diagonal terms of balanced gramian and dc gain'])
        xlabel('state')
```

```
          ylabel('diagonal of gramian')
          legend('balanced','dc gain',3)
          grid off

          disp('execution paused to display figure, "enter" to continue'); pause

          num_oscil_states_used = 2*num_modes_used - 2;

%         use modred to reduce states from balanced system using both "del" and "mdc"

          bsys_delo = modred(sysob,num_oscil_states_used+1:2*num_modes_total-2,'del');

          bsys_mdco = modred(sysob,num_oscil_states_used+1:2*num_modes_total-2,'mdc');

%         rebuild system by appending balanced realization of oscillatory modes to
%                          rigid body mode

          [a_delo_bal,b_delo_bal,c_delo_bal,d_delo_bal] = ssdata(bsys_delo);

          a_del_bal = [   a(1:2,1:2)     zeros(2,num_oscil_states_used)
                          zeros(num_oscil_states_used,2)     a_delo_bal     ];

          b_del_bal = [b(1:2,:)
                          b_delo_bal];

          c_del_bal = [c_disp(index_out+6,1:2) c_delo_bal];

          bsys_del = ss(a_del_bal,b_del_bal,c_del_bal,d);

          [a_mdco_bal,b_mdco_bal,c_mdco_bal,d_mdco_bal] = ssdata(bsys_mdco);

          a_mdc_bal = [   a(1:2,1:2)     zeros(2,num_oscil_states_used)
                          zeros(num_oscil_states_used,2)     a_mdco_bal     ];

          b_mdc_bal = [b(1:2,:)
                          b_mdco_bal];

          c_mdc_bal = [c_disp(index_out+6,1:2) c_mdco_bal];

          bsys_mdc = ss(a_mdc_bal,b_mdc_bal,c_mdc_bal,d);

          [magr_del,phsr_del] = bode(bsys_del,frad);

          [magr_mdc,phsr_mdc] = bode(bsys_mdc,frad);

%         compare frequency responses for all four reduction methods

          loglog(f,mag(index_out,:),'k--',f,mag_sort_red(index_out,:),'k-', ...
                          f,magr_del(1,:),'k.-')
          title([headstr ', results comparison, ',num2str(num_modes_used),' modes, ', ...
                          num2str(num_oscil_states_used),' oscillatory balanced states'])
          xlabel('Frequency, hz')
          ylabel('Magnitude, mm')
          axis([500 25000 1e-8 1e-4])
          legend('all modes','sorted truncated','balreal modred del',3)
```

```
grid off

disp('execution paused to display figure, "enter" to continue'); pause

loglog(f,mag(index_out,:),'k--',f,mag_mdc(index_out,:),'k-',f,magr_mdc(1,:),'k.-')
title([headstr ', results comparison, ',num2str(num_modes_used),' modes, ', ...
              num2str(num_oscil_states_used),' oscillatory balanced states'])
xlabel('Frequency, hz')
ylabel('Magnitude, mm')
axis([500 25000 1e-8 1e-4])
legend('all modes','sorted mdc','balreal modred mdc',3)
grid off

disp('execution paused to display figure, "enter" to continue'); pause
```

Figure 18.12 shows the Hankel singular values and sorted dc gains versus number of states. At this point it is interesting to compare frequency responses for the two ranking techniques to see how each decides which modes/states to eliminate.

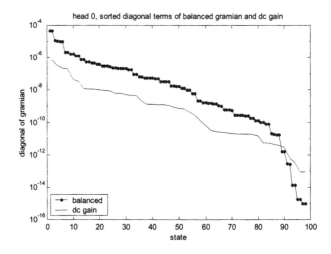

Figure 18.12: Balanced gramian diagonal terms (Hankel singular values) and sorted dc gain.

Figures 18.13 to 18.18 show frequency response plots for different numbers of retained modes, from two to seven modes, including the rigid body mode.

While the code above calculates "sorted truncated" and "balreal modred del" responses, we will only show the following in the figures below:

1) "sorted mdc" – uses dc gain ranking and modred "mdc" to reduce

2) "balreal modred mdc" – uses balreal for ranking and modred "mdc" to reduce

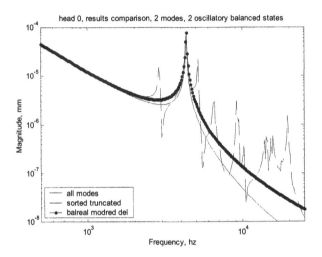

Figure 18.13: Two modes included.

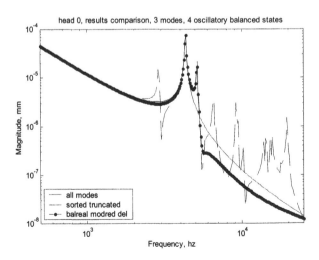

Figure 18.14: Three modes included.

Note that the two ranking methods chose different modes for the three reduced modes. The dc gain method chose the two system modes in the 4.2 khz range (almost coincident) while the balanced method chose one mode at 4.2 khz and another at 5.1 khz.

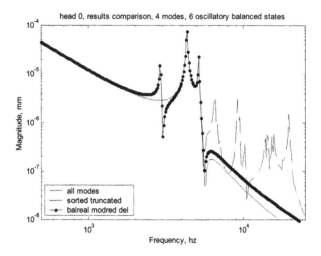

Figure 18.15: Four modes included.

For the four reduced mode case, the dc gain method picked up the 5.1 khz mode, while the balanced method chose the suspension torsion mode at 2.9 khz.

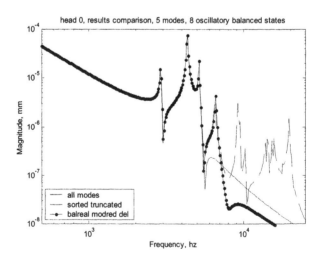

Figure 18.16: Five modes included.

For the five reduced mode case the dc gain method included the torsion mode but missed the mode at 5.5 khz which was picked up by the balanced method.

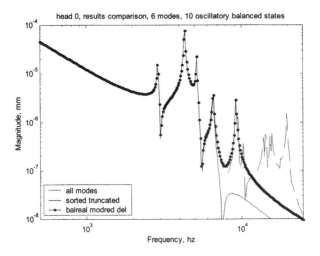

Figure 18.17: Six modes included.

With six reduced modes the balanced method includes the mode at 9 khz, but the dc gain method missed it.

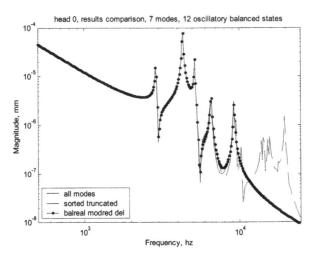

Figure 18.18: Seven modes included.

With seven or higher modes the balanced and dc gain results are very similar. We will see later when analyzing impulse responses of the oscillatory system that the two methods give results which are within a few percent of each other when seven or more modes are included in the reduced model.

18.8 Balanced and dc Gain Ranking Impulse Response Comparison

This section will compare the impulse responses for four different reduced systems, using from 2 through 15 modes. Only the matched dc gain (mdc) methods will be compared as there are minimal differences between the mdc method and the truncation or "del" method of reducing, as can be seen from the eight reduced mode results below.

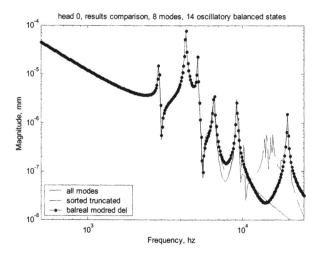

Figure 18.19: Frequency response for eight-mode reduced models, sorted truncated and balreal modred "del."

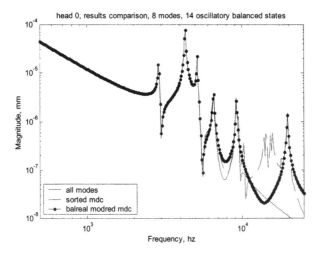

Figure 18.20: Frequency response for 8-mode reduced models, sorted "mdc" and balreal modred "mdc."

In studying the impulse response, we will use only the oscillatory modes. The final model will of course include the rigid body mode, but to study the effects of the various reduced models on transient response it is useful to include only the oscillatory modes.   The reason this is useful is that a typical forcing function applied to a rigid body mode will move the system from one position to another, with rigid body displacements quite large relative to the displacements of the oscillatory modes, creating roundoff errors that mask the oscillatory mode responses.

The code below calculates the impulse response using the "lsim" function for five oscillatory systems, the original "all modes included" system and the four reduced systems. The impulse responses are then plotted and the normalized reduction index, δ (Gawronski 1998), is calculated, where the index is defined as:

$$\delta = \frac{\|\text{disp(all mode model)} - \text{disp(reduced model)}\|}{\|\text{disp(all mode model)}\|} \qquad (18.28)$$

A table of results for δ from earlier runs with different numbers of retained modes is included in the code listing below. Information in the table is also shown graphically in Figures 18.25 and 18.26.

```
%        calculate impulse responses of all four oscillatory systems for comparison

         ttotal = 0.0025;
```

```
        t = linspace(0,ttotal,400);
%       define oscillatory systems for models
%       sorted reduced system
        red_size = 2*num_modes_used;
        [a_sys_sort_red,b_sys_sort_red,c_sys_sort_red,d_sys_sort_red] = ...
                            ssdata(sys_sort_red);
        a_sys_sort_redo = a_sys_sort_red(3:red_size,3:red_size);
        b_sys_sort_redo = b_sys_sort_red(3:red_size);
        c_sys_sort_redo = c_sys_sort_red(index_out,3:red_size);
        sys_sort_redo = ss(a_sys_sort_redo,b_sys_sort_redo,c_sys_sort_redo,d);
%       sorted mdc reduced system
        [a_sys_sort_mdc,b_sys_sort_mdc,c_sys_sort_mdc,d_sys_sort_mdc] = ...
                            ssdata(sys_mdc);
        a_sys_sort_mdc = a_sys_sort_red(3:red_size,3:red_size);
        b_sys_sort_mdc = b_sys_sort_red(3:red_size);
        c_sys_sort_mdc = c_sys_sort_red(index_out,3:red_size);
        sys_mdco = ss(a_sys_sort_mdc,b_sys_sort_mdc,c_sys_sort_mdc,d);
%       use lsim to calculate transient response
        [disp_syso,t_syso] = impulse(syso,t);
        [disp_sys_sort_redo,t_sys_sort_redo] = impulse(sys_sort_redo,t);
        [disp_sys_sort_mdco,t_sys_sort_mdco] = impulse(sys_mdco,t);
        [disp_bsys_delo,t_bsys_delo] = impulse(bsys_delo,t);
        [disp_bsys_mdco,t_bsys_mdco] = impulse(bsys_mdco,t);
%       build matrix of results
        dispo = [disp_syso(:,1) disp_sys_sort_redo(:,1) ...
                    disp_sys_sort_mdco(:,1) disp_bsys_delo(:,1) ...
                    disp_bsys_mdco(:,1)];
        sort_redo_del = dispo(:,1) - dispo(:,2);
        sort_mdco_del = dispo(:,1) - dispo(:,3);
```

```
          delo_del = dispo(:,1) - dispo(:,4);

          mdco_del = dispo(:,1) - dispo(:,5);

%         calculate normalized reduction index

          index_sort_redo = ...
              sqrt(sum(sort_redo_del.*sort_redo_del))/sqrt(sum(dispo(:,1).*dispo(:,1)))

          index_sort_mdco = ...
             sqrt(sum(sort_mdco_del.*sort_mdco_del))/sqrt(sum(dispo(:,1).*dispo(:,1)))

          index_delo = ...
              sqrt(sum(delo_del.*delo_del))/sqrt(sum(dispo(:,1).*dispo(:,1)))

          index_mdco = ...
              sqrt(sum(mdco_del.*mdco_del))/sqrt(sum(dispo(:,1).*dispo(:,1)))


          [num_modes_used index_sort_redo index_sort_mdco index_delo index_mdco]

          plot(t_syso,disp_syso(:,1),'k-',t_sys_sort_redo,disp_sys_sort_redo(:,1),'k.-')
          title([headstr ', displacement vs time, ',num2str(num_modes_used-1), ...
                          ' oscillatory modes'])
          xlabel('time, sec')
          ylabel('displacement, mm')
          legend('all modes','sorted reduced system',4)
          grid off

          disp('execution paused to display figure, "enter" to continue'); pause

          plot(t_syso,disp_syso(:,1),'k-',t_sys_sort_mdco,disp_sys_sort_mdco(:,1),'k.-')
          title([headstr ', displacement vs time, ',num2str(num_modes_used-1), ...
                          ' oscillatory modes'])
          xlabel('time, sec')
          ylabel('displacement, mm')
          legend('all modes','sorted modred mdc',4)
          grid off

          disp('execution paused to display figure, "enter" to continue'); pause

          plot(t_syso,disp_syso(:,1),'k-',t_bsys_delo,disp_bsys_delo(:,1),'k.-')
          title([headstr ', displacement vs time, ',num2str(num_oscil_states_used), ...
                          ' oscillatory balanced states'])
          xlabel('time, sec')
          ylabel('displacement, mm')
          legend('all modes','balreal modred del',4)
          grid off

          disp('execution paused to display figure, "enter" to continue'); pause

          plot(t_syso,disp_syso(:,1),'k-',t_bsys_mdco,disp_bsys_mdco(:,1),'k.-')
          title([headstr ', displacement vs time, ',num2str(num_oscil_states_used), ...
                          ' oscillatory balanced states'])
          xlabel('time, sec')
          ylabel('displacement, mm')
```

```
            legend('all modes','balreal modred mdc',4)
            grid off

            disp('execution paused to display figure, "enter" to continue'); pause

%           plot results of oscillatory impulse response normalized error index versus
%           number of modes used

            error_norm = [     2       .4332     .4332     0.3007     0.3008
                               3       .3041     .3041     0.1777     0.1823
                               4       .1759     .1759     0.1135     0.1137
                               5       .1134     .1134     0.0845     0.0841
                               6       .0851     .0851     0.0598     0.0603
                               7       .0637     .0637     0.0582     0.0583
                               8       .0599     .0599     0.0383     0.0401
                               9       .0594     .0594     0.0343     0.0356
                              10       .0572     .0572     0.0338     0.0347
                              11       .0555     .0555     0.0258     0.0264
                              12       .0392     .0392     0.0280     0.0268
                              13       .0327     .0327     0.0167     0.0168
                              14       .0270     .0270     0.0162     0.0158
                              15       .0209     .0209     0.0162     0.0156];

        nmode = error_norm(:,1);

        error_sort_red = error_norm(:,2);

        error_sort_mdc = error_norm(:,3);

        error_bal_del = error_norm(:,4);

        error_bal_mdc = error_norm(:,5);

        plot(nmode,error_sort_red,'k.-',nmode,error_bal_del,'ko-')
        title([headstr ', normalized reduction index versus number of modes included'])
        xlabel('number of modes included')
        ylabel('normalized reduction index')
        legend('sorted reduced','balanced del')
        axis([0 15 0 0.5])
        grid off

        disp('execution paused to display figure, "enter" to continue'); pause

        plot(nmode,error_sort_mdc,'k.-',nmode,error_bal_mdc,'ko-')
        title([headstr ', normalized reduction index versus number of modes included'])
        xlabel('number of modes included')
        ylabel('normalized reduction index')
        legend('sorted mdc','balanced mdc')
        axis([0 15 0 0.5])
        grid off

        disp('execution paused to display figure, "enter" to continue'); pause

        save balred_data;
```

The impulse response comparisons for the same four reduced methods are shown in the four figures below.

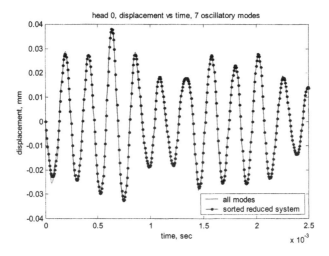

Figure 18.21: Impulse response comparisons for oscillatory system, full model (all oscillatory modes) and sorted reduced system with seven oscillatory modes.

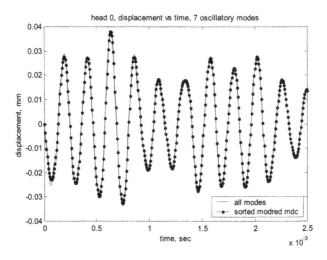

Figure 18.22: Impulse response comparisons for oscillatory system, full model (all oscillatory modes) and sorted modred with "mdc" option with seven oscillatory modes.

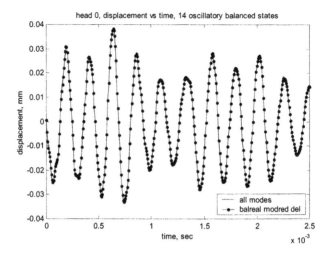

Figure 18.23: Impulse response comparisons for oscillatory system, full model (all oscillatory modes) and balreal modred "del" reduced system with seven oscillatory modes.

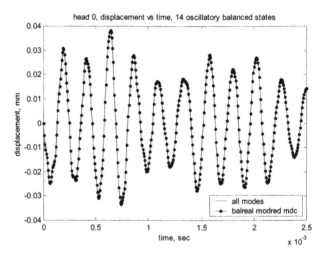

Figure 18.24:   Impulse response comparisons for oscillatory system, full model (all oscillatory modes) and balreal modred "mdc" reduced system with seven oscillatory modes.

The two figures below compare the normalized reduction index, δ, as a function of the number of modes included in the various reduced model methods.

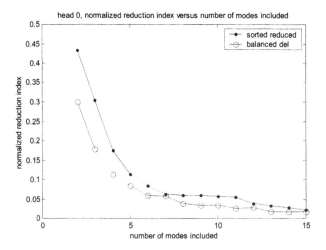

Figure 18.25: Impulse response normalized reduction index versus number of modes included in reduction for sorted reduced and balanced modred "del" option reductions.

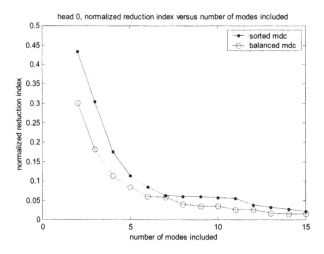

Figure 18.26: Impulse response normalized reduction index versus number of modes included in reduction for sorted modred "mdc" and balanced modred "mdc" options reductions.

As mentioned in the frequency response section, when five or more modes are included, the impulse responses are almost identical for all reduction techniques, with small differences in normalized reduction indices. For less than five modes, it is better to use the balanced technique because it picks up an additional mode in addition to the system mode, whereas the dc gain method assigns the first two modes to the almost coincident two modes near the system mode.

CHAPTER 19

MIMO TWO-STAGE ACTUATOR MODEL

19.1 Introduction

In this chapter we will use an ANSYS model of a two-stage disk drive actuator/suspension system to illustrate the creation of a reduced model for a Multiple Input, Multiple Output (MIMO) system using the balanced reduction method. The results will seem somewhat anticlimactic since the previous chapter covered most aspects of how to use the balanced reduction method. However, understanding the mechanics of setting up a MIMO system should prove useful.

As the track density (tracks per inch, tpi) of disk drives continues to increase, it will be necessary to add a second stage of actuation to the system in order to have the high servo bandwidths required to accurately follow the closely spaced tracks. Many different types of two-stage actuator architectures are being explored. The actuator architecture used for this example is not meant to represent a practical embodiment but will serve to illustrate a two-input, two-output system.

We will begin with descriptions of the actuator system and ANSYS model. Then, ANSYS output, mode shape plots, frequency responses and a partial eigenvector listing will be discussed.   The pertinent eigenvector and eigenvalue information will be extracted into a .mat file for input to MATLAB.

The MATLAB code will calculate either dc or peak gains, depending on whether uniform or non-uniform damping is defined. There are four gains to be plotted for this two-input, two-output MIMO system. While dc and peak gains are not required for the "balreal" and "modred" model reduction, they will serve to bridge our understanding from SISO models to MIMO models. We will see the difficulty of choosing which modes to include in a MIMO model using dc or peak gain sorting by discussing the ranking of modes for the four input/output combinations.

In order to perform a balanced reduction, the system is partitioned into rigid body and oscillatory modes, similar to the method used in Chapter 18. The oscillatory modes are balanced and "modred" is used with both the "del" and "mdc" options to reduce the model. Frequency responses for head 0 for both coil and piezo inputs for "del" reduction are shown for various numbers of reduced modes, from 6 oscillatory states to 20 oscillatory states included. The 20-state case shows both "del" and "mdc" for comparison.

Impulse responses are calculated for oscillatory systems with various numbers of reduced modes retained. The error is plotted as a function of number of modes retained.

19.2 Actuator Description

Figure 19.1 shows top and cross-sectioned side views of the two-stage actuator used for the analysis.

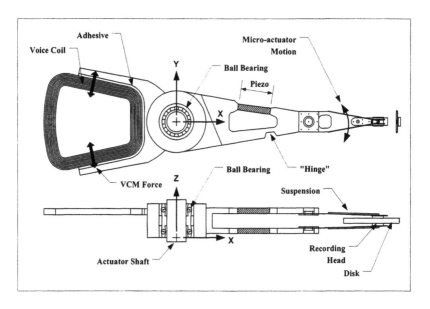

Figure 19.1: Drawing of actuator/suspension system.

The model is similar to the actuator used in Chapters 17 and 18 except that the arms are now the same thickness and are symmetrically located with respect to the pivot bearing z axis centerline. Also, there is now a piezo-actuator bonded into one side of each of the arms. The piezo actuator consists of a ceramic element that changes size when a voltage is applied. In this case, the voltage would be applied to the piezo element so that it changes length, creating a rotation about the "hinge" section in the other side of the arm. This rotation translates the recording head in the circumferential direction. When this "fine positioning" motion is used in conjunction with the VCM's "coarse positioning" motion, higher servo bandwidths and consequently higher tpi are possible.

The actuator example in the last two chapters had a coil forcing function applied at four nodes in the coil body. Even though there were multiple points at which the force was applied, the fact that the same force was applied to all nodes defined a Single Input system.

Instead of applying voltage as the input into the piezo element, we will assume that we have calculated an equivalent set of forces which can be applied at the ends of the element that will replicate the voltage forcing function. In this model, we will be applying forces to multiple nodes at the ends of both piezo elements. Since the same forces are being applied to both piezo elements, they represent the second input to the now Multi Input system, the first input being the coil force. We will apply equal and opposite forces to the two ends of each piezo actuator, and reverse the signs of the forces applied to the two separate elements. If the same forcing function were applied to both elements, an inertial moment arises which would tend to rotate the entire actuator about the pivot. By using opposite signs for the two arms, this moment is largely eliminated, generating less cross-coupling between the coarse and fine actuator inputs.

In order to make this example a "Multiple Output" system, we will output the displacements of both lower and upper heads, head 0 and head 1.

19.3 ANSYS Model Description

The model description is the same as for the model in Chapter 17. The ANSYS model is shown below, along with a drawing showing the node locations for the coil, piezo elements and heads.

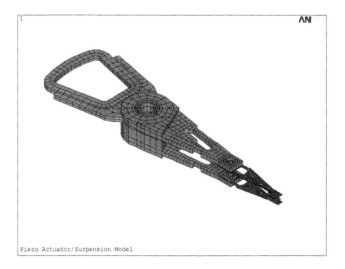

Figure 19.2: Complete piezo actuator/suspension model.

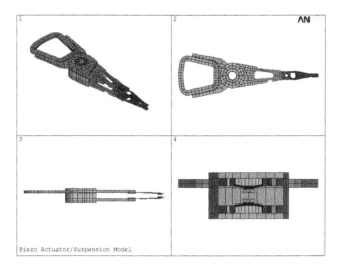

Figure 19.3: Piezo actuator/suspension model, four views.

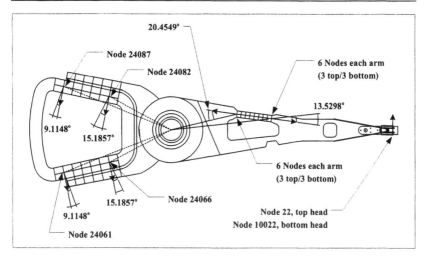

Figure 19.4: Nodes used for reduced MATLAB model, shown with partial mesh at coil and piezo element.

Since the model uses cylindrical coordinates, the coil and piezo forces are at an angle to the radial line joining the pivot bearing centerline to the node location. Both coil and piezo element forces are decomposed into radial and circumferential elements using the angles shown for each in Figure 19.4.

19.4 ANSYS Piezo Actuator/Suspension Model Results

19.4.1 Eigenvalues, Frequency Response

The first 50 modes were extracted using the Block Lanczos method. Frequency versus mode number is plotted in Figure 19.5.

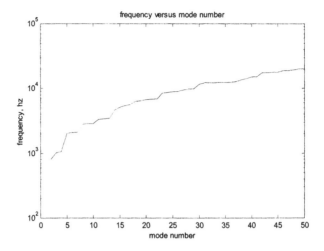

Figure 19.5: Frequencies versus mode number.

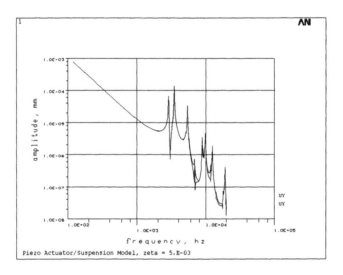

**Figure 19.6: Coil input frequency responses for head 0 and head 1 from ANSYS, zeta =
0.005.**

Figure 19.6 is the frequency response from ANSYS for coil input for both
heads. The same frequency response from the 50-mode MATLAB model is
shown in Figure 19.7. Figure 19.8 plots the frequency response for the two
piezo inputs.

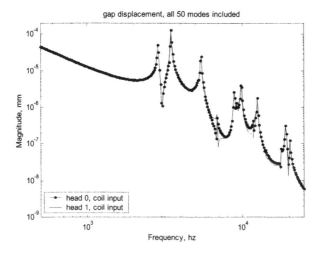

Figure 19.7: Coil input frequency response from MATLAB, zeta = 0.005.

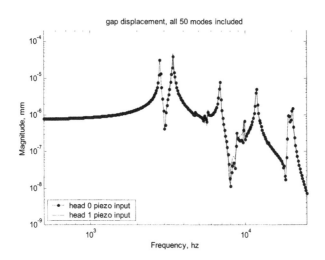

Figure 19.8: Piezo input frequency response from MATLAB, zeta = 0.005.

19.4.2 Mode Shape Plots

Selected mode shape plots are shown below, with a brief discussion of each in the following section.

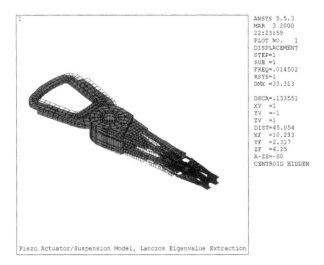

Figure 19.9: Mode 1 undeformed/deformed plot, 0.014 hz, rigid body rotation.

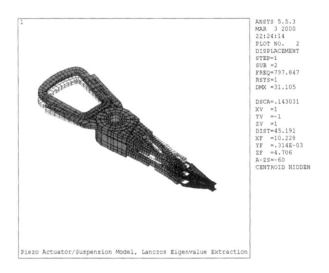

Figure 19.10: Mode 2, 798 hz, actuator pitching mode.

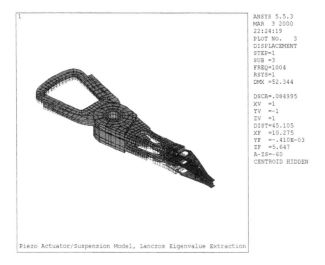

```
1                                                ANSYS 5.5.3
                                                 MAR  3 2000
                                                 22:24:19
                                                 PLOT NO.    3
                                                 DISPLACEMENT
                                                 STEP=1
                                                 SUB =3
                                                 FREQ=1004
                                                 RSYS=1
                                                 DMX =52.344

                                                 DSCA=.084995
                                                 XV  =1
                                                 YV  =-1
                                                 ZV  =1
                                                 DIST=45.105
                                                 XF  =10.275
                                                 YF  =-.410E-03
                                                 ZF  =5.647
                                                 A-ZS=-60
                                                 CENTROID HIDDEN




Piezo Actuator/Suspension Model, Lanczos Eigenvalue Extraction
```

Figure 19.11: Mode 3, 1004 hz, arm/coil bending in phase.

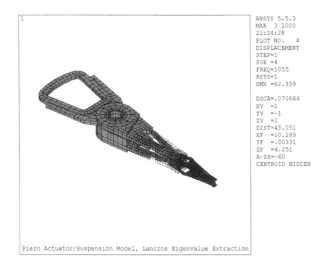

```
1                                                ANSYS 5.5.3
                                                 MAR  3 2000
                                                 22:24:28
                                                 PLOT NO.    4
                                                 DISPLACEMENT
                                                 STEP=1
                                                 SUB =4
                                                 FREQ=1055
                                                 RSYS=1
                                                 DMX =62.959

                                                 DSCA=.070664
                                                 XV  =1
                                                 YV  =-1
                                                 ZV  =1
                                                 DIST=45.051
                                                 XF  =10.289
                                                 YF  =.00331
                                                 ZF  =4.251
                                                 A-ZS=-60
                                                 CENTROID HIDDEN




Piezo Actuator/Suspension Model, Lanczos Eigenvalue Extraction
```

Figure 19.12: Mode 4, 1055 hz, arms bending out of phase.

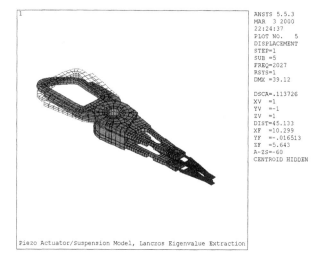

Figure 19.13: Mode 5, 2027 hz, actuator/coil torsion about x axis.

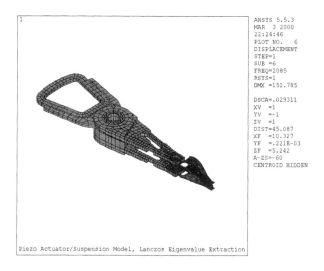

Figure 19.14: Mode 6, 2085 hz, suspension bending mode, some arm interaction.

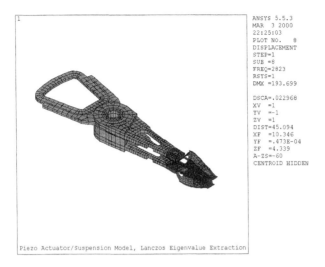

Figure 19.15: Mode 8, 2823 hz, suspension torsion, in phase, arm tip interaction.

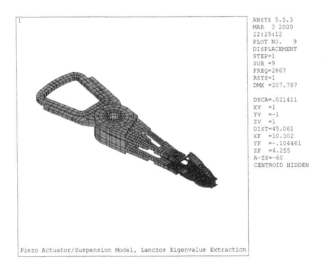

Figure 19.16: Mode 9, 2867 hz, suspension torsion, out of phase.

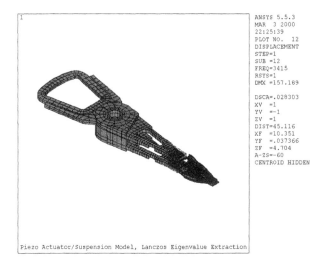

Figure 19.17: Mode 12, 3415 hz, suspension torsion, arm tip lateral.

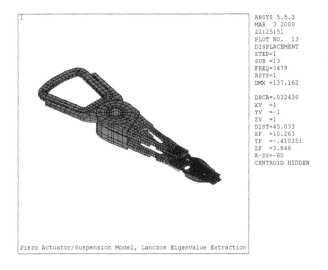

Figure 19.18: Mode 13, 3479 hz, coil/arm/suspension lateral mode.

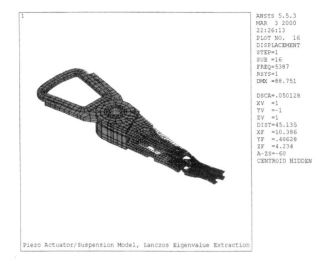

Figure 19.19: Mode 16, 5387 hz, suspension sway, arm tip lateral.

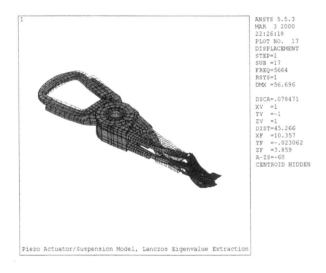

Figure 19.20: Mode 17, 5664 hz, piezo bending, arm tip torsion, coil bending.

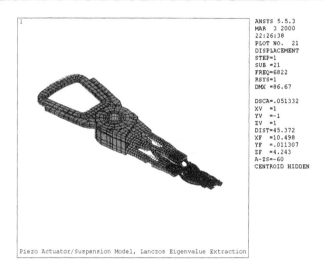

Figure 19.21: Mode 21, 6822 hz, suspension/arm lateral out of phase.

19.4.3 Mode Shape Discussion

As in Chapter 17, we will now describe the major modes which couple into the frequency response as well as several that do not couple, associating them with the frequency responses in Figures 19.7 and 19.8.

Mode 1 is the rigid body rotation mode, which ANSYS again does not calculate at zero hz because of slight geometric and numerical roundoff issues. The frequency for the rigid body mode is set to zero in the MATLAB code.

Modes 2, 3 and 4 are all modes which involve motion only in the x-z plane, bending type motions. Since the motions are perpendicular, or orthogonal, to the direction of input forces and output displacements, they do not couple into any of the frequency responses.

Mode 5 is an actuator/coil torsion mode, rotating about the x axis. A similar mode can be seen on the model in Chapter 17 as a small pole/zero pair on head 1. A torsional mode like this can be excited by: (1) coil forces, since the coil is offset from both the mass center and bearing stiffness center, and (2) inertial forces, because of the asymmetry of the structure about the mass center location in the z direction. Because the arms are more symmetric on this model than the model in Chapter 17, the pole/zero mode does not appear on the frequency response plot of either head. We will see in the dc gain ranking that mode 5 is two orders of magnitude less important than the major modes of the system for coil input, and is almost three orders of magnitude less important for piezo input.

Mode 6 is a suspension bending mode, once again a bending-only mode with no coupling into the circumferential direction.

Mode 8 is a suspension torsion, arm-tip interaction mode. It is the second most important mode for piezo input, but is unimportant for coil input.

Mode 9 is a suspension torsion mode. It is the second most important mode for coil input, but is unimportant for piezo input. The peak on the two frequency responses, just below 3 khz, is in fact two different frequencies and two different modes for the two different forcing functions. For the coil input the peak is at 2867 hz, mode 9. For piezo input, the peak is at 2823 hz, mode 8.

Modes 12 and 13 are the most important modes for piezo and coil inputs, respectively. Mode 12 involves arm tip lateral motion which the piezo can easily excite. Mode 13 is the "system" lateral mode with all components moving laterally, in phase.

Mode 16, another mode involving the tips of the arms and this time the suspension sway mode, is the third most important mode for coil input.

Mode 17 is the fifth most important piezo excitation mode, involving piezo bending, arm tip torsion and coil bending.

Mode 21 is the third most important mode for piezo excitation, with the suspensions and arms moving laterally, out of phase.

19.4.4 ANSYS Output Listing

The ANSYS output listing for input and output nodes for modes 1, 2 and 13 are listed below. These three modes were selected for discussion in order to highlight different aspects of the eigenvectors. Compared with the ANSYS output listing in Chapter 17, there are significantly more nodes in the output, with the additional nodes representing the six nodes at each end of the bottom and top piezo elements.

The rigid body mode, mode 1, should have only UY displacements (circumferential motion in the cylindrical coordinate system). Mode 2, an actuator pitching mode has its most significant motion in the UZ direction, with some slight coupling into the UX and UY directions. Mode 13 is a highly coupled mode, with significant displacements in all three directions for some nodes. The UY direction displacements are significant with respect to the UY displacements of mode 2.

```
***** POST1 NODAL DEGREE OF FREEDOM LISTING *****

LOAD STEP=  1 SUBSTEP=   1
FREQ=  0.14502E-01 LOAD CASE=  0

THE FOLLOWING DEGREE OF FREEDOM RESULTS ARE IN COORDINATE SYSTEM  1

    NODE    UX         UY         UZ         ROTX     ROTY     ROTZ
      22 0.30584E-06  32.618   0.11285E-11  0.0000   0.0000   0.0000
   10022 0.30627E-06  32.618  -0.46777E-10  0.0000   0.0000   0.0000
   21538 0.85322E-07   9.7742  0.21745E-08  0.0000   0.0000   0.0000
   21546 0.82634E-07  14.735   0.36557E-08  0.0000   0.0000   0.0000
   21576 0.10309E-06   9.9634  0.21924E-08  0.0000   0.0000   0.0000
   21584 0.16887E-06  14.883   0.37407E-08  0.0000   0.0000   0.0000
   21617 0.10951E-06  10.147   0.22079E-08  0.0000   0.0000   0.0000
   21625 0.11092E-06  14.978   0.37980E-08  0.0000   0.0000   0.0000
   22538 0.85184E-07   9.7742  0.21706E-08  0.0000   0.0000   0.0000
   22546 0.82327E-07  14.735   0.36546E-08  0.0000   0.0000   0.0000
   22576 0.10295E-06   9.9634  0.21900E-08  0.0000   0.0000   0.0000
   22584 0.16856E-06  14.883   0.37381E-08  0.0000   0.0000   0.0000
   22617 0.10937E-06  10.147   0.22067E-08  0.0000   0.0000   0.0000
   22625 0.11061E-06  14.978   0.37940E-08  0.0000   0.0000   0.0000
   24061 0.11911E-06  16.888  -0.95894E-09  0.0000   0.0000   0.0000
   24066 0.77030E-07  10.226  -0.53758E-09  0.0000   0.0000   0.0000
   24082 0.68150E-07  10.226  -0.48785E-09  0.0000   0.0000   0.0000
   24087 0.10037E-06  16.888  -0.86954E-09  0.0000   0.0000   0.0000
   24538 0.84850E-07   9.7742  0.20872E-08  0.0000   0.0000   0.0000
   24546 0.81937E-07  14.735   0.18321E-08  0.0000   0.0000   0.0000
   24576 0.10262E-06   9.9634  0.20998E-08  0.0000   0.0000   0.0000
   24584 0.16817E-06  14.883   0.17648E-08  0.0000   0.0000   0.0000
   24617 0.10904E-06  10.147   0.21122E-08  0.0000   0.0000   0.0000
   24625 0.11021E-06  14.978   0.17139E-08  0.0000   0.0000   0.0000
   25538 0.84745E-07   9.7742  0.20835E-08  0.0000   0.0000   0.0000
   25546 0.82082E-07  14.735   0.18310E-08  0.0000   0.0000   0.0000
   25576 0.10251E-06   9.9634  0.20975E-08  0.0000   0.0000   0.0000
   25584 0.16832E-06  14.883   0.17623E-08  0.0000   0.0000   0.0000
   25617 0.10894E-06  10.147   0.21110E-08  0.0000   0.0000   0.0000
   25625 0.11036E-06  14.978   0.17100E-08  0.0000   0.0000   0.0000

MAXIMUM ABSOLUTE VALUES
NODE    10022       22        21625       0        0        0
VALUE   0.30627E-06 32.618    0.37980E-08 0.0000   0.0000   0.0000

*ENDDO  INDEX= I

***** POST1 NODAL DEGREE OF FREEDOM LISTING *****

LOAD STEP=  1 SUBSTEP=   2
 FREQ=  797.85    LOAD CASE=  0
```

THE FOLLOWING DEGREE OF FREEDOM RESULTS ARE IN COORDINATE SYSTEM   1

NODE	UX	UY	UZ	ROTX	ROTY	ROTZ
22	0.49229	-0.14022	-0.10321E-03	0.0000	0.0000	0.0000
10022	-0.89140	0.14245	-0.83465E-03	0.0000	0.0000	0.0000
21538	-1.0283	0.18631	-4.0091	0.0000	0.0000	0.0000
21546	-1.5471	0.23464E-01	-10.200	0.0000	0.0000	0.0000
21576	-1.0204	0.23663	-4.0561	0.0000	0.0000	0.0000
21584	-1.5459	0.72962E-01	-10.473	0.0000	0.0000	0.0000
21617	-1.0084	0.27685	-4.0950	0.0000	0.0000	0.0000
21625	-1.5436	0.11594	-10.631	0.0000	0.0000	0.0000
22538	-0.61275	0.10972	-4.0090	0.0000	0.0000	0.0000
22546	-0.12481	0.83127E-01	-10.200	0.0000	0.0000	0.0000
22576	-0.60478	0.13415	-4.0560	0.0000	0.0000	0.0000
22584	-0.12184	0.86554E-01	-10.473	0.0000	0.0000	0.0000
22617	-0.60100	0.15502	-4.0950	0.0000	0.0000	0.0000
22625	-0.11925	0.89513E-01	-10.631	0.0000	0.0000	0.0000
24061	-0.35220	0.13939	19.652	0.0000	0.0000	0.0000
24066	-0.33572	0.17431	7.3143	0.0000	0.0000	0.0000
24082	-0.33512	-0.17241	7.3089	0.0000	0.0000	0.0000
24087	-0.35171	-0.13563	19.644	0.0000	0.0000	0.0000
24538	0.22023	-0.36868E-01	-4.0205	0.0000	0.0000	0.0000
24546	-0.27795	-0.52244E-01	-10.250	0.0000	0.0000	0.0000
24576	0.21597	-0.43317E-01	-4.0680	0.0000	0.0000	0.0000
24584	-0.27997	-0.42854E-01	-10.524	0.0000	0.0000	0.0000
24617	0.21591	-0.49478E-01	-4.1074	0.0000	0.0000	0.0000
24625	-0.28139	-0.34705E-01	-10.683	0.0000	0.0000	0.0000
25538	0.63806	-0.11349	-4.0206	0.0000	0.0000	0.0000
25546	1.1532	0.79337E-02	-10.250	0.0000	0.0000	0.0000
25576	0.63387	-0.14598	-4.0680	0.0000	0.0000	0.0000
25584	1.1531	-0.29036E-01	-10.524	0.0000	0.0000	0.0000
25617	0.62557	-0.17161	-4.1074	0.0000	0.0000	0.0000
25625	1.1519	-0.61159E-01	-10.683	0.0000	0.0000	0.0000

MAXIMUM ABSOLUTE VALUES

NODE	21546	21617	24061	0	0	0
VALUE	-1.5471	0.27685	19.652	0.0000	0.0000	0.0000

***** POST1 NODAL DEGREE OF FREEDOM LISTING *****

LOAD STEP=   1 **SUBSTEP=   13**
FREQ=   3479.3   LOAD CASE= 0

THE FOLLOWING DEGREE OF FREEDOM RESULTS ARE IN COORDINATE SYSTEM   1

NODE	UX	UY	UZ	ROTX	ROTY	ROTZ
22	-2.1984	60.376	-0.14239E-02	0.0000	0.0000	0.0000
10022	-1.9960	77.045	0.31840E-01	0.0000	0.0000	0.0000
21538	0.80764E-01	0.40397E-01	0.49848	0.0000	0.0000	0.0000
21546	-6.4836	3.9912	-1.2673	0.0000	0.0000	0.0000
21576	0.72358E-01	0.63009E-01	0.42663	0.0000	0.0000	0.0000
21584	-7.6689	4.6553	-1.8884	0.0000	0.0000	0.0000
21617	0.12273	0.57379E-01	0.37047	0.0000	0.0000	0.0000

21625	-8.7016	5.1772	-2.4325	0.0000	0.0000	0.0000
22538	0.87706E-01	0.17543	0.56748	0.0000	0.0000	0.0000
22546	-6.2831	5.0182	-1.2225	0.0000	0.0000	0.0000
22576	0.92974E-01	0.18824	0.48659	0.0000	0.0000	0.0000
22584	-7.4322	5.6835	-1.8299	0.0000	0.0000	0.0000
22617	0.14368	0.17617	0.42076	0.0000	0.0000	0.0000
22625	-8.4357	6.2048	-2.3541	0.0000	0.0000	0.0000
24061	-1.9369	-12.670	-0.95604	0.0000	0.0000	0.0000
24066	-1.0801	-4.7937	-1.0649	0.0000	0.0000	0.0000
24082	1.5007	-4.5559	-1.4595	0.0000	0.0000	0.0000
24087	2.3829	-12.467	0.10330	0.0000	0.0000	0.0000
24538	-0.93404E-01	0.37757	1.0909	0.0000	0.0000	0.0000
24546	-5.5118	4.1576	2.6594	0.0000	0.0000	0.0000
24576	-0.66009E-01	0.38853	1.0874	0.0000	0.0000	0.0000
24584	-6.3981	4.6967	3.0133	0.0000	0.0000	0.0000
24617	-0.78948E-02	0.37908	1.0812	0.0000	0.0000	0.0000
24625	-7.1715	5.1206	3.3430	0.0000	0.0000	0.0000
25538	-0.30931	0.47682	1.1451	0.0000	0.0000	0.0000
25546	-5.2283	3.4392	2.6949	0.0000	0.0000	0.0000
25576	-0.28463	0.50756	1.1349	0.0000	0.0000	0.0000
25584	-6.1405	3.9607	3.0595	0.0000	0.0000	0.0000
25617	-0.21671	0.51131	1.1213	0.0000	0.0000	0.0000
25625	-6.9354	4.3710	3.4049	0.0000	0.0000	0.0000

```
MAXIMUM ABSOLUTE VALUES
NODE    21625    10022    25625      0        0        0
VALUE  -8.7016   77.045   3.4049   0.0000   0.0000   0.0000
```

The eigenvalues and eigenvectors are stripped out of the ANSYS actrlpz.eig file and are stored in the MATLAB .mat file actrlpz_eig.mat.

19.5 MATLAB Model, MATLAB Code act8pz.m Listing and Results

19.5.1 Input, dof Definition

The **act8pz.m** MATLAB code starts by defining the degrees of freedom, nodes, directions and locations for the problem for reference in building the model.   The degrees of freedom are extracted from the ANSYS eigenvalue/eigenvector listing and are ordered by node number, first the UX direction and then the UY direction.   Once again, the UX direction information is required to transform the coil and piezo forces into cylindrical coordinates.   The eigenvalue/eigenvector information is then loaded by reading the .mat file **actrlpz_eig.mat** and the rigid body mode is set to zero frequency.

```
%      act8pz.m

       clear all;
```

```
            hold off;

            clf;

%           load the Block Lanczos .mat file actrl_eig.mat, containing evr – the
%           modal matrix, freqvec -the frequency vector and node_numbers - the
%           vector of node numbers for the modal matrix

%           the output for the ANSYS run is the following dof's

% dof      node    dir        where
%
%  1          22   ux - radial, top head gap
%  2       10022   ux - radial, bottom head gap
%  3       21538   ux - radial, bottom arm piezo, hub end
%  4       21546   ux - radial, bottom arm piezo, head end
%  5       21576   ux - radial, bottom arm piezo, hub end
%  6       21584   ux - radial, bottom arm piezo, head end
%  7       21617   ux - radial, bottom arm piezo, hub end
%  8       21625   ux - radial, bottom arm piezo, head end
%  9       22538   ux - radial, bottom arm piezo, hub end
% 10       22546   ux - radial, bottom arm piezo, head end
% 11       22576   ux - radial, bottom arm piezo, hub end
% 12       22584   ux - radial, bottom arm piezo, head end
% 13       22617   ux - radial, bottom arm piezo, hub end
% 14       22625   ux - radial, bottom arm piezo, head end
% 15       24061   ux - radial, bottom arm piezo, coil
% 16       24066   ux - radial, bottom arm piezo, coil
% 17       24082   ux - radial, bottom arm piezo, coil
% 18       24087   ux - radial, bottom arm piezo, coil
% 19       24538   ux - radial, top arm piezo, hub end
% 20       24546   ux - radial, top arm piezo, head end
% 21       24576   ux - radial, top arm piezo, hub end
% 22       24584   ux - radial, top arm piezo, head end
% 23       24617   ux - radial, top arm piezo, hub end
% 24       24625   ux - radial, top arm piezo, head end
% 25       25538   ux - radial, top arm piezo, hub end
% 26       25546   ux - radial, top arm piezo, head end
% 27       25576   ux - radial, top arm piezo, hub end
% 28       25584   ux - radial, top arm piezo, head end
% 29       25617   ux - radial, top arm piezo, hub end
% 30       25625   ux - radial, top arm piezo, head end
% 31          22   uy - circumferential, top head gap
% 32       10022   uy - circumferential, bottom head gap
% 33       21538   uy - circumferential, bottom arm piezo, hub end
% 34       21546   uy - circumferential, bottom arm piezo, head end
% 35       21576   uy - circumferential, bottom arm piezo, hub end
% 36       21584   uy - circumferential, bottom arm piezo, head end
% 37       21617   uy - circumferential, bottom arm piezo, hub end
% 38       21625   uy - circumferential, bottom arm piezo, head end
% 39       22538   uy - circumferential, bottom arm piezo, hub end
% 40       22546   uy - circumferential, bottom arm piezo, head end
% 41       22576   uy - circumferential, bottom arm piezo, hub end
% 42       22584   uy - circumferential, bottom arm piezo, head end
% 43       22617   uy - circumferential, bottom arm piezo, hub end
```

```
% 44      22625      uy - circumferential, bottom arm piezo, head end
% 45      24061      uy - circumferential, bottom arm piezo, coil
% 46      24066      uy - circumferential, bottom arm piezo, coil
% 47      24082      uy - circumferential, bottom arm piezo, coil
% 48      24087      uy - circumferential, bottom arm piezo, coil
% 49      24538      uy - circumferential, top arm piezo, hub end
% 50      24546      uy - circumferential, top arm piezo, head end
% 51      24576      uy - circumferential, top arm piezo, hub end
% 52      24584      uy - circumferential, top arm piezo, head end
% 53      24617      uy - circumferential, top arm piezo, hub end
% 54      24625      uy - circumferential, top arm piezo, head end
% 55      25538      uy - circumferential, top arm piezo, hub end
% 56      25546      uy - circumferential, top arm piezo, head end
% 57      25576      uy - circumferential, top arm piezo, hub end
% 58      25584      uy - circumferential, top arm piezo, head end
% 59      25617      uy - circumferential, top arm piezo, hub end
% 60      25625      uy - circumferential, top arm piezo, head end

        load actrlpz_eig;

        [numdof,num_modes_total] = size(evr);

        freqvec(1) = 0;          % set rigid body mode to zero frequency

        xn = evr;
```

19.5.2 Forcing Function Definition, dc Gain Calculations

The unity coil force is equally divided between the four coil nodes. For this model, the piezo force, "fpz," is arbitrarily set at 0.2, to be applied with equal magnitudes and with opposite signs to the two ends of each piezo element. For an actual system, the piezo force would be related to the coil force by the appropriate force constants for the VCM and the appropriate voltage/force relationships for the piezo, and would not be arbitrarily chosen.

Given the directions of the coil and piezo forces in Figure 19.4, the forces are transformed to cylindrical coordinates and two forcing function vectors are formed, one for the coil and one for the piezo.

The user is prompted for whether uniform or non-uniform damping is to be used and then dc or peak gains are calculated, respectively.

For a SISO system, we can rank the relative importance of modes using two methods, by using dc or peak gains and by using balancing. For a MIMO system, balancing is the only practical option. However, we will still calculate the dc gains for this MIMO system to get a feel for the relative importance of each of the modes for both forcing functions. This will require calculating dc gains for the four combinations possible for the two-input, two-output system.

The four dc gains are calculated, sorted and plotted in the code below.

```
%          define radial and circumferential forces applied at four coil force nodes
%          "x" is radial, "y" is circumferential, total force is unity

           fcoil = 0.25;

           n24061fx = fcoil*sin(9.1148*pi/180);
           n24061fy = fcoil*cos(9.1148*pi/180);

           n24066fx = fcoil*sin(15.1657*pi/180);
           n24066fy = fcoil*cos(15.1657*pi/180);

           n24082fx = -fcoil*sin(15.1657*pi/180);
           n24082fy = fcoil*cos(15.1657*pi/180);

           n24087fx = -fcoil*sin(9.1148*pi/180);
           n24087fy = fcoil*cos(9.1148*pi/180);

%          define radial and circumferential forces applied at ends of piezo element
%          "x" is radial, "y" is circumferential, total force is unity

           fpz = 0.2/6;          % six nodes at each end of the piezo

%          bottom arm radial force

           n21538fx = fpz*cos(20.4549*pi/180);
           n21546fx = -fpz*cos(13.5298*pi/180);
           n21576fx = fpz*cos(20.4549*pi/180);
           n21584fx = -fpz*cos(13.5298*pi/180);
           n21617fx = fpz*cos(20.4549*pi/180);
           n21625fx = -fpz*cos(13.5298*pi/180);
           n22538fx = fpz*cos(20.4549*pi/180);
           n22546fx = -fpz*cos(13.5298*pi/180);
           n22576fx = fpz*cos(20.4549*pi/180);
           n22584fx = -fpz*cos(13.5298*pi/180);
           n22617fx = fpz*cos(20.4549*pi/180);
           n22625fx = -fpz*cos(13.5298*pi/180);

%          top arm radial force

           n24538fx = -fpz*cos(20.4549*pi/180);
           n24546fx = fpz*cos(13.5298*pi/180);
           n24576fx = -fpz*cos(20.4549*pi/180);
           n24584fx = fpz*cos(13.5298*pi/180);
           n24617fx = -fpz*cos(20.4549*pi/180);
           n24625fx = fpz*cos(13.5298*pi/180);
           n25538fx = -fpz*cos(20.4549*pi/180);
           n25546fx = fpz*cos(13.5298*pi/180);
           n25576fx = -fpz*cos(20.4549*pi/180);
```

```
        n25584fx = fpz*cos(13.5298*pi/180);
        n25617fx = -fpz*cos(20.4549*pi/180);
        n25625fx = fpz*cos(13.5298*pi/180);

%       bottom arm circumferential force

        n21538fy = -fpz*sin(20.4549*pi/180);
        n21546fy = fpz*sin(13.5298*pi/180);
        n21576fy = -fpz*sin(20.4549*pi/180);
        n21584fy = fpz*sin(13.5298*pi/180);
        n21617fy = -fpz*sin(20.4549*pi/180);
        n21625fy = fpz*sin(13.5298*pi/180);
        n22538fy = -fpz*sin(20.4549*pi/180);
        n22546fy = fpz*sin(13.5298*pi/180);
        n22576fy = -fpz*sin(20.4549*pi/180);
        n22584fy = fpz*sin(13.5298*pi/180);
        n22617fy = -fpz*sin(20.4549*pi/180);
        n22625fy = fpz*sin(13.5298*pi/180);

%       top arm circumferential force

        n24538fy = fpz*sin(20.4549*pi/180);
        n24546fy = -fpz*sin(13.5298*pi/180);
        n24576fy = fpz*sin(20.4549*pi/180);
        n24584fy = -fpz*sin(13.5298*pi/180);
        n24617fy = fpz*sin(20.4549*pi/180);
        n24625fy = -fpz*sin(13.5298*pi/180);
        n25538fy = fpz*sin(20.4549*pi/180);
        n25546fy = -fpz*sin(13.5298*pi/180);
        n25576fy = fpz*sin(20.4549*pi/180);
        n25584fy = -fpz*sin(13.5298*pi/180);
        n25617fy = fpz*sin(20.4549*pi/180);
        n25625fy = -fpz*sin(13.5298*pi/180);

%       two-input system

%               first input is coil force
%               second input is excitation of both piezo elements with opposite polarity

%       f_coil is the vector of forces applied to coil

        f_coil = [zeros(14,1)
                  n24061fx
                  n24066fx
                  n24082fx
                  n24087fx
                  zeros(26,1)
                  n24061fy
                  n24066fy
                  n24082fy
                  n24087fy
                  zeros(12,1)];

%       f_piezo is vector of forces applied to piezo ends
```

```
f_piezo = [        0
                   0
                   n21538fx  %        bottom arm radial force
                   n21546fx
                   n21576fx
                   n21584fx
                   n21617fx
                   n21625fx
                   n22538fx
                   n22546fx
                   n22576fx
                   n22584fx
                   n22617fx
                   n22625fx
                   0
                   0
                   0
                   0
                   n24538fx  %        top arm radial force
                   n24546fx
                   n24576fx
                   n24584fx
                   n24617fx
                   n24625fx
                   n25538fx
                   n25546fx
                   n25576fx
                   n25584fx
                   n25617fx
                   n25625fx
                   0
                   0
                   n21538fy  %        bottom arm circumferential force
                   n21546fy
                   n21576fy
                   n21584fy
                   n21617fy
                   n21625fy
                   n22538fy
                   n22546fy
                   n22576fy
                   n22584fy
                   n22617fy
                   n22625fy
                   0
                   0
                   0
                   0
                   n24538fy  %        top arm circumferential force
                   n24546fy
                   n24576fy
                   n24584fy
                   n24617fy
                   n24625fy
                   n25538fy
```

```
                                    n25546fy
                                    n25576fy
                                    n25584fy
                                    n25617fy
                                    n25625fy ];
```

%   define composite forcing function, force applied to each node times
%   eigenvector value for that node

```
        force_coil = f_coil'*xn;

        force_piezo = f_piezo'*xn;
```

%   prompt for uniform or variable zeta

```
        zeta_type = input('enter "1" to read in damping vector (zetain.m) ...
                    or "enter" for uniform damping ... ');

        if  (isempty(zeta_type))

                zeta_type = 0;

                zeta_uniform = input('enter value for uniform damping, ...
                    .005 is 0.5% of critical (default) ... ');

                if (isempty(zeta_uniform))
                        zeta_uniform = 0.005;
                    end

                zeta_unsort = zeta_uniform*ones(num_modes_total,1);

        else

                zetain;      % read in zeta_unsort damping vector from zetain.m file

        end

        if  length(zeta_unsort) ~= num_modes_total

                    error(['error - zetain vector has ',num2str(length(zeta_unsort)), ...
                        ' entries instead of ',num2str(num_modes_total)]);

        end
```

%   define dc gains, 31 is head 1, 32 is head 0

```
        omega2 = (2*pi*freqvec)'.^2;     % convert to radians and square
```

%   define frequency range for frequency response

```
        freqlo = 501;

        freqhi = 25000;

        flo=log10(freqlo) ;
```

```
        fhi=log10(freqhi) ;

        f=logspace(flo,fhi,300) ;
        frad=f*2*pi ;

%       calculate dc gains if uniform damping, peak gains if non-uniform

        if  zeta_type == 0              % dc gain

                gain_h0_coil = abs([force_coil(1)*xn(32,1)/frad(1) ...
                force_coil(2:num_modes_total).*xn(32,2:num_modes_total) ...
                        ./omega2(2:num_modes_total)]);

                gain_h1_coil = abs([force_coil(1)*xn(31,1)/frad(1) ...
                force_coil(2:num_modes_total).*xn(31,2:num_modes_total) ...
                ./omega2(2:num_modes_total)]);

                gain_h0_piezo = abs([force_piezo(1)*xn(31,1)/frad(1) ...
                force_piezo(2:num_modes_total).*xn(32,2:num_modes_total) ...
                ./omega2(2:num_modes_total)]);

                gain_h1_piezo = abs([force_piezo(1)*xn(31,1)/frad(1) ...
                force_piezo(2:num_modes_total).*xn(31,2:num_modes_total) ...
                ./omega2(2:num_modes_total)]);

        elseif  zeta_type == 1          % peak gain

                gain_h0_coil = abs([force_coil(1)*xn(32,1)/frad(1) ...
                force_coil(2:num_modes_total).*xn(32,2:num_modes_total) ...
                ./((2*zeta_unsort(2:num_modes_total))'.*omega2(2:num_modes_total))]);

                gain_h1_coil = abs([force_coil(1)*xn(31,1)/frad(1) ...
                force_coil(2:num_modes_total).*xn(31,2:num_modes_total) ...
                ./((2*zeta_unsort(2:num_modes_total))'.*omega2(2:num_modes_total))]);

                gain_h0_piezo = abs([force_piezo(1)*xn(31,1)/frad(1) ...
                force_piezo(2:num_modes_total).*xn(32,2:num_modes_total) ...
                ./((2*zeta_unsort(2:num_modes_total))'.*omega2(2:num_modes_total))]);

                gain_h1_piezo = abs([force_piezo(1)*xn(31,1)/frad(1) ...
                force_piezo(2:num_modes_total).*xn(31,2:num_modes_total) ...
                ./((2*zeta_unsort(2:num_modes_total))'.*omega2(2:num_modes_total))]);

        end

%       sort gains, keeping track of original and new indices so can rearrange
%       eigenvalues and eigenvectors

        [gain_h0_coil_sort,index_h0_coil_sort] = sort(gain_h0_coil);

        [gain_h1_coil_sort,index_h1_coil_sort] = sort(gain_h1_coil);

        [gain_h0_piezo_sort,index_h0_piezo_sort] = sort(gain_h0_piezo);

        [gain_h1_piezo_sort,index_h1_piezo_sort] = sort(gain_h1_piezo);
```

```
            gain_h0_coil_sort = fliplr(gain_h0_coil_sort);        % max to min

            gain_h1_coil_sort = fliplr(gain_h1_coil_sort);        % max to min

            gain_h0_piezo_sort = fliplr(gain_h0_piezo_sort);      % max to min

            gain_h1_piezo_sort = fliplr(gain_h1_piezo_sort);      % max to min

            index_h0_coil_sort = fliplr(index_h0_coil_sort)       % max to min indices

            index_h1_coil_sort = fliplr(index_h1_coil_sort)       % max to min indices

            index_h0_piez_sort = fliplr(index_h0_piezo_sort)      % max to min indices

            index_h1_piez_sort = fliplr(index_h1_piezo_sort)      % max to min indices

            index_orig = 1:num_modes_total;

            [index_h0_coil_sort' index_h1_coil_sort' index_h0_piez_sort' index_h1_piez_sort']

%        plot results

            semilogy(index_orig(2:num_modes_total),freqvec(2:num_modes_total),'k-');
            title(['frequency versus mode number'])
            xlabel('mode number')
            ylabel('frequency, hz')
            grid off
            disp('execution paused to display figure, "enter" to continue');%pause

            semilogy(index_orig,gain_h0_coil,'k.-',index_orig,gain_h1_coil,'k-')
            title(['coil input:  dc value of each mode contribution versus mode number'])
            xlabel('mode number')
            ylabel('dc value')
            legend('h0 coil input','h1 coil input')
            grid off
            disp('execution paused to display figure, "enter" to continue');%pause

            semilogy(index_orig,gain_h0_piezo,'k.-',index_orig,gain_h1_piezo,'k-')
            title(['piezo input:  dc value of each mode contribution versus mode number'])
            xlabel('mode number')
            ylabel('dc value')
            legend('h0 piezo input','h1 piezo input')
            grid off
            disp('execution paused to display figure, "enter" to continue');%pause

            loglog(freqvec(2:num_modes_total),gain_h0_coil(2:num_modes_total),'k.-', ...
                     freqvec(2:num_modes_total),gain_h1_coil(2:num_modes_total),'k-')
            title(['coil input:  dc value of each mode contribution versus frequency'])
            xlabel('frequency, hz')
            ylabel('dc value')
            axis([500 25000 -inf inf])
            legend('h0 coil input','h1 coil input')
            grid off
            disp('execution paused to display figure, "enter" to continue');%pause
```

```
loglog(freqvec(2:num_modes_total),gain_h0_piezo(2:num_modes_total),'k.-', ...
               freqvec(2:num_modes_total),gain_h1_piezo(2:num_modes_total),'k-')
title(['piezo input:  dc value of each mode contribution versus frequency'])
xlabel('frequency, hz')
ylabel('dc value')
axis([500 25000 -inf inf])
legend('h0 piezo input','h1 piezo input')
grid off
disp('execution paused to display figure, "enter" to continue');%pause

semilogy(index_orig,gain_h0_coil_sort,'k.-',index_orig,gain_h1_coil_sort,'k-')
title(['coil input:  sorted dc value of each mode versus number of modes included'])
xlabel('modes included')
ylabel('sorted dc value')
legend('h0 coil input','h1 coil input')
grid off
disp('execution paused to display figure, "enter" to continue');%pause

semilogy(index_orig,gain_h0_piezo_sort,'k.-',index_orig,gain_h1_piezo_sort,'k-')
title(['piezo input:  sorted dc value of each mode versus number of modes included'])
xlabel('modes included')
ylabel('sorted dc value')
legend('h0 piezo input','h1 piezo input')
grid off
disp('execution paused to display figure, "enter" to continue');%pause

semilogy(index_orig,gain_h0_coil_sort,'k.-',index_orig,gain_h1_coil_sort,'k.-', ...
                    index_orig,gain_h0_piezo_sort, ...
                    'k-',index_orig,gain_h1_piezo_sort,'k-')
title(['coil and piezo input:  sorted dc value of each mode versus number ...
                    of modes included'])
xlabel('modes included')
ylabel('sorted dc value')
legend('h0 coil input','h1 coil input','h0 piezo input','h1 piezo input')
grid off
disp('execution paused to display figure, "enter" to continue');%pause
```

Figure 19.22 repeats Figure 19.5, plotting resonant frequency versus mode number. Note that there are several "jumps" in the curve, the most significant between mode 4 and mode 5. As indicated in Section 17.6, "jumps" in the frequency plot can indicate the system transitioning from one type of characteristic motion to another. In this case modes 2, 3 and 4 involve bending motions of the system, while mode 5 involves coil torsion.

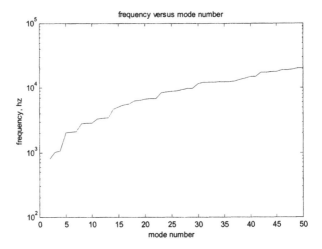

Figure 19.22: Resonant frequencies versus mode number.

The dc gains for head 0 and head 1 for coil input are shown in Figure 19.23. Because the actuator is nearly symmetrical in design the gains of the two heads are quite similar.

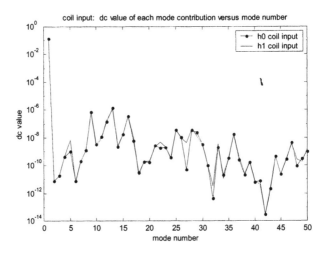

Figure 19.23: dc gain versus mode for both heads for coil input.

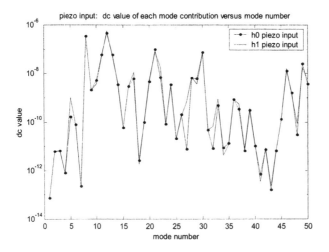

Figure 19.24: dc gain versus mode number for both heads for piezo input.

The gains for both heads for piezo inputs are shown in Figure 19.24.

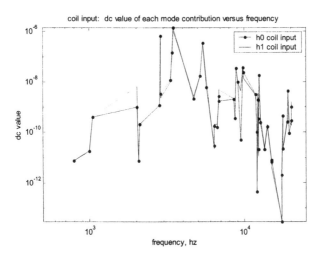

Figure 19.25: dc gain versus frequency for both heads for coil input.

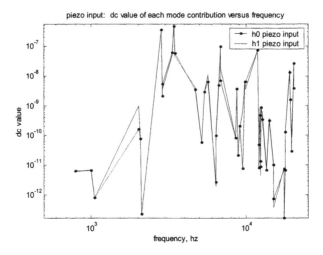

Figure 19.26: dc gain versus frequency for both heads for piezo input.

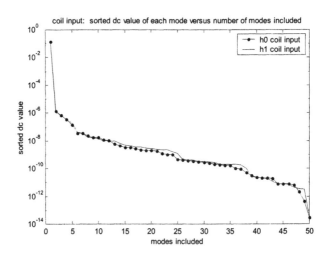

Figure 19.27: Sorted dc gain for both heads for coil input.

The sorted dc gains of the two heads, Figure 19.27, are very similar because the actuator design is so symmetrical.

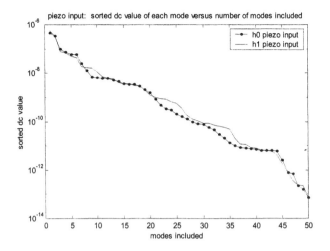

Figure 19.28: Sorted dc gain for both heads for piezo input.

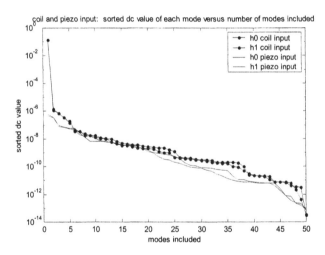

Figure 19.29: Sorted dc gain for both heads for both coil and piezo inputs.

The sorted gains of head 0 and head 1 for both coil and piezo inputs can be seen in Figure 19.29. They are of similar magnitude because the piezo force "fpz" in Section 19.5.2 was chosen to be 0.2.

With the partial listing of mode ranking for both heads and both inputs shown in Table 19.1, we can start looking at the difficulties of using dc and peak gains for ranking MIMO systems.

Table 19.1 lists the mode ranking for the first 15 modes for:

Column 1: head 0, coil input

Column 2: head 1, coil input

Column 3: head 0, piezo input

Column 4: head 1, piezo input

1	1	12	12
13	13	8	8
9	9	21	21
16	16	30	30
12	12	11	11
28	28	13	13
25	25	49	22
29	36	46	49
36	15	22	46
15	29	28	17
11	17	17	28
26	11	29	20
17	26	10	50
47	5	20	14
10	22	50	29

Table 19.1: Ranking for first 15 modes for head 0 and head 1 for coil and piezo inputs.

The first two columns in Table 19.1 show that for coil input, head 0 and head 1 have the same ranking through the first seven modes, then their rankings change. The second two columns show that for piezo input, head 0 and head 1 have the same ranking through the first six modes, then their rankings change.

If one were to choose a single ranking for the model which would take into account both inputs and both outputs, it is difficult to see how to do it given the rankings in the table. Thus the necessity of balanced reduction for MIMO models. (See Problem P19.1 for using dc gain to rank for reduction.)

19.5.3 Building State Space Matrices

In this section of code the system matrices are assembled and the four frequency responses are plotted. For all previous SISO models in the book we have built the system matrices using dc gain ordering of modes. Here, for the MIMO model, we will assemble the system using the original, unsorted ordering and will let "balreal" do all the work of sorting in the next section.

```
%       create five state space systems with all modes included, differing in the ordering
%       of the modes, the unsorted system will be used for all reductions, letting balreal do all
```

```
%           the ordering, the sorted systems will be used to show how the dc gain ordering
%           compares with the balanced ordering

%                    1) unsorted
%                    2) sorted, head 0, coil input
%                    3) sorted, head 1, coil input
%                    4) sorted, head 0, piezo input
%                    5) sorted, head 1, piezo input

       for num_model = 1:5

       if num_model == 1                      % unsorted

              xnnew = xn;

              freqnew = freqvec;

       elseif num_model == 2                         % sorted, head 0, coil input

              xnnew = xn(:,index_h0_coil_sort);

              freqnew = freqvec(index_h0_coil_sort);

       elseif num_model == 3                   % sorted, head 1, coil input

              xnnew = xn(:,index_h1_coil_sort);

              freqnew = freqvec(index_h1_coil_sort);

       elseif num_model == 4                          % sorted, head 0, piezo input

              xnnew = xn(:,index_h0_piezo_sort);

              freqnew = freqvec(index_h0_piezo_sort);

       elseif num_model == 5                          % sorted, head 1, piezo input

              xnnew = xn(:,index_h1_piezo_sort);

              freqnew = freqvec(index_h1_piezo_sort);

       end

%      define variables for all modes included system matrix, a

       w = freqnew*2*pi;              %          frequencies in rad/sec

       w2 = w.^2;

       zw = 2*zeta_unsort.*w;

%      define size of system matrix

       asize = 2*num_modes_total;
```

```
            disp(' ');
            disp(' ');
            disp(['size of system matrix a is ',num2str(asize)]);

%           setup system matrix for all modes included model

            a = zeros(asize);

            for  col = 2:2:asize

            row = col-1;

            a(row,col) = 1;

            end

            for  col = 1:2:asize

            row = col+1;

            a(row,col) = -w2((col+1)/2);

            end

            for  col = 2:2:asize

            row = col;

            a(row,col) = -zw(col/2);

            end

%           setup input matrix b, state space forcing function in principal coordinates

%           two-input system

%                   first input is coil force
%                   second input is excitation of both piezo elements with opposite polarity

            f_physical = [f_coil f_piezo];

%           f_principal is the matrix of forces in principal coordinates

            f_principal = xnnew'*f_physical;

%           b is the matrix of forces in principal coordinates, state space form

            b = zeros(2*num_modes_total,2);

            for  cnt = 1:num_modes_total

                    b(2*cnt,:) = f_principal(cnt,:);

            end
```

```
%          setup cdisp and cvel, padded xn matrices to give the displacement and velocity
%          vectors in physical coordinates cdisp and cvel each have numdof rows
%          and alternating columns consisting of columns of xnnew and zeros to give total
%          columns equal to the number of states

%          all modes included cdisp and cvel

           for  col = 1:2:2*length(freqnew)

           for  row = 1:numdof

           c_disp(row,col) = xnnew(row,ceil(col/2));

           cvel(row,col) = 0;

           end

           end

           for  col = 2:2:2*length(freqnew)

           for  row = 1:numdof

           c_disp(row,col) = 0;

                cvel(row,col) = xnnew(row,col/2);

           end

           end

%          define output

           d = [0];  %

           if  num_model == 1                          % unsorted

                     sys = ss(a,b,c_disp(31:32,:),d);

           elseif  num_model == 2                       % sorted, head 0, coil input

                     sys_h0_coil = ss(a,b,c_disp(31:32,:),d);

           elseif  num_model == 3                       % sorted, head 1, coil input

                     sys_h1_coil = ss(a,b,c_disp(31:32,:),d);

           elseif  num_model == 4                       % sorted, head 0, piezo input

                     sys_h0_piezo = ss(a,b,c_disp(31:32,:),d);

           elseif  num_model == 5                       % sorted, head 1, piezo input

                     sys_h1_piezo = ss(a,b,c_disp(31:32,:),d);
```

```
           end

           end                    % end of for loop for creating system matrices
```

19.5.4 Balancing, Reduction

Balancing the system involves calculating gramians, which are only defined for negative definite systems. This requires separating the rigid body mode from the oscillatory modes and balancing the oscillatory modes. The system matrices are partitioned and a model of only oscillatory modes is created and balanced. Plotting the diagonal gramian terms (squares of the Hankel singular values) reveals the relative importance of the states.

Modred is used to reduce the states using both the "del" and "mdc" options. The complete system is rebuilt by augmenting the rigid body mode (states) with the reduced oscillatory modes (states). Frequency responses are then plotted, comparing the two reducing methods with the original 50-mode model.

```
%        partition system matrices into rigid body mode and oscillatory modes, can't use balreal
%        with rigid body mode so will reduce the oscillatory modes and then augment the
%        resulting system with the rigid body mode

%        define oscillatory system, where output 31 is head 1, output 32 is head 0

         [a,b,c_disp,d] = ssdata(sys);

         a_syso = a(3:asize,3:asize);

         b_syso = b(3:asize,:);

         c_syso = c_disp(1:2,3:asize);

         syso = ss(a_syso,b_syso,c_syso,d);

%        define controllability and observability gramians for oscillatory system, syso

         wc = gram(syso,'c');

         wo = gram(syso,'o');

         [row_syso,col_syso] = size(a_syso);

         statevec = 1:row_syso;

%        plot controllability and observability gramians

         meshz(wc);
         view(60,30);
```

```
            title(['controllability gramian for oscillatory system'])
            xlabel('state')
            ylabel('state')
            grid on

            disp('execution paused to display figure, "enter" to continue');%pause

            meshz(wo);
            view(60,30);
            title(['observability gramian for oscillatory system'])
            xlabel('state')
            ylabel('state')
            grid on

            disp('execution paused to display figure, "enter" to continue');%pause

%           pull out diagonal elements

            wc_diag = diag(wc);

            wo_diag = diag(wo);

%           plot diagonal terms of controllability and observability gramians

            semilogy(statevec,wc_diag,'k.-')
            title(['controllability gramian diagonal terms'])
            xlabel('states')
            ylabel('diagonal')
            grid off

            disp('execution paused to display figure, "enter" to continue');%pause

            semilogy(statevec,wo_diag,'k.-')
            title(['observability gramian diagonal terms'])
            xlabel('states')
            ylabel('diagonal')
            grid off

            disp('execution paused to display figure, "enter" to continue');%pause

%           position and velocity states plotted separately

            semilogy(statevec(1:2:row_syso),wc_diag(1:2:row_syso),'k.-', ...

            statevec(2:2:row_syso),wc_diag(2:2:row_syso),'k-')
            title(['controllability gramian diagonal terms'])
            xlabel('states')
            ylabel('diagonal')
            legend('position states','velocity states',3)
            grid off

            disp('execution paused to display figure, "enter" to continue');%pause

            semilogy(statevec(1:2:row_syso),wo_diag(1:2:row_syso),'k.-', ...
```

```
            statevec(2:2:row_syso),wo_diag(2:2:row_syso),'k-')
            title(['observability gramian diagonal terms'])
            xlabel('states')
            ylabel('diagonal')
            legend('position states','velocity states',3)
            grid off

            disp('execution paused to display figure, "enter" to continue');%pause

%           use balreal to rank oscillatory states and modred to reduce for comparison

            [sysob,g,T,Ti] = balreal(syso);

            [ao_bal,bo_bal,cdispo_bal,do_bal] = ssdata(sysob);

            semilogy(g,'k.-')
            title('diagonal of balanced gramian versus number of states')
            xlabel('state number')
            ylabel('diagonal of balanced gramian')
            grid off

            osc_states_used = input(['enter number of oscillatory states to use, default 20 ... ']);

            if  isempty(osc_states_used)

                    osc_states_used = 20;

            end

            num_modes_used = 1 + osc_states_used/2;      % number of modes for overlaid plots

%           use modred to order oscillatory states from balreal to define reduced order
%           oscillatory system using both "del" and "mdc"

            rsys_delo = modred(sysob,osc_states_used+1:2*num_modes_total-2,'del');

            rsys_mdco = modred(sysob,osc_states_used+1:2*num_modes_total-2,'mdc');

%           rebuild system by appending balanced realization of oscillatory modes to rigid
%           body mode

            [a_delo_bal,b_delo_bal,c_delo_bal,d_delo_bal] = ssdata(rsys_delo);

            a_del_bal = [  a(1:2,1:2)    zeros(2,osc_states_used)
                                    zeros(osc_states_used,2)    a_delo_bal   ];

            b_del_bal = [b(1:2,:)
                                    b_delo_bal];

            c_del_bal = [c_disp(1:2,1:2) c_delo_bal];

            rsys_del = ss(a_del_bal,b_del_bal,c_del_bal,d);

            [a_mdco_bal,b_mdco_bal,c_mdco_bal,d_mdco_bal] = ssdata(rsys_mdco);
```

```
        a_mdc_bal = [   a(1:2,1:2)     zeros(2,osc_states_used)
                             zeros(osc_states_used,2)     a_mdco_bal    ];

        b_mdc_bal = [b(1:2,:)
                             b_mdco_bal];

        c_mdc_bal = [c_disp(1:2,1:2) c_mdco_bal];

        rsys_mdc = ss(a_mdc_bal,b_mdc_bal,c_mdc_bal,d);

%       frequency response for unsorted system

        [mag,phs] = bode(sys,frad);

%       plot original system response, output of bode command has dimensions
%       of "i" x "j" x "k" where "i" is output row, "j" is input column and "k" is the
%       vector of frequencies

        magh0coil = mag(2,1,:);
        magh1coil = mag(1,1,:);
        magh0pz  = mag(2,2,:);
        magh1pz  = mag(1,2,:);

        loglog(f,magh0coil(1,:),'k.-',f,magh1coil(1,:),'k-')
        title(['gap displacement, all ',num2str(num_modes_total),' modes included'])
        xlabel('Frequency, hz')
        ylabel('Magnitude, mm')
        axis([500 25000 1e-9 2e-4])
        legend('head 0, coil input','head 1, coil input',3)
        grid off
        disp('execution paused to display figure, "enter" to continue');%pause

        loglog(f,magh0pz(1,:),'k.-',f,magh1pz(1,:),'k-')
        title(['gap displacement, all ',num2str(num_modes_total),' modes included'])
        xlabel('Frequency, hz')
        ylabel('Magnitude, mm')
        axis([500 25000 1e-9 2e-4])
        legend('head 0 piezo input','head 1 piezo input',3)
        grid off
        disp('execution paused to display figure, "enter" to continue');%pause

        loglog(f,magh0coil(1,:),'k.-',f,magh1coil(1,:),'k.-',f,magh0pz(1,:),'k-',f,magh1pz(1,:),'k-
')
        title(['gap displacement, all ',num2str(num_modes_total),' modes included'])
        xlabel('Frequency, hz')
        ylabel('Magnitude, mm')
        axis([500 25000 1e-9 2e-4])
        legend('head 0, coil input','head 1, coil input','head 0 piezo input','head 1 piezo ...
                             input',3)
        grid off
        disp('execution paused to display figure, "enter" to continue');%pause

%       frequency response for balanced reduced modred "del"
```

```
[magr_del,phsr_del] = bode(rsys_del,frad);

magr_delh0coil = magr_del(2,1,:);
magr_delh1coil = magr_del(1,1,:);
magr_delh0pz  = magr_del(2,2,:);
magr_delh1pz  = magr_del(1,2,:);

loglog(f,magr_delh0coil(1,:),'k-',f,magr_delh1coil(1,:),'k.-',f,magr_delh0pz(1,:), ...
                   'k.-',f,magr_delh1pz(1,:),'k-')
title(['gap displacement, modred "del", ',num2str(osc_states_used), ...
                ' oscillatory states included'])
xlabel('Frequency, hz')
ylabel('Magnitude, mm')
axis([500 25000 1e-9 2e-4])
legend('head 0, coil input','head 1, coil input','head 0 piezo input' ...
                ,'head 1 piezo input',3)
grid off
disp('execution paused to display figure, "enter" to continue');%pause

loglog(f,magh0coil(1,:),'k-',f,magr_delh0coil(1,:),'k.-')
title(['gap displacement, modred "del", ',num2str(osc_states_used), ...
                ' oscillatory states included'])
xlabel('Frequency, hz')
ylabel('Magnitude, mm')
axis([500 25000 1e-9 2e-4])
legend('head 0, coil input','"del" reduced head 0, coil input',3)
grid off
disp('execution paused to display figure, "enter" to continue'); pause

loglog(f,magh1coil(1,:),'k-',f,magr_delh1coil(1,:),'k.-')
title(['gap displacement, modred "del", ',num2str(osc_states_used), ...
                ' oscillatory states included'])
xlabel('Frequency, hz')
ylabel('Magnitude, mm')
axis([500 25000 1e-9 2e-4])
legend('head 1, coil input','"del" reduced head 1, coil input',3)
grid off
disp('execution paused to display figure, "enter" to continue');%pause

loglog(f,magh0pz(1,:),'k-',f,magr_delh0pz(1,:),'k.-')
title(['gap displacement, modred "del", ',num2str(osc_states_used), ...
                ' oscillatory states included'])
xlabel('Frequency, hz')
ylabel('Magnitude, mm')
axis([500 25000 1e-9 2e-4])
legend('head 0, piezo input','"del" reduced head 0, piezo input',3)
grid off
disp('execution paused to display figure, "enter" to continue'); pause

loglog(f,magh1pz(1,:),'k-',f,magr_delh1pz(1,:),'k.-')
title(['gap displacement, modred "del", ',num2str(osc_states_used), ...
                ' oscillatory states included'])
xlabel('Frequency, hz')
ylabel('Magnitude, mm')
axis([500 25000 1e-9 2e-4])
```

```
          legend('head 1, piezo input','"del" reduced head 1, piezo input',3)
          grid off
          disp('execution paused to display figure, "enter" to continue');%pause

%         frequency response for balanced reduced modred "mdc"

          [magr_mdc,phsr_mdc] = bode(rsys_mdc,frad);

          magr_mdch0coil = magr_mdc(2,1,:);
          magr_mdch1coil = magr_mdc(1,1,:);
          magr_mdch0pz  = magr_mdc(2,2,:);
          magr_mdch1pz   = magr_mdc(1,2,:);

          loglog(f,magr_mdch0coil(1,:),'k-',f,magr_mdch1coil(1,:),'k.-', ...
                  f,magr_mdch0pz(1,:),'k.-',f,magr_mdch1pz(1,:),'k-')
          title(['gap displacement, modred "mdc", ',num2str(osc_states_used), ...
                      ' oscillatory states included'])
          xlabel('Frequency, hz')
          ylabel('Magnitude, mm')
          axis([500 25000 1e-9 2e-4])
          legend('head 0, coil input','head 1, coil input','head 0 piezo input','head 1 piezo ...
                              input',3)
          grid off
          disp('execution paused to display figure, "enter" to continue');%pause

          loglog(f,magh0coil(1,:),'k-',f,magr_mdch0coil(1,:),'k.-')
          title(['gap displacement, modred "mdc", ',num2str(osc_states_used), ...
                          ' oscillatory states included'])
          xlabel('Frequency, hz')
          ylabel('Magnitude, mm')
          axis([500 25000 1e-9 2e-4])
          legend('head 0, coil input','"mdc" reduced head 0, coil input',3)
          grid off
          disp('execution paused to display figure, "enter" to continue'); pause

          loglog(f,magh1coil(1,:),'k-',f,magr_mdch1coil(1,:),'k.-')
          title(['gap displacement, modred "mdc", ',num2str(osc_states_used), ...
                          ' oscillatory states included'])
          xlabel('Frequency, hz')
          ylabel('Magnitude, mm')
          axis([500 25000 1e-9 2e-4])
          legend('head 1, coil input','"mdc" reduced head 1, coil input',3)
          grid off
          disp('execution paused to display figure, "enter" to continue');%pause

          loglog(f,magh0pz(1,:),'k-',f,magr_mdch0pz(1,:),'k.-')
          title(['gap displacement, modred "mdc", ',num2str(osc_states_used), ...
                          ' oscillatory states included'])
          xlabel('Frequency, hz')
          ylabel('Magnitude, mm')
          axis([500 25000 1e-9 2e-4])
          legend('head 0, piezo input','"mdc" reduced head 0, piezo input',3)
          grid off
          disp('execution paused to display figure, "enter" to continue'); pause
```

```
loglog(f,magh1pz(1,:),'k-',f,magr_mdch1pz(1,:),'k.-')
title(['gap displacement, modred "mdc", ',num2str(osc_states_used), ...
                ' oscillatory states included'])
xlabel('Frequency, hz')
ylabel('Magnitude, mm')
axis([500 25000 1e-9 2e-4])
legend('head 1, piezo input','"mdc" reduced head 1, piezo input',3)
grid off
disp('execution paused to display figure, "enter" to continue');%pause
```

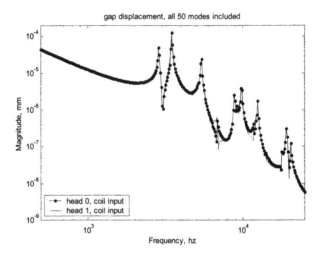

Figure 19.30: Frequency response for coil input for both heads, all modes included.

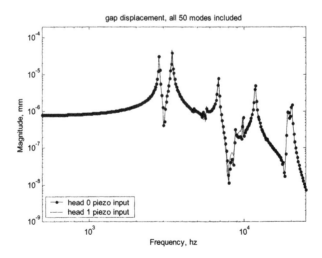

Figure 19.31: Frequency response for piezo input for both heads, all modes included.

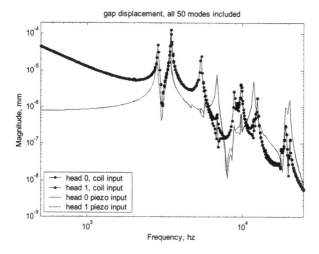

Figure 19.32: Frequency response for both coil and piezo inputs for both heads, all modes included.

The frequency response plots for both inputs and both outputs are shown above for reference.

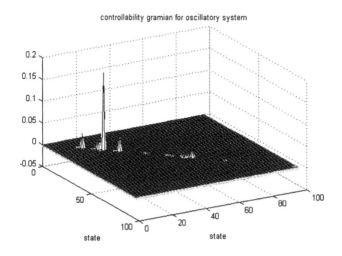

Figure 19.33: Controllability gramian values.

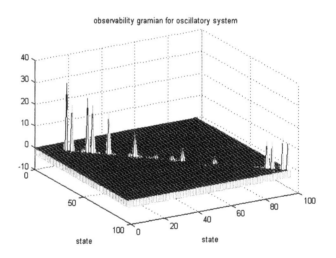

Figure 19.34: Observability gramian values.

Graphically, Figures 19.33 and 19.34 show the two gramians for this MIMO system. The gramians are nearly diagonal. The controllability gramian displays a predominance of lower frequency states, while the observability gramian has some higher frequency states included.

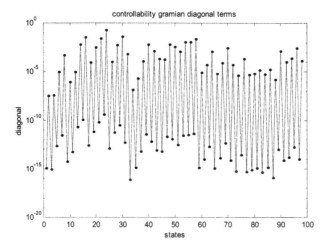

Figure 19.35: Controllability gramian diagonal terms versus states.

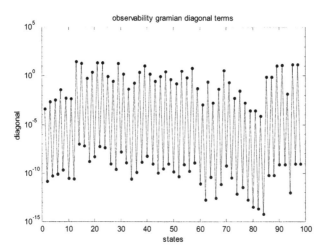

Figure 19.36: Observability gramian diagonal terms versus states.

Plotting the diagonal elements of the two gramians reveals the same pattern as for the SISO model. The maximum and minimum values for each mode are related by the square of the eigenvalue for that mode.

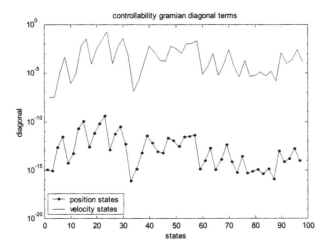

Figure 19.37: Controllability gramian diagonal position and velocity state terms.

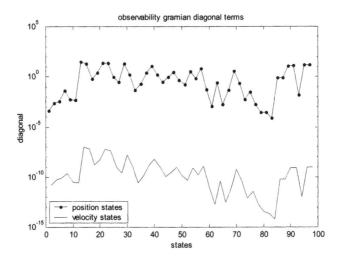

Figure 19.38: Observability gramian diagonal position and velocity state terms.

Plotting the position and velocity terms for each gramian separately displays their character on a mode-by-mode basis.

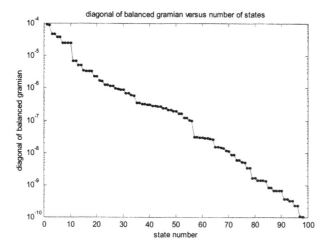

Figure 19.39: Balanced gramian diagonal terms (Hankel singular values) versus state number.

The balanced gramian shows several sharp drops in magnitude, one at 10 states and one at 56 states. We will see in Section 19.5.7 that 10 oscillatory modes (20 oscillatory states) are required for a normalized reduction index of less than 5% for coil input, and that 16 oscillatory modes (32 oscillatory states) are required for a normalized reduction index of less than 5% for piezo input.

19.5.5 Frequency Responses for Different Numbers of Retained States

This section displays pairs of frequency responses, one for head 0 for coil input and one for head 0 for piezo input. Each pair of plots represents an increasing number of oscillatory modes included in the reduced model. The original 50 mode model is overlaid to show the error in the reduced model. Note how the balanced method adds modes and which modes it chooses.

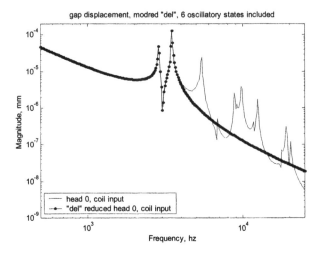

Figure 19.40: Head 0, coil input, six reduced oscillatory states included.

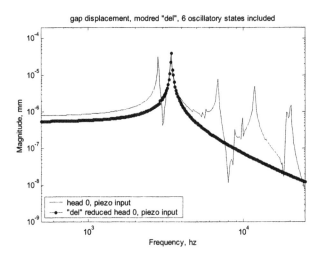

Figure 19.41: Head 0, piezo input, six reduced oscillatory states included.

With only six oscillatory states included the coil input captures the first two resonances but the piezo input misses the first resonance.

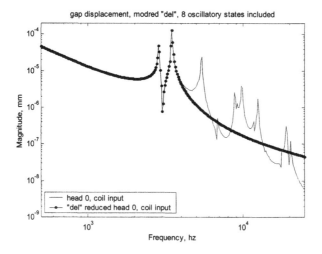

Figure 19.42: Head 0, coil input, eight reduced oscillatory states included.

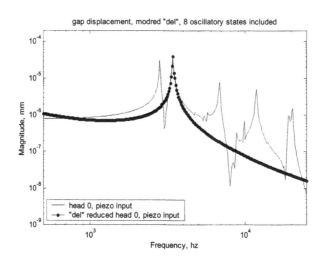

Figure 19.43: Head 0, piezo input, eight reduced oscillatory states included.

With 8 oscillatory states included the coil input captures the first two
resonances but the piezo input again misses the first resonance.

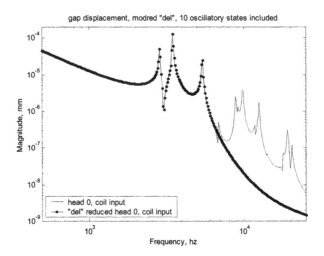

Figure 19.44: Head 0, coil input, 10 reduced oscillatory states included.

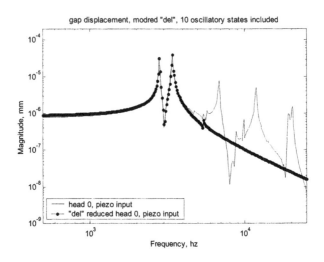

Figure 19.45: Head 0, piezo input, 10 reduced oscillatory states included.

With 10 oscillatory states included the first three coil input modes are fit well and also the first two piezo input modes.

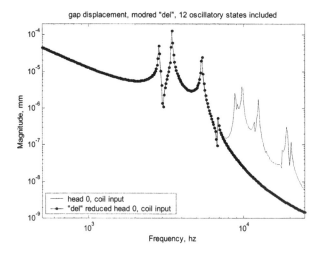

Figure 19.46:   Head 0, coil input, 12 reduced oscillatory states included.

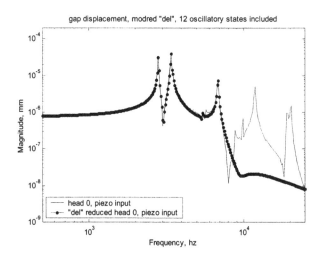

Figure 19.47:   Head 0, piezo input, 12 reduced oscillatory states included.

With 12 oscillatory states included the first three major modes are fitted for
both coil and piezo inputs.

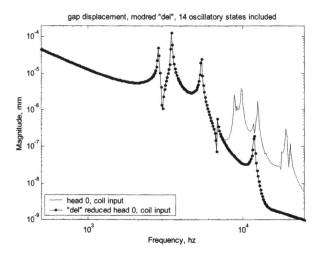

Figure 19.48: Head 0, coil input, 14 reduced oscillatory states included.

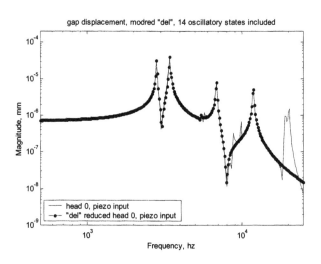

Figure 19.49: Head 0, piezo input, 14 reduced oscillatory states included.

For 14 oscillatory states included now the first four major piezo modes are fitted while the coil input starts missing some modes in the 10khz range.

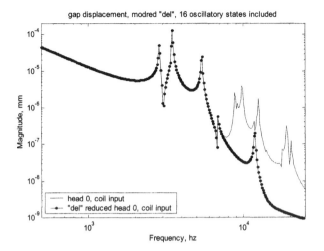

Figure 19.50: Head 0, coil input, 16 reduced oscillatory states included.

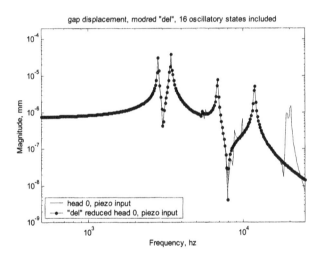

Figure 19.51: Head 0, piezo input, 16 reduced oscillatory states included.

For 16 oscillatory states included the only visible effect of the extra two states is in the piezo input zero in the 8khz range.

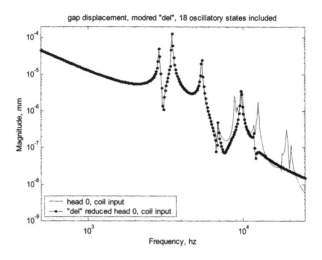

Figure 19.52: Head 0, coil input, 18 reduced oscillatory states included.

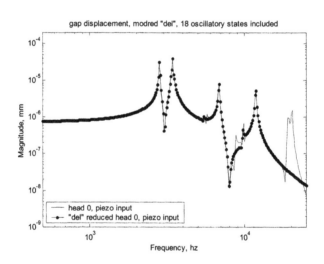

Figure 19.53: Head 0, piezo input, 18 reduced oscillatory states included.

For 18 oscillatory states included the coil input response picks up an additional mode in the 10khz range.

19.5.6 "del" and "mdc" Frequency Response Comparison

This section compares the "del" and "mdc" reduced models for the case of 20 included oscillatory states.

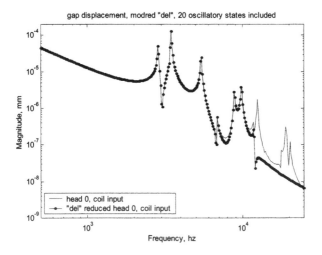

Figure 19.54: Head 0, coil input, 20 reduced oscillatory states included, modred "del."

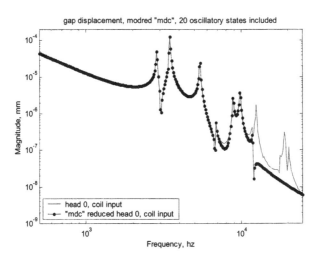

Figure 19.55: Head 0, coil input, 20 reduced oscillatory states included, modred "mdc."

There is virtually no difference between the "del" and "mdc" reductions in the two figures above for coil input.

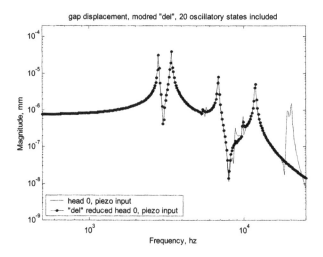

Figure 19.56: Head 0, piezo input, 20 reduced oscillatory states included, modred "del."

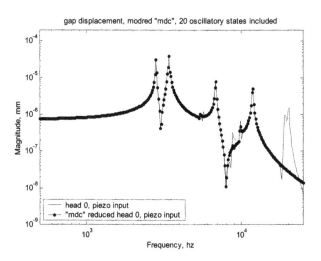

Figure 19.57: Head 0, piezo input, 20 reduced oscillatory states included, modred "mdc."

Similarly, there is no difference between the "del" and "mdc" reductions for piezo input.

19.5.7 Impulse Response

Oscillatory system impulse responses due to both coil and piezo forcing functions are calculated.   Previously calculated results for normalized reduction index (18.28) versus number of modes included are shown.

```
%          calculate impulse responses

           ttotal = 0.0025;

           t = linspace(0,ttotal,400)';

           [disp_syso,t_syso] = impulse(syso,t);

           [disp_rsys_delo,t_rsys_delo] = impulse(rsys_delo,t);

           [disp_rsys_mdco,t_rsys_mdco] = impulse(rsys_mdco,t);

           disph0coil = disp_syso(:,2,1);
           disph1coil = disp_syso(:,1,1);
           disph0pz  = disp_syso(:,2,2);
           disph1pz  = disp_syso(:,1,2);

           dispr_delh0coil = disp_rsys_delo(:,2,1);
           dispr_delh1coil = disp_rsys_delo(:,1,1);
           dispr_delh0pz  = disp_rsys_delo(:,2,2);
           dispr_delh1pz  = disp_rsys_delo(:,1,2);

           dispr_mdch0coil = disp_rsys_mdco(:,2,1);
           dispr_mdch1coil = disp_rsys_mdco(:,1,1);
           dispr_mdch0pz  = disp_rsys_mdco(:,2,2);
           dispr_mdch1pz  = disp_rsys_mdco(:,1,2);

%          build matrix of results

           dispo = [disph0coil disph1coil disph0pz disph1pz ...
                       dispr_delh0coil dispr_delh1coil dispr_delh0pz dispr_delh1pz ...
                       dispr_mdch0coil dispr_mdch1coil dispr_mdch0pz dispr_mdch1pz];

           h0coil_del_del =  dispo(:,1) - dispo(:,5);

           h1coil_del_del =  dispo(:,2) - dispo(:,6);

           h0piezo_del_del = dispo(:,3) - dispo(:,7);

           h1piezo_del_del = dispo(:,4) - dispo(:,8);

           h0coil_mdc_del =  dispo(:,1) - dispo(:,9);

           h1coil_mdc_del =  dispo(:,2) - dispo(:,10);

           h0piezo_mdc_del = dispo(:,3) - dispo(:,11);

           h1piezo_mdc_del = dispo(:,4) - dispo(:,12);

           index_h0coil_del = ...
                   sqrt(sum(h0coil_del_del.*h0coil_del_del))/sqrt(sum(dispo(:,1).*dispo(:,1)));

           index_h1coil_del = ...
                   sqrt(sum(h1coil_del_del.*h1coil_del_del))/sqrt(sum(dispo(:,2).*dispo(:,2)));
```

```
index_h0piezo_del = ...
    sqrt(sum(h0piezo_del_del.*h0piezo_del_del))/sqrt(sum(dispo(:,3).*dispo(:,3)));

index_h1piezo_del = ...
    sqrt(sum(h1piezo_del_del.*h1piezo_del_del))/sqrt(sum(dispo(:,4).*dispo(:,4)));

index_h0coil_mdc = ...
    sqrt(sum(h0coil_mdc_del.*h0coil_mdc_del))/sqrt(sum(dispo(:,1).*dispo(:,1)));

index_h1coil_mdc = ...
    sqrt(sum(h1coil_mdc_del.*h1coil_mdc_del))/sqrt(sum(dispo(:,2).*dispo(:,2)));

index_h0piezo_mdc = ...
    sqrt(sum(h0piezo_mdc_del.*h0piezo_mdc_del))/sqrt(sum(dispo(:,3).*dispo(:,3)));

index_h1piezo_mdc = ...
    sqrt(sum(h1piezo_mdc_del.*h1piezo_mdc_del))/sqrt(sum(dispo(:,4).*dispo(:,4)));

[index_h0coil_del index_h1coil_del index_h0piezo_del index_h1piezo_del ...
index_h0coil_mdc index_h1coil_mdc index_h0piezo_mdc index_h1piezo_mdc]

plot(t_syso,disph0coil,'k.-',t_rsys_delo,dispr_delh0coil, ...
            'k-',t_rsys_mdco,dispr_mdch0coil,'k--')
title(['head 0, displacement vs time, coil impulse input, ', ...
        num2str(osc_states_used),' oscillatory states included'])
xlabel('time, sec')
ylabel('displacement, mm')
legend('all modes','modred del','modred mdc',4)
grid off

disp('execution paused to display figure, "enter" to continue');%pause

plot(t_syso,disph1coil,'k.-',t_rsys_delo,dispr_delh1coil, ...
        'k-',t_rsys_mdco,dispr_mdch1coil,'k--')
title(['head 1, displacement vs time, coil impulse input, ', ...
        num2str(osc_states_used),' oscillatory states included'])
xlabel('time, sec')
ylabel('displacement, mm')
legend('all modes','modred del','modred mdc',4)
grid off

disp('execution paused to display figure, "enter" to continue');%pause

plot(t_syso,disph0pz,'k.-',t_rsys_delo,dispr_delh0pz, ...
            'k-',t_rsys_mdco,dispr_mdch0pz,'k--')
title(['head 0, displacement vs time, piezo impulse input, ', ...
            num2str(osc_states_used),' oscillatory states included'])
xlabel('time, sec')
ylabel('displacement, mm')
legend('all modes','modred del','modred mdc',4)
grid off

disp('execution paused to display figure, "enter" to continue');%pause
```

```
            plot(t_syso,disph1pz,'k.-',t_rsys_delo,dispr_delh1pz, ...
                    'k-',t_rsys_mdco,dispr_mdch1pz,'k--')
            title(['head 1, displacement vs time, piezo impulse input, ', ...
                    num2str(osc_states_used),' oscillatory states included'])
            xlabel('time, sec')
            ylabel('displacement, mm')
            legend('all modes','modred del','modred mdc',4)
            grid off

            disp('execution paused to display figure, "enter" to continue');%pause

%               states h0cd   h1cd   h0pd   h1pd   h0cm   h1cm   h0pm   h1pm

            error = [   10 0.1081 0.1075 0.4162 0.3963 0.1081 0.1075 0.4165 0.3964

                        12 0.1079 0.1072 0.3154 0.3058 0.1079 0.1073 0.3157 0.3061

                        16 0.1075 0.1070 0.1393 0.1421 0.1074 0.1070 0.1393 0.1419

                        20 0.0395 0.0425 0.1391 0.1410 0.0397 0.0425 0.1391 0.1411

                        24 0.0363 0.0374 0.0839 0.0873 0.0463 0.0473 0.0841 0.0875

                        28 0.0161 0.0178 0.0469 0.0495 0.0160 0.0191 0.0791 0.0794

                        32 0.0140 0.0142 0.0145 0.0160 0.0142 0.0143 0.0146 0.0163];

            nmode = error(:,1)/2;

            error_h0coil_del = error(:,2);

            error_h1coil_del = error(:,3);

            error_h0piezo_del = error(:,4);

            error_h1piezo_del = error(:,5);

            error_h0coil_mdc = error(:,6);

            error_h1coil_mdc = error(:,7);

            error_h0piezo_mdc = error(:,8);

            error_h1piezo_mdc = error(:,9);

            plot(nmode,error_h0coil_del,'k.-',nmode,error_h0coil_mdc,'k-')
            title('head 0, coil input normalized reduction index')
            xlabel('number of modes included')
            ylabel('normalized reduction index')
            legend('modred del','modred mdc')
            axis([0 20 0 0.5])
            grid off

            disp('execution paused to display figure, "enter" to continue');%pause
```

```
plot(nmode,error_h1coil_del,'k.-',nmode,error_h1coil_mdc,'k-')
title('head 1, coil input normalized reduction index')
xlabel('number of modes included')
ylabel('normalized reduction index')
legend('modred del','modred mdc')
axis([0 20 0 0.5])
grid off

disp('execution paused to display figure, "enter" to continue');%pause

plot(nmode,error_h0piezo_del,'k.-',nmode,error_h0piezo_mdc,'k-')
title('head 0, piezo input normalized reduction index')
xlabel('number of modes included')
ylabel('normalized reduction index')
legend('modred del','modred mdc')
axis([0 20 0 0.5])
grid off

disp('execution paused to display figure, "enter" to continue');%pause

plot(nmode,error_h1piezo_del,'k.-',nmode,error_h1piezo_mdc,'k-')
title('head 1, piezo input normalized reduction index')
xlabel('number of modes included')
ylabel('normalized reduction index')
legend('modred del','modred mdc')
axis([0 20 0 0.5])
grid off

disp('execution paused to display figure, "enter" to continue');%pause
```

The pages following will show impulse responses for head 0 for both coil and piezo inputs and for both "del" and "mdc" reduced models. Following the impulse responses, the normalized reduction index versus number of reduced modes is plotted. It shows very little difference between the two reduction methods.

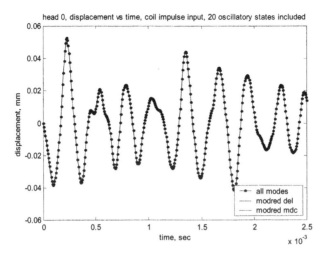

Figure 19.58: Impulse response comparison for head 0 for coil input for oscillatory system, full model (all oscillatory modes) and balreal modred "del" and "mdc" reduced systems with 20 oscillatory modes.

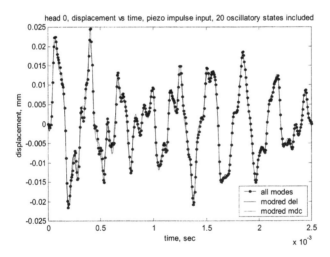

Figure 19.59: Impulse response comparison for head 0 for piezo input for oscillatory system, full model (all oscillatory modes) and balreal modred "del" and "mdc" reduced systems with 20 oscillatory modes.

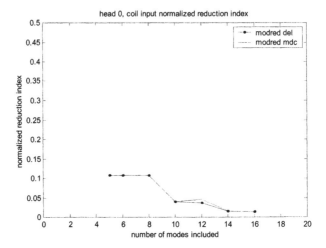

Figure 19.60: Head 0 impulse response normalized error index comparison for reduced modred models using "del" and "mdc" methods, coil input.

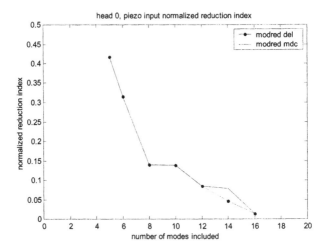

Figure 19.61: Head 0 impulse response normalized error index comparison for reduced modred models using "del" and "mdc" methods, piezo input.

19.6 MIMO Summary

We started the chapter with a description of key mode shapes for the two-stage actuator/suspension system. ANSYS eigenvector listings for several modes allowed comparing the numeric values in the eigenvector to the visual interpretation from the mode shape plot. Small displacements in the deformed mode shape plot correlate to small numerical values in the eigenvector. If the small numerical values in the eigenvector occur in the input and/or output degrees of freedom, the mode will have a "small" dc gain and is relatively unimportant.

In the next section we calculated and plotted the dc gains for all four input/output combinations.   In Table 19.1 we listed the modes for the input/output combinations, sorted by dc gain. We found that head 0 and head 1 dc gain sorted modes for coil input are the same for the first seven modes. For piezo input, both heads have the same mode ranking for the first six modes. This similarity in the most important modes for both heads for the coil and piezo inputs is brought about by the physical symmetry of the actuator/suspension system, and in general will not be the case.

As in the previous chapter, we used balancing to define the system for reduction and used the "modred" "del" and "mdc" options to reduce. Frequency responses for different number of states were plotted and compared for both coil and piezo inputs, overlaying the non-reduced transfer function.

Visually comparing the reduced and non-reduced frequency response magnitudes, we found that including 20 oscillatory states (plus the states from the one rigid body mode) gave a "good" fit through the 10khz range.

The MATLAB model was then used to calculate the impulse responses for the oscillatory reduced and non-reduced systems, where we found that 10 oscillatory modes (20 oscillatory states) were required to have a normalized error index of less than 5% for coil inputs. For piezo inputs, 16 oscillatory modes (32 oscillatory states) were required for less than 5% normalized error index. There was little difference in normalized error index between the "del" and "mdc" reduction options.

Problems

P19.1 Modify the MATLAB code **act8pz.m** to reduce the piezo force "fpz" (Section 19.5.2) from the 0.2 value used in the text to 0.02 and 0.002. In both cases, examine the frequency and impulse responses for different number of oscillatory states used.   Does the balanced reduction method technique continue to choose roughly equal number of modes for both coil and piezo inputs even when there are large differences in dc gain values between the two inputs?

P19.2 For the piezo force "fpz" of 0.2, choose the first five oscillatory modes from the coil input and the first five oscillatory modes from the piezo input (Table 19.1). Assemble the state equations from the rigid body mode and the 10 oscillatory modes and solve for the frequency and impulse responses. Compare the responses to the 20 oscillatory state balanced reduction. Comment on the similarities/differences.

APPENDIX 1

MATLAB AND ANSYS PROGRAMS

This appendix lists all the MATLAB and ANSYS codes used in each chapter, along with a short description of the purpose of each.

MATLAB codes have the suffix ".m" and the ANSYS codes have the suffix ".inp." Additional output files from previous runs are stored as ".grp" or other suffixes and will be used from time to time.

Coding format: All the MATLAB code available from downloading and shown in the book starts over one tab, allowing comment lines to stand out. The code also includes a lot of blank lines for readability (my apologies to tight "c" code programmers).

In most MATLAB code, critical definitions and calculations are only a few lines of code, while plotting and annotating are the bulk of the space. For this reason, some code listings in the book do not show all the plotting commands.

ANSYS eigenvalue/eigenvector results are converted to MATLAB input form using the following MATLAB extraction codes:

ext56ux.m	extracts the ANSYS UX degree of freedom
ext56uy.m	extracts the ANSYS UY degree of freedom
ext56uz.m	extracts the ANSYS UZ degree of freedom
ext56uxuy.m	extracts the ANSYS UX and UY degrees of freedom
ext56uxuz.m	extracts the ANSYS UX and UZ degrees of freedom
ext56uyuz.m	extracts the ANSYS UY and UZ degrees of freedom
ext56uxuyuz.m	extracts the ANSYS UX, UY and UZ degrees of freedom

The codes above all call a supporting MATLAB code **ext56chk.m**. All the codes should be installed in the same directory as the ANSYS output code which is to be extracted or should be installed in a directory which is in the MATLAB path. To use the extraction code, just rename the ANSYS eigenvector output file to have a ".eig" extension and open MATLAB in the same directory. MATLAB will then open a window showing all the ".eig" files in the directory. Double-click on the file to extract and MATLAB will output a file with the "ext56xx.mat" name. If several files are to be extracted in the same directory, rename the "ext56xx.mat" name to a unique name with the ".mat" extension.

The ".mat" extracted MATLAB file contains the following information:

evr, the modal matrix, with rows consisting of degrees of freedom and each column representing a mode. The numbering of degrees of freedom is the same as the ANSYS listing, which is in ascending order of the selected node numbers. Where multiple directions are extracted, for instance UX and UY degrees of freedom, the degrees of freedom are listed in that order, first the UX degrees of freedom and then the UY degrees of freedom. The extracted modal matrix is of size: (total dof) x (modes).

freqvec, a vector listing the eigenvalues (resonant frequencies), in hz values. The size of the frequency vector is (modes) x (1).

node_numbers, a vector listing the node numbers for the extracted data, of size (dof) x (1).

The extracted data can then be loaded and used to develop state space models of the system.

Chapter 2: Transfer Function Analysis

sdofxfer.m: Calculates and plots magnitude and phase for a single degree of freedom system over a range of damping values.

tdofpz3x3.m: Uses the "num/den" form of the transfer function, calculates and plots all nine pole/zero combinations for the nine different transfer functions for tdof model. It prompts for values of the two dampers, $c1$ and $c2$, where the default (hitting the "enter" key) values are set to zero to match the hand calculated values in (2.82). The "transfer function" forms of the transfer functions are then converted to "zpk - zero/pole/gain" form to enable graphical construction of frequency response in the next chapter.

tdofpz3x3_rlocus.m: Plots pole and zero values for z11 transfer function for a range of damping values.

Chapter 3: Frequency Response Analysis

tdofxfer.m: Plots tdof model poles and zeros in complex plane, user choice of damping values. Uses several different model descriptions and frequency response calculating techniques.   The model is described in polynomial, transfer function and zpk forms.   Magnitude and phase versus frequency are calculated using a scalar frequency "for loop," vector frequency, automatic bode plotting and bode with magnitude and frequency outputs.

Chapter 4: Zeros in SISO Mechanical Systems

ndof_numzeros.m: Calculates and plots poles/zeros and transfer functions for user selected input/output locations on a "n" dof series spring/mass model. Shows that poles of "constrained" structures to left and right of input/output degrees of freedom are the zeros of the unconstrained structure.

cantfem.inp: ANSYS code for resonant frequencies of cantilever and tip driving point transfer function.   Used to identify zero locations to compare with poles of "constrained" system in cantzero.inp.

cantzero.inp: ANSYS code for resonant frequencies of cantilever with simple support at tip. Used to identify poles of "constrained" structure.

cantzero.m: Uses eigenvalues and eigenvectors from cantfem.inp and cantzero.inp to plot overlay of zeros of cantilever with poles of tip supported cantilever,   showing   the   correspondence.   Calls   **cantzero_freq.m**, **cantfem_magphs.m**.

Chapter 5: State Space Analysis

tdof_non_prop_damped.m: This code is used to develop an understanding of the results of MATLAB's eigenvalue analysis and complex modes.

Chapter 6: State Space: Frequency Response, Time Domain

tdofss.m: Calculates and plots the four distinct frequency responses for the tdof model.

tdof_ss_time_ode45_slnk.m: Solves for time domain response of tdof problem using MATLAB's ODE45 solver, a Runga-Kutta method of solving differential equations, as well as, MATLAB's Simulink block-diagram simulation tool.

tdof_ss_time_slnk_plot.m: Plots results from tdof_ss_time_ode45_slnk.m.

tdofssfun.m: Function code called by tdof_ss_time_ode45_slnk.m, contains state equations.

tdofss_simulink.mdl: Simulink   model   called   by
tdof_ss_time_ode45_slnk.m, defines state equations.

Chapter 8: Frequency Response: Modal Form

tdof_modal_xfer.m: Calculates and plots the four distinct frequency responses and the individual modal contributions.

threedof.inp: ANSYS code that builds the undamped tdof model, calculates eigenvalues and eigenvectors, outputs the frequency listing and eigenvectors, plots the mode shapes. Calculates and plots all three transfer functions for a force applied to mass 1.

Chapter 9: Transient Response: Modal Form

tdof_modal_time.m: Plots displacements versus time in principal and physical coordinates.

Chapter 10: Modal Analysis: State Space Form

tdofss_eig.m: Solves for the eigenvalues and eigenvectors in the state space form of the tdof system.

tdof_prop_damped.m: Calculates poles and zeros of proportionally damped tdof system. Plots initial condition responses for modes 2 and 3 in physical and principal coordinate systems.

Chapter 11: Frequency Response: Modal State Space Form

tdofss_modal_xfer_modes.m: Solves for and plots frequency responses for individual modal contributions and overall responses. Has code for plotting frequency responses in different forms.

Chapter 12: Time Domain: Modal State Space Form

tdofss_modal_time_ode45.m: Plots tdof transient responses for overall and individual modal contributions. Calls the function files below, which define the state space system and individual modes.

tdofssmodalfun.m,   tdofssmodal1fun.m,   tdofssmodal2fun.m, tdofssmodal3fun.m: Function files called by tdofss_modal_time_ode45.m.

Chapter 14: Finite Elements: Dynamics

cant_2el_guyan.m: Solves for the eigenvalues and eigenvectors of a two-element cantilever beam.

cantbeam_guyan.m: Solves for eigenvalues and eigenvectors of a cantilever with user-defined dimensions, material properties, number of elements and number of mode shapes to plot. Guyan Reduction is an option. A 10-element beam is used as an example.

cantbeam.inp: ANSYS code solves for the eigenvalues and eigenvectors of a 10 element cantilever, the same beam as the cantbeam_guyan.m example.

Chapter 15: SISO State Space MATLAB Model from ANSYS Model

cantbeam_ss.inp: ANSYS code for cantilever beam, allows the user to change the number of elements and the eigenvalue extraction technique. The two variables "num_elem" and "eigext" can be easily changed to see their effects.

cantbeam_ss_freq.m: Compares theoretical frequencies for the first 16 modes for a cantilever beam with MATLAB finite element and ANSYS finite element results.

cantbeam_ss_modred.m: Creates a MATLAB state space model using the eigenvalue and eigenvector results from previous ANSYS runs. Modes are ranked for importance and several reduction techniques are used.

Chapter 16: Ground Acceleration MATLAB Model from ANSYS Model

cantbeam_ss_spring_shkr.inp: ANSYS model of shaker mounted cantilever with tip mass and tip spring to shaker. Outputs mode shape plot file **cantbeam16red.grp**.

cantbeam_ss_tip_con.inp: ANSYS model of shaker mounted constrained tip cantilever. Outputs mode shape file **tipcon16red.grp**.

cantbeam_shkr_modeshape.m: Plots mode shapes from ANSYS modal analysis results for any of the tip spring models, with 2, 4, 8, 10, 12, 16, 32 and 64 beam elements.

cantbeam_ss_shkr_modred.m: Creates a MATLAB state space model using the results from ANSYS model **cantbeam_ss_spring_shkr.inp**. Ranks modes, then uses several reduction techniques to define smaller model.

Chapter 17: SISO Disk Drive Actuator Model

srun.inp: ANSYS model of suspension.

arun.inp: ANSYS model of actuator/suspension system.

act8.m: MATLAB code for dc and peak gain ranking and reduction of actuator/suspension model. Output from program is used for some input to **balred.m** in Chapter 18.

Chapter 18: Balanced Reduction

balred.m: MATLAB code for balanced reduction of actuator/suspension model from **act8.m**.

Chapter 19: MIMO Two-Stage Actuator Model

arunpz.inp: ANSYS model of two-stage actuator/suspension system.

act8pz.m: MATLAB model of two-stage actuator/suspension system, balanced reduction.

Downloading

All the programs listed can be downloaded from the MathWorks FTP site at www.mathworks.com or from the author's site at www.hatchcon.com.

APPENDIX 2

LAPLACE TRANSFORMS

This appendix presents a short introduction to Laplace transforms, the basic tool used in analyzing continuous systems in the frequency domain. The Laplace transform converts linear ordinary differential equations (LODE's) into algebraic equations, making them easy to solve for their frequency and time-domain behavior. There are many excellent presentations of the Laplace transform, as in Oppenheim [1997], for those who would like more information.

A2.1 Definitions

The Laplace transform is a generalized Fourier transform, where given any function f(t), the Fourier transform $F(\omega)$ is defined as:

$$F(\omega) = \mathcal{F}\{f(\cdot)\}(\omega) = \int_{-\infty}^{\infty} f(t)\, e^{j\omega t} dt \qquad (A2.1)$$

where $\omega = 2\pi f$ and f is frequency, in hz.

In the same spirit, we can define the Laplace transform as:

$$F(s) = \mathcal{L}\{f(\cdot)\}(s) = \int_{0^-}^{+\infty} f(t)\, e^{-st} dt \qquad (A2.2)$$

where s is complex:

$$s = \sigma + j\omega, \qquad (A2.3)$$

σ and ω are real numbers which define the locations of "s" in the complex plane, see Figure A2.1 below. Also, $\omega = 2\pi f$ as above.

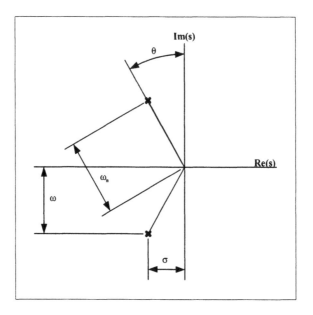

Figure A2.1: σ and ω **definitions in complex plane.**

Remarks:

1) if $f(t) \equiv 0$ for $t < 0$, then

$$\mathcal{F}\{f(\cdot)\}(\omega) = \mathcal{L}\{f(\cdot)\}(j\omega) \qquad (A2.4)$$

2) The " 0^- " limit in the Laplace transform definition takes care of $f(t)$'s which contain the δ function.

3) The integral in the definition of the Laplace transform need not be finite, i.e. $\mathcal{L}\{f\}(s)$ may not exist for all $s \in \square$. However, if $f(t)$ is bounded by some exponential:

$$|f(t)| \leq M e^{\sigma_0 t} \qquad (A2.5)$$

then $\mathcal{L}\{f\}(s)$ will make sense for $s \in \square$ such that $Re\{s\} > \sigma_0$.

4) The Laplace transform is linear:

$$\mathcal{L}\{a_1f_1 + a_2f_2\} = a_1\mathcal{L}\{f_1\} + a_2\mathcal{L}\{f_2\} \qquad (A2.6)$$

A2.2 Examples, Laplace Transform Table

1) Exponential

$$f(t) = e^{-at}1(t)$$

$$F(s) = \int_{0^-}^{\infty} e^{-at}1(t)e^{-st}dt = \int_{0^-}^{\infty} e^{-(s+a)t}\, dt = \frac{1}{s+a} \qquad [s > a] \qquad (A2.7a,b)$$

2) Impulse

$$f(t) = \delta(t)$$

$$F(s) = \int_{0^-}^{\infty} \delta(t)e^{-st}dt = e^{-0} = 1 \qquad [\text{for any } s] \qquad (A2.8a,b)$$

3) Step

$$f(t) = 1(t)$$

$$F(s) = \int_{0^-}^{\infty} e^{-st}dt = \frac{-\left[e^{-s(\infty)} - e^{-s(0)}\right]}{s} = \frac{1}{s} \qquad [s > 0] \qquad (A2.9a,b)$$

Table A2.1 below contains Laplace transforms for a few selected functions in the time domain. The "Region of Convergence" or "ROC" is defined as the range of values of "s" for which the integral in the definition of the Laplace transform (A2.2) converges (Oppenheim 1997).

	f(t)	Laplace Transform	Region of Convergence
1)	$\delta(t)$	1	all s
2)	$\delta(t-T)$	e^{-sT}	all s
3)	$1(t)$	$\dfrac{1}{s}$	$\mathrm{Re}\{s\} > 0$
4)	$\dfrac{1}{m!}t^m 1(t)$	$\dfrac{1}{s^{m+1}}$	$\mathrm{Re}\{s\} > 0$
5)	$e^{-at}1(t)$	$\dfrac{1}{s+a}$	$\mathrm{Re}\{s\} > \mathrm{Re}\{a\}$
6)	$\dfrac{1}{(m-1)!}t^{m-1}e^{-at}1(t)$	$\dfrac{1}{(s+a)^m}$	$\mathrm{Re}\{s\} > \mathrm{Re}\{a\}$
7)	$(1-e^{-at})1(t)$	$\dfrac{a}{s(s+a)}$	$\mathrm{Re}\{s\} > \max\{0, \mathrm{Re}\{a\}\}$
8)	$(e^{-at} - a^{-bt})1(t)$	$\dfrac{b-a}{(s+a)(s+b)}$	$\mathrm{Re}\{s\} > \max\{\mathrm{Re}\{a\}, \mathrm{Re}\{b\}\}$
9)	$\sin(at)\,1(t)$	$\dfrac{a}{s^2+a^2}$	$\mathrm{Re}\{s\} > 0$
10)	$\cos(at)1(t)$	$\dfrac{s}{s^2+a^2}$	$\mathrm{Re}\{s\} > 0$
11)	$e^{-at}\sin(bt)1(t)$	$\dfrac{b}{(s+a)^2+b^2}$	$\mathrm{Re}\{s\} > a$
12)	$e^{-at}\cos(bt)1(t)$	$\dfrac{s+a}{(s+a)^2+b^2}$	$\mathrm{Re}\{s\} > a$

Table A2.1: Laplace transform table.

A2.3 Duality

The following duality conditions exist:

$$t\, f(t) \iff -\frac{d}{ds}\, F(s)$$

(A2.10a,b)

$$\frac{d}{dt} f(t) \iff s\, F(s)$$

A2.4 Differentiation and Integration

Differentiation and the Laplace transform: Suppose

$$\mathcal{L}\{x\}(s) = X(s) \tag{A2.11}$$

then

$$\mathcal{L}\{\dot{x}\}(s) = sX(s) - x(0^-)\,, \tag{A2.12}$$

so we can interpret "s" as a differentiation operator:

$$\frac{d}{dt} \leftrightarrow s \tag{A2.13}$$

Integration and the Laplace transform: Suppose

$$\mathcal{L}\{x\}(s) = X(s)\,, \tag{A2.14}$$

then

$$L\left\{\int_0^t x(\tau)d\tau\right\}(s) = \frac{1}{s}\, X(s)\,, \tag{A2.15}$$

and we can interpret "1/s" as an integration operator:

$$\frac{1}{s} \leftrightarrow \int_0^t dt \tag{A2.16}$$

A2.5 Applying Laplace Transforms to LODE's with Zero Initial Conditions

Assume we have a linear ordinary differential equation as shown in (A2.17):

$$\dddot{y}(t) + a_1\ddot{y}(t) + a_2\dot{y}(t) + a_3 y(t) = b_1\ddot{u}(t) + b_2\dot{u}(t) + b_3 u(t) \qquad (A2.17)$$

Assume $\ddot{y}(t) = 0$, $\dot{y}(t) = 0$, $y(t) = 0$ and take the Laplace transform of both sides, using the linearity property (A2.6):

$$\mathcal{L}\{\dddot{y}\}(s) + a_1\mathcal{L}\{\ddot{y}\}(s) + a_2\mathcal{L}\{\dot{y}\}(s) + a_3\mathcal{L}\{y\}(s) = \\ b_1\mathcal{L}\{\ddot{u}\}(s) + b_2\mathcal{L}\{\dot{u}\}(s) + b_3\mathcal{L}\{u\}(s) \qquad (A2.18)$$

Recalling that "s" is the differentiation operator, replace "dots" with "s":

$$s^3 Y(s) + a_1 s^2 Y(s) + a_2 s Y(s) + a_3 Y(s) = b_1 s^2 U(s) + b_2 s U(s) + b_3 U(s) \quad (A2.19)$$

We are now left with a polynomial equation in "s" that can be factored into terms multiplying Y(s) and U(s):

$$\left[s^3 + a_1 s^2 + a_2 s + a_3 \right] Y(s) = \left[b_1 s^2 + b_2 s + b_3 \right] U(s) \qquad (A2.20)$$

Solving for Y(s):

$$Y(s) = \frac{\left[b_1 s^2 + b_2 s + b_3 \right]}{\left[s^3 + a_1 s^2 + a_2 s + a_3 \right]} U(s) \qquad (A2.21)$$

It can be shown that the terms in the numerator and denominator above are the Laplace transform of the impulse response, H(s):

$$Y(s) = H(s)U(s), \qquad (A2.22)$$

$$H(s) = \mathcal{L}\left[h(\cdot) \right](s), \qquad (A2.23)$$

and $h(\cdot)$ is the impulse response. For the example LODE (A2.17) the Laplace transform of the impulse response is:

$$H(s) = \frac{\left[b_1 s^2 + b_2 s + b_3 \right]}{\left[s^3 + a_1 s^2 + a_2 s + a_3 \right]} \qquad (A2.24)$$

A2.6 Transfer Function Definition

It can be shown that the transfer function of a system described by a LODE is the Laplace transform of its impulse response, H(s), (A2.23).

Taking the Laplace transform of the LODE has provided the Laplace transform of the impulse response. If we could inverse-transform H(s) we could get the impulse response h(t) without having to integrate the differential equation. Typically the inverse transform is found by simplifying/expanding H(s) into terms which can be found in tables, such as Table A2.1, and than inverting "by inspection."

A2.7 Frequency Response Definition

Having obtained H(s) directly from the LODE by replacing "dots" by "s," we can obtain the frequency response of the system (the Fourier transform of the impulse response) by substituting " $j\omega$ " for "s" in H(s).

$$H(j\omega) = H(s)\big|_{s=j\omega} \qquad\qquad (A2.25)$$

A2.8 Applying Laplace Transforms to LODE's with Initial Conditions

In A2.5 we looked at applying Laplace transforms to LODE's with zero initial conditions, which led to transfer function and frequency response definitions. Since transfer functions and frequency responses deal with steady state sinusoidal excitation response of the system, initial conditions are of no significance, as it is assumed that all measurements of the system undergoing sinusoidal excitation are taken over a long enough period of time that transients have died out.

On the other hand, if we are solving for the transient response of a system defined by a LODE that has initial conditions, obviously the initial conditions will not be zero. We will use the basic definition of the differentiation operation from (A2.12) to define the Laplace transform of 1st and 2nd order differential equations with initial conditions $x(0)$ and $\dot{x}(0)$:

1st Order: $\qquad\qquad \mathcal{L}\{\dot{x}(t)\} = sX(s) - x(0) \qquad\qquad (A2.26)$

2nd Order: $\qquad\qquad \mathcal{L}\{\ddot{x}(t)\} = s^2X(s) - sx(0) - \dot{x}(0) \qquad\qquad (A2.27)$

A2.9 Applying Laplace Transform to State Space

We defined the form of state space equations in Chapter 5 as below:

$$\dot{\mathbf{x}}(t) = \mathbf{A}\mathbf{x}(t) + \mathbf{B}u(t) \tag{A2.28}$$

$$\mathbf{y}(t) = \mathbf{C}\mathbf{x}(t) + \mathbf{D}u(t) \tag{A2.29}$$

where the initial conditions are set by $\mathbf{x}(0) = \mathbf{x}_o$. The general block diagram for a SISO state space system is shown in Figure A2.1.

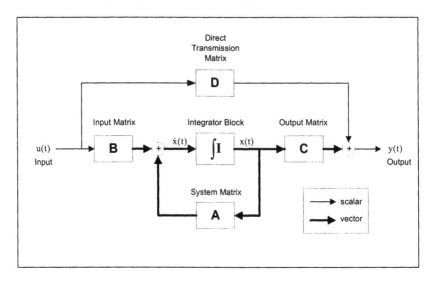

Figure A2.1: State space block diagram.

Taking Laplace transform of (A2.28):

$$\begin{aligned}
\mathcal{L}\{\dot{\mathbf{x}}\}(s) &= \mathcal{L}\{\mathbf{A}\mathbf{x}\}(s) + \mathcal{L}\{\mathbf{B}u\}(s) \\
s\mathbf{X}(s) - \mathbf{x}(0^-) &= \mathbf{A}\mathcal{L}\{\mathbf{x}\}(s) + \mathbf{B}\mathcal{L}\{u\}(s) \\
&= \mathbf{A}\mathbf{X}(s) + \mathbf{B}U(s)
\end{aligned} \tag{A2.30a,b}$$

Solving for $\mathbf{X}(s)$:

$$\begin{aligned}
s\mathbf{X}(s) - \mathbf{A}\mathbf{X}(s) &= \mathbf{x}(0^-) + \mathbf{B}U(s) \\
(s\mathbf{I} - \mathbf{A})\mathbf{X}(s) &= \mathbf{x}(0^-) + \mathbf{B}U(s) \\
\mathbf{X}(s) &= (s\mathbf{I} - \mathbf{A})^{-1}\mathbf{x}(0^-) + (s\mathbf{I} - \mathbf{A})^{-1}\mathbf{B}U(s)
\end{aligned} \tag{A2.31a,b,c}$$

The two terms on the right-hand side of (A2.31c) have special significance:

1) $(s\mathbf{I} - \mathbf{A})^{-1}\mathbf{x}(0^-)$ is the Laplace transform of the homogeneous solution, the initial condition response.

2) $(s\mathbf{I} - \mathbf{A})^{-1}\mathbf{B}U(s)$ is the Laplace transform of the particular solution, the forced response.

Taking the Laplace transform of (A2.29), the output equation:

$$\mathbf{Y}(s) = \mathbf{C}\mathbf{X}(s) + \mathbf{D}U(s) \tag{A2.32}$$

Knowing $X(s)$ from (A2.31c) and substituting in (A2.32):

$$\mathbf{Y}(s) = \mathbf{C}(s\mathbf{I} - \mathbf{A})^{-1}\mathbf{x}(0^-) + \left[\mathbf{C}(s\mathbf{I} - \mathbf{A})^{-1}\mathbf{B} + \mathbf{D}\right]U(s) \tag{A2.33}$$

If the initial conditions are zero, $x(0^-) = 0$, then

$$\mathbf{Y}(s) = \left[\mathbf{C}(s\mathbf{I} - \mathbf{A})^{-1}\mathbf{B} + \mathbf{D}\right]U(s), \tag{A2.34}$$

with the transfer function for the system being defined by $H(s)$:

$$\mathbf{H}(s) = \left[\mathbf{C}(s\mathbf{I} - \mathbf{A})^{-1}\mathbf{B} + \mathbf{D}\right] \tag{A2.35}$$

When the terms in $H(s)$ above are multiplied out, they will result in the following polynomial form:

$$\mathbf{H}(s) = \frac{b(s)}{a(s)} + \mathbf{D} \tag{A2.36}$$

REFERENCES

Archer, John S., Consistent Mass Matrix for Distributed Mass Systems, *Journal of the Structural Division, Proceedings of the American Society of Civil Engineers*, ST4, August, 1963, p. 161.

Bay, J.S., *Fundamentals of Linear State Space Systems*, McGraw-Hill, Boston, MA, 1999.

Chang, Tish-Chun and Craig, Roy R., Jr., Normal Modes of Uniform Beams, *Journal of Engineering Mechanics Division, Proceedings of the American Society of Civil Engineers*, Vol. 95, No. EM4, August, 1969, p. 1027.

Chen, C.T., *Linear System Theory and Design,* Third Edition, Oxford University Press, New York, 1999.

Craig, R.R., Jr., *Structural Dynamics, An Introduction to Computer Methods*, John Wiley & Sons, New York, 1981.

Evans, W.R., Graphical Analysis of Control Systems, *Trans. AIEE*, vol. 68, 1949, pp. 765–777.

Franklin, G.F., Powell, J.D., and Emami-Naeini, A., *Feedback Control of Dynamic Systems*, Third Edition, Addison-Wesley, Menlo Park, CA, 1994.

Franklin, G.F., Powell, J.D., and Workman, M., *Digital Control of Dynamic Systems*, Third Edition, Addison-Wesley, Menlo Park, CA, 1998.

Gawronski, W.K., *Balanced Control of Flexible Structures,* Springer, New York, 1996.

Gawronski, W.K., *Dynamics and Control of Structures, A Modal Approach*, Springer, New York, 1998.

Johnson, C.D. and Kienholz, D.A., Finite Element Prediction of Damping in Structures with Constrained Viscoelastic Layers, *AIAA Journal*, 20(9), September 1982, p. 1284.

Kailath, T., *Linear Systems*, Prentice-Hall, Englewood Cliffs, NJ, 1980.

Laub, A.J., Heath, M.T., Paige, C.C., and Ward, R.C., "Computations of System Balancing Transformations and Other Applications of Simultaneous Diagonalization Algorithms," *IEEE Transactions on Automatic Control*, AC-32 (1987), pp. 115–122.

Maia, N.M.M. and Silva, J.M.M., *Theoretical and Experimental Modal Analysis*, Research Studies Press LTD, Taunton, Somerset, U.K., 1997.

Miu, D.K., Poles and Zeros, *Mechatronics,* Springer-Verlag, New York, 1993.

Moore, B., "Principal Component Analysis in Linear Systems: Controllability, Observability and Model Reduction," *IEEE Transsactions on Automatic Control,* AC-26 (1981), pp. 17–31.

Newland, D.E., *Mechanical Vibration Analysis and Computation*, John Wiley & Sons, Inc., New York, 1989.

Oppenheim, A.V., Willsky, A.S., and Nawab, S.H., *Signals and Systems*, Prentice-Hall, Upper Saddle River, NJ, 1997.

Pilkey, Walter D., *Formulas for Stress, Strain, and Structural Matrices*, John Wiley & Sons, New York, 1994.

Strang, G., *Introduction to Linear Algebra*, 2nd Edition, Wellesley-Cambridge Press, Wellesley, MA, 1998.

Weaver, W. Jr., Timoshenko, S.P., and Young, D.H., *Vibration Problems in Engineering*, 5th Edition, John Wiley & Sons, New York, 1990.

Zhou, K., Doyle, J.C., and Glover, K., *Robust and Optimal Control*, Prentice-Hall, Upper Saddle River, NJ, 1996.

Zhou, K. and Doyle, J.C., *Essentials of Robust Control*, Prentice-Hall, Upper Saddle River, NJ, 1998.

INDEX

H

G

L

M

Milton Keynes UK
Ingram Content Group UK Ltd.
UKHW031123141024
449569UK00006B/467